The Anthropocene: Politik—Economics—Society—Science

Volume 18

Series editor

Hans Günter Brauch, Mosbach, Germany

More information about this series at http://www.springer.com/series/15232
http://www.afes-press-books.de/html/APESS.htm
http://www.afes-press-books.de/html/APESS_18.htm

Tsugihiro Watanabe · Selim Kapur
Mehmet Aydın · Rıza Kanber
Erhan Akça
Editors

Climate Change Impacts on Basin Agro-ecosystems

 Springer

Editors
Tsugihiro Watanabe
Graduate School of Global
  Environmental Studies
Kyoto University
Kyoto, Japan

Rıza Kanber
Department of Agricultural Structures
  and Irrigation Engineering
Çukurova University
Adana, Turkey

Selim Kapur
Department of Soil Science and Plant
  Nutrition, Faculty of Agriculture
Çukurova University
Adana, Turkey

Erhan Akça
School of Technical Sciences
Adıyaman University
Adıyaman, Turkey

Mehmet Aydın
Department of Soil Science, Faculty
  of Agriculture
Mustafa Kemal University
Antakya, Turkey

The cover photograph was provided by Adana Museum and shows Teşup or Tarhunda, the god of atmospheric events of rain, thunder and storm during the late Hittite Period (BC 700). This statue of Tarhunda (limestone) on a basalt oxen-driven chariot (2700 BP) was discovered by the Adana Regional Museum in the Lower Seyhan Irrigation Plain (the ICCAP site) in the late 1990s. To illustrate the drought theme, the photograph on the internal title page (iii) shows an abandoned well on the Seyhan Plain. Both photographs are used with the kind permission of the photographers.

For more information on this book, see: http://afes-press-books.de/html/APESS_18.htm.

ISSN 2367-4024   ISSN 2367-4032   (electronic)
The Anthropocene: Politik—Economics—Society—Science
ISBN 978-3-030-01035-5   ISBN 978-3-030-01036-2   (eBook)
https://doi.org/10.1007/978-3-030-01036-2

Library of Congress Control Number: 2018955450

Editor: PD Dr. Hans Günter Brauch, AFES-PRESS e.V., Mosbach, Germany.
English Language Editor: Dr. Vanessa Greatorex, England.

This Springer imprint is published by the registered company Springer Nature Switzerland AG
The registered company address is: Gewerbestrasse 11, 6330 Cham, Switzerland

Foreword by Fujio Kimura

I was involved in the Impact of Climate Change on Agricultural Production System in Arid Area (ICCAP) Project, with its outcomes reported in this book, from its very beginning. My career as a meteorologist and climatologist was influenced by this project where I learned a lot on assessing the basin-wide impact of climate change on hydrological systems and agricultural production. During the collaboration with the leader, Dr. Watanabe, and the other researchers of the ICCAP, I tried to develop the future climate projection model for the target region. While this was quite a tough task for me, as I had mainly worked on the climate change in Japan, I remember being very happy when I finally achieved satisfactory results. We had lively discussions with many researchers and practitioners of different fields from Japan and abroad about the latest scientific issues and developments in methodology.

In the project, from the beginning, the collaborators in the fields of hydrology and agronomy had requested the establishment of a reliable climate change scenario with considerably higher horizontal resolution. Long discussions and exchanges of knowledge went on until a deep mutual understanding of "projection" and "prediction" was reached. I had primarily tried to convince them on the basis of the climate model and also how to use its outputs, and finally share the common comprehension concerning the mechanism of climate change and its prediction as well as the limitations on the prediction.

Climate change prediction with higher resolution and accuracy was the general requirement in the studies related to the ICCAP. During the ICCAP Project, across the world, higher-resolution general circulation models (GCMs) were being developed, and methods to downscale the GCM outputs were also being developed with improved regional climate models (RCM) nesting the GCM.

The future regional climate scenarios were generated when the ICCAP introduced the pseudo-global warming experiment, with the future climate change predictions by the GCM, the detailed climate distributions currently observed, and the RCM downscaling the GCM outputs. At the time, this method was the most advanced, offering higher accuracy and resolution. This technique is being still used in other countries, such as the USA.

The researchers in various fields, who had developed well-coordinated models to assess climate change utilising the generated regional climate scenarios, really seemed to feel that they provided exact and meaningful future predictions. In addition, with the sophisticated prediction of changes resulting from agricultural production and land and water use, an understanding of the necessary feedback of changes in agriculture and water use to the climate system was extended and resulted in new developments. While I regret that I myself did not record the development of the method related to the generation of the climate scenario in this book, it is my great pleasure that some of the chapters describe the method and use their outcomes, and refer to the journal paper that summarises the research results.

As mentioned above, related to the main target of the ICCAP, close collaborations and cooperation between the researchers of various academic fields were achieved with regard to climate change. The "warm and close cooperation" coordinated by the leader, Dr. Watanabe, seems to underpin each chapter in this book and appears explicitly at the surface of some chapters.

After the ICCAP Project, I have been involved in implementing research programmes on the development of the adaptation measures to climate change impacts on Japan, where my experiences and learning during the international ICCAP Project helped me solve domestic research issues and supported my research capability.

Such a cross-disciplinary and international "warm collaboration" as of the ICCAP is to be further enhanced. I hope the integrated approach to climate change impact assessment and the establishment of adaptation measures will be promoted extensively, leading to the basis for a sustainable society all over the world, where people are not subjected to the adverse effects of climate change.

Tokyo, Japan   Dr. Fujio Kimura
March 2018   Professor Emeritus of Tsukuba University
                                  Programme Director, Advanced Atmosphere-Ocean-Land
                               Modelling Program; Research Institute for Global Change
                               Japan Agency for Marine-Earth Science and Technology
                                                                                (JAMSTEC)

Foreword by Eiichi Nakakita

As a researcher working in the relevant field, I am very pleased about the publication of this book on the integrated assessment of climate change impacts on hydrological processes and water resource management, including agricultural water use.

Needless to say, climate change due to global warming has become a major urgent issue for humanity. It is therefore vital to further accelerate assessment of its impacts, including the exact possible changes and problems, and establish adaptation measures based on the impacts that have already been observed. Under such circumstances, with the development of science and technology supported by the expansion of observation data and their availability, the accuracy and reliability of the projection of future climate change have certainly increased. On the other hand, integrating various aspects of impact assessment with adaptation scenarios is still developing, since it has many constraints, including its complex processes with diverse factors of relevant target fields, such as the natural physical and chemical phenomena, biosystems and ecosystems, as well as the lifestyle and production activities of the various societies.

In Japan, in view of this situation, the National Government Ministry of Education, Culture, Sports, Science, and Technology has been leading large-scale research programs, such as the Innovative Program of Climate Change Projection for the twenty-first century (KAKUSHIN) (2007–2012) and the Program for Risk Information on Climate Change (SOUSEI) (2012–2017), aiming at increased accuracy of projecting the future climate and precise impact assessment. The Integrated Research Program for Advancing Climate Models (TOUGOU) (2017–2022) was launched as their successor. In these programmes, I have been coordinating the team to more precisely forecast and assess the climate change impacts on hydrological systems, water resource management, and the eco-environmental system of basins. Drs. Kenji Tanaka, Tsugihiro Watanabe, Takanori Nagano, and Yoichi Fujiwara are the key researchers of the ICCAP Project whose methodologies and results are introduced in this book, and who have participated in the consecutive programmes mentioned above. This means that the outcomes of the ICCAP Project are partly reflected in the framework and methodologies of both the programmes of the internationally challenging research programmes. Additionally, the development and outcomes of these programs have served as a reference in the process of compiling this book in terms of academic sense with financial support. I am honoured that our programme has contributed to this publication.

In view of the urgency of the climate change issues, and considering the advancement history mentioned above, I have a strong interest in this publication. I believe that this book will be a major force in response to climate change, with the progress of application of its outcomes and further improvements.

Kyoto, Japan   Dr. Eiichi Nakakita
March 2018   Professor, Division of Atmospheric and Hydrospheric Disasters
                              Disaster Prevention Research Institute, Kyoto University
                              The Principal Investigator of the Research Theme D
                              "Integrated Hazard Projection" of the Integrated Research
                              Programme for Advancing Climate Models of MEXT
                              Ministry of Education, Culture, Sports
                              Science and Technology, Japan

Foreword by Cemal Saydam

Back in 2001, I was informed that a scientist from Japan named Tsugihiro Watanabe wished to visit me. I thought that he would be yet another scientist who was willing to establish a collaboration with us and as always I accepted with great pleasure. During a long conversation, we mutually agreed on a collaboration. However, it was initially difficult for me to assign a specific research grant group to further extend a possible cooperation. Having the quick approval of The Science and Technology Promotion Agency of Turkey's (TÜBİTAK) president, I was personally appointed to be in charge of the project since I realised that the proposed project encompassed the research interests of several grant committees.

It was challenging to set up everything, since we had to decide on a river basin in order to initiate this multidisciplinary project. Cross-boundary rivers certainly offer ideal basins to study, but in this part of the world, as can be seen today, it is hard—if not impossible—to plan scientific projects since there are so many other factors that scientists are not empowered to resolve. Considering all these factors and to eliminate obstacles that we were not in a position to handle, we mutually agreed on the Seyhan Basin as a case study area. Such a multidisciplinary project is not easy to coordinate and run smoothly in its rather complicated integrity, but we were lucky to have a gentleman named Prof. Dr. Rıza Kanber from the Çukurova University to fulfill this task. We are also deeply indebted to Prof. Dr. Mehmet Aydın for his valuable efforts in creating the initial link between Turkey and Japan for the ICCAP and selection of the Çukurova plain as the project site via discussions that started in Japan with Profs. Tsugihiro Watanabe and Tomohisa Yano, and followed in Turkey. Professors Aytekin Berkman, Osman Tekinel (deceased), and Neşet Kılınçer should be acknowledged for their invaluable efforts in developing the relationships with other universities, Government establishments, and TÜBİTAK. We also extend our thanks to Prof. Dr. Selim Kapur for his efforts regarding the decision given on the implementation site of the ICCAP.

Thus with the support of the Research Institute for Humanity and Nature (RIHN) and TÜBİTAK project, the study was set to roll out in 2002 and was finalised during 2007 with great success. This was one of the perfect examples of excellent scientific cooperation between different cultures and proved that such things can be achieved. It is nice to see that the termination of the project at the end of its expected lifetime was, in fact, the start of so many collaborations that are still running in a perfect manner. This book is the compilation of the outcomes and successes of the programme and a testament to the excellent collaboration.

Ankara, Turkey   Dr. Cemal Saydam
March 2018   Professor of Hacettepe University, Turkey
                                                  Former Vice-President of TÜBİTAK
        The Scientific and Technological Research Council of Turkey

Preface

I am delighted that the latest research results from the integrated assessment of climate change impacts on basin hydrological systems and agricultural production have been collated for publication as a volume. As editor, I would like to express my sincere and profound gratitude to all contributors and those who have supported the research projects and publication of the findings.

The contents of this book consist of the outcomes of the large-scale research project Impact of Climate Change on Agricultural Production System in Arid Areas (ICCAP) of the Research Institute for Humanity and Nature (RIHN) in Kyoto, Japan. This ICCAP Project was implemented for five years, from April 2002 to March 2007. Its aim was to assess the impacts of global-warming-induced climate change on agriculture. The principal investigator of the project was Dr. Tsugihiro Watanabe. In the project, the Seyhan River Basin of the Mediterranean region of Turkey was selected as the main case study region. Scientists from various relevant academic disciplines, of Japan, Turkey, and other countries participated in the projects. At that time, it was a very ambitious project, which addressed the very urgent and complicated issues by introducing state-of-the-art technology and knowledge.

The RIHN, which launched and implemented the ICCAP Project, was founded by the Japanese Government's Ministry of Education, Culture, Sports, Science, and Technology (MEXT), with the aim of establishing new academic fields related to the global environment, with the recognition of "global environmental issues" as a major concern in the relationship between nature and human beings. This relationship is "culture" in its wider sense, and in this sense, the environmental problem is the cultural issue. The institute tried to promote exact cross-disciplinary approaches to the issues to identify their deep implication and propose practical solutions, beyond their collateral nature. In this policy and context, and as a matter of urgency, the project was carried out to assess climate change and its likely impact on agricultural production. The meaning and management of natural resources, including climate, in agricultural production were re-studied during the project, and the existing management knowledge was reviewed to re-tailor the wisdom for "*futurability*". Thus, the context of the ICCAP Project was a uniquely ambitious challenge rarely encountered in the studies of the other relevant research institutes.

The project initially targeted a different region of the Middle East. However, because of the security and social situation of the region as well as the appropriate research environment, the Seyhan River Basin of Turkey was selected as the main case study area. The project of the RIHN was implemented with the financial support of The Science and Technology Promotion Agency of Turkey (TÜBİTAK), with the participation of many Turkish researchers, principally from Çukurova University, and the engineers of the 6th Regional Directorate of the State Hydraulic Works (DSİ) in Adana, located in the centre of the case study. In this collaborative work, Dr. Rıza Kanber, Professor of Çukurova University, served as the leader of the Turkish Team.

Six sub-teams were established: climate, hydrology, vegetation, agriculture, irrigation, and economics based on the generated future climate change predicted to occur by the 2070s. This was the date each of the teams used when assessing future impacts on the basin hydrology and agriculture. Finally, possible and practical methods were developed and applied despite the difficulties encountered concerning the research approach and context of the project. These challenges and developments deserve to be highly appreciated.

The challenges and outcomes are summarised in the following chapters of this book. At the beginning, the basic ideas and techniques are introduced, and the outcomes of each sub-team are grouped in the following parts. In addition, because some years have passed since the completion of the project and further progress on the methodology has been made, the chapters have been updated and the relevant research outcomes of other researchers have been incorporated in the book. Although each chapter concerns an independent study so could theoretically be read in isolation, I respectfully suggest you pay attention to all the findings obtained from the mutual relationships and accumulated knowledge developed in multi-authored chapters.

Again, looking back on the ICCAP, while some years have passed, its perspectives and approaches still address ongoing challenges and remain at the forefront of research in the field of climate change impact assessment and adaptation strategy development. I hope this book will contribute to further development of strategies to address the issues, inspired by the methods and results summarised here, and applications corresponding to the exact case situation. In conclusion, I would like to express my gratitude for this publication, the fruitful outcomes of the project and the meticulous and dedicated input of my co-editors, Selim Kapur, Mehmet Aydın, Rıza Kanber, and Erhan Akça.

Kyoto, Japan   Dr. Tsugihiro Watanabe
March 2018   Guest Editor of *Climate Change*
Impacts on Basin Agro-ecosystems
Professor, Graduate School of Global Environmental
Studies, Kyoto University
The Principal Investigator of the ICCAP

Acknowledgements

The guest editors would like to express their utmost gratitude to PD Dr. Hans Günter Brauch, an independent editor of five book series published by Springer Nature, for his everlasting support and the unlimited hours of work he devoted to enhancing the quality of this book. Dr. Vanessa Greatorex is highly acknowledged for helping to make each chapter understandable to learners of all levels by refining the English.

Contents

Chapter 1
An Integrated Approach to Climate Change Impact Assessment on Basin Hydrology and Agriculture

Tsugihiro Watanabe, Takanori Nagano, Rıza Kanber and Selim Kapur

Abstract Climate change, including changes in air temperature and precipitation, would affect the basin hydrological regime, and the change in the hydrological system might have some impacts on agriculture. To assess the impacts of climate change on the hydrology and agriculture of a basin, the relationship between climate and basin hydrology, and between hydrology and the agriculture of the basin need to be analysed systemically and integrally. In this study, an integrated approach to these analyses or diagnoses is developed for a better projection and evaluation of the climate change impacts as the foundation for better adaptation. It takes into account various complicated factors in these relationships, which are often uncertain and affect each other and must therefore be treated in a particular way.

Consequently, an integrated approach to the issues in question was developed and applied to a large-scale research project, which is the ICCAP Project of RIHN, Japan. The primary aim of the approach and the project was to integrate the concept and processes of the integration approach, which are outlined together with their application in the case study. The methodologies and major objectives of the integrated approach as the core task of the project were:

(1) To diagnose the structure of land and water management in agricultural production systems in a basin, and especially to evaluate quantitatively the relationship between cropping systems, the hydrological cycle, and water balance in farmland and its environs.

(2) To develop the methodology or model for integrated assessment of impacts of climate change and adaptations to it on agricultural production systems, mainly agricultural land and water management.

T. Watanabe, Professor, Kyoto University, Graduate School of Global Environmental Studies, Yoshida-Honmachi, Sakyo-ku, Kyoto 606-8501, Japan; e-mail: nabe@kais.kyoto-u.ac.jp.

T. Nagano, Associate Professor, Kobe University, Graduate School of Agricultural Science, 1-1 Rokkodai, Nada-ku, Kobe, Japan; e-mail: naganot@ruby.kobe-u.ac.jp.

R. Kanber, Retired Professor, Çukurova University, Department of Agricultural Structures and Irrigation Engineering, Adana, Turkey; e-mail: kanber@cu.edu.tr.

S. Kapur, Retired Professor, Çukurova University, Department of Soil Science and Plant Nutrition, Adana, Turkey; e-mail: kapurs@cu.edu.tr.

© Springer Nature Switzerland AG 2019
T. Watanabe et al. (eds.), *Climate Change Impacts on Basin Agro-ecosystems*,
The Anthropocene: Politik—Economics—Society—Science 18,
https://doi.org/10.1007/978-3-030-01036-2_1

(3) To develop and improve the Regional Climate Model (RCM) for more accurate projection with higher resolution of future changes in a regional climate, as the base for better assessment of climate change impacts.
(4) To assess the vulnerability of agricultural production systems to natural change and to suggest possible and effective measures for enhancing the sustainability of agriculture through the integrated assessment of climate change impacts.

Keywords Adaptation · Agricultural production · Climate change Impact assessment · Regional climate · Turkey

1.1 Introduction

The Assessment Report of IPCC WGII released in 2014 clearly identified the future risk of "loss of rural livelihood and income due to insufficient access to drinking and irrigation water and reduced agricultural productivity, particularly for farmers and pastoralists with minimal capital in semi-arid regions" (IPCC 2014). This is one of eight key risks that have been identified with a high degree of confidence, spanning sectors and regions. Some fundamental questions are, however, raised and they include what exact impacts of climate change may manifest on irrigation and agricultural systems, how the systems can adapt to the changes, and what measures should be applied to sustain productivity.

Identifying the direction and dimension of potential impacts on irrigation and agricultural production systems, based on the projection of future regional climate change and consequent hydrological changes, is an essential process to adapt to them. The current structure and problems of the systems are also elucidated through analysing the impacts of climate change and developing adaptation measures.

To assess the climate change impacts on basin agriculture and its water use as its base, the analyses of the relationship between the climate and basin hydrology regime with water resources availability, and between water management and agricultural production of the basin are essential. These analyses or diagnoses involve various factors, which are often uncertain and inter-dependent. Therefore, integrated assessment needed to be developed for better projection and evaluation of climate change impacts as the basis for better adaptation.

While the "integrated assessment" is easier to establish in theory, it is, however, actually very difficult to develop and implement, since the behaviour and future statuses of the involved factors and players affecting each other are difficult to project in detail. Since this is a very implicit system, step-wise approaches with scenario-based projections are acceptable. These scenario-generating processes are useful when assessing the vulnerability of the present system.

In this chapter, first the general flow of this stepwise approach with scenario-based projection is outlined. Then, by way of example, a case study is introduced, this being the research project ICCAP that the authors of this chapter coordinated. Although the ICCAP was completed in 2007 and the authors have introduced this approach and outcomes of the project in several instances in the past (RIHN 2007; Watanabe 2012; Watanabe/Nagano 2014), its challenges are still considered novel and have not been elaborated in the context above. In this chapter, some additional information and modified points are highlight further developments in the study area. The main contents of this chapter are quoted from the presentation paper published in the proceedings of the 22nd ICID Congress (Watanabe/Nagano 2014).

Most of the following chapters of this book explain the details of its challenges in each part of the integrated approach overviewed here.

1.2 Step-Wise Integrated Approach to Climate Change Impacts Assessment

Figure 1.1 depicts the main flow of the step-wise integrated approach to climate change impacts assessment, including the adaptation measures evaluation. Hereafter each step, shown in the boxes of 'A' to 'I', is introduced briefly. The exact detailed processes or procedures are explained in the following case study section.

1.2.1 Step A

To project and assess future changes in water use and agriculture caused by climate change due to global warming, the target period or year in the future needs to be decided first and the future climate should be projected thereafter. Generally, future climate scenarios are best generated by outputs of *General Circulation Models* (GCMs) and/or outputs of Regional Climate Models (RCMs) nested with or downscaled from the GCM outputs. Then, future climate factors, including air temperature, humidity, radiation, wind, and so on, are provided as the basis of impact assessment. Taking the uncertainty of climate models into account, it is recommended that at least two models are used.

1.2.2 Step B

The basin hydrology model can simulate the future hydrological regime of the basin, with meteorological parameters provided by the climate scenarios. The outputs of the distributed process model include distribution of precipitation, evapotranspiration, seepage and other hydrological processes in each part of the basin, and finally river discharge at any point in the water system. Given the

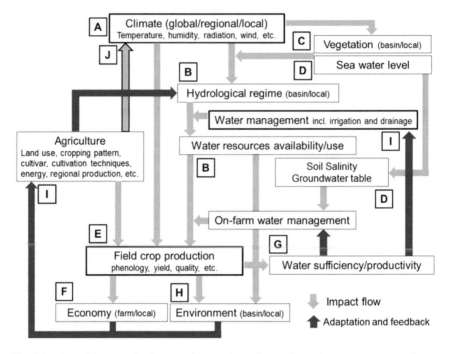

Fig. 1.1 Flow of the step-wise integrated approach to climate change impacts assessment. *Source* The authors

infrastructures for the management of water resources, such as reservoirs, ponds, diversion works and canals, and their operation rules, changes in the availability of water resources are predicted.

1.2.3 Step C

The vegetation model enables future land cover, especially vegetation, to be predicted by the projected meteorological parameters, which include future changes in the distribution of forests and species of trees and grasses. The transition of the vegetation, especially forest vegetation, takes longer and it should be noted that the future possible vegetation with future climate parameters is unlikely to be realised during the target period for which the future climate is generated. The change in vegetation should be reflected in the basin model.

1.2.4 Step D

When the target area is located in a coastal zone, the future sea level is generally projected simultaneously by the future climate. The projected sea level's impacts on the basin hydrological regime, including groundwater dynamics with seawater intrusion into the coastal region, are estimated. At this point the special groundwater dynamics model can be utilised in conjunction with the surface hydrology model. These models can provide predictions on groundwater availability and salinity.

1.2.5 Step E

Changes in crop growth and production in the fields are projected by many available crop models developed by many research organisations. In addition to the parameters of crop phenology, these projections require meteorological parameters and conditions of soil and water in the field, which might be provided in by steps A to D. In the initial phase this step assumes that a single crop is under cultivation in the current situation, since the future cropping pattern, which is affected by many un-redictable factors beyond the basin or region, cannot be projected.

1.2.6 Step F

Changes in crop production in the field affect farm economy and local/regional economics, leading to the alteration of cultivars, crops and cropping patterns. The changes in economics are estimated by specific micro and macro models developed in the field of agro-economics.

1.2.7 Step G

Comparing the future water availability provided by step B and the future water requirements calculated in step E may indicate possible water shortages in the water balance in the fields. For this estimation, any suitable model on field-level soil and water dynamics can be introduced. These changes in water productivity might be the basis for modified water management.

1.2.8 Step H

Future changes in, basin hydrology and water resources management, field conditions of soil, water and crop, and basin vegetation, might cause some environmental problems, including degradation of water quality, the eco-system and biodiversity. These changes need to be predicted. And, not only future changes but also present problems need to be diagnosed, since some measures to solve them need to be introduced when the adaptation measures to climate change are developed.

1.2.9 Step I

This step is for impact assessment with adaptation measures. With the outputs of steps A to H, the impacts of future climate change are projected. Here, the impacts do not mean just changes, but imply adverse effects with some damage. Assessment also implies not only prediction but also evaluation of the nature and extent of the problems.

To improve the future possible degraded situation, many alternatives will be proposed and compared before selection, since limited time and resources make it unfeasible to test all. Both modest and challenging adaptations should be used to predict the most likely or serious cases; in addition even undesirable cases could be tested, to identify what the stakeholders should be prepared for.

For example, in the case of a future water shortage prediction, expansion of the irrigation scheme may be proposed, and it would be considered undesirable for farmers wanting increased profits to introduce high water-consuming cash crops in the predicted conditions. In such a scenario, the construction of additional reservoirs or the introduction of improved management practices for efficient use are other options.

In this step, changes in land use and cropping patterns are the most usual options. Through these sets of counter measures and options, some adaptation scenarios are generated, as the assumption of future basin and agriculture.

1.2.10 Step J

This is not an adaptation process that can be recognised as the feedback of the basin to the climate system. Future conditions of the basin land use and cover, and the hydrological regime including the water content of the soil profile and evapotranspiration, affect the regional climate system. In most of the past and on-going impact assessments, these feedbacks are not involved, and are expected in the future with feasible conditions.

1.3   Trial Approach of the Integrated Impacts Assessment

1.3.1   General Overview

The step-wise integrated approach to climate change impacts assessment was tested in one of the research projects of Japan's Research Institute for Humanity and Nature (RIHN). The project was titled "The Impact of Climate Change on Agricultural Production System in Arid Areas (ICCAP)" and implemented from 2002 to 2007 in the case study area of the Seyhan River Basin in Turkey. Overviews of its concept, methodology and outputs have been previously published (RIHN 2007; Watanabe 2012), and in this chapter the processes of approach explained above are summarised according to their steps.

The main objectives of the project were:

i. To examine and diagnose the structure of land and water management in agricultural production systems in arid areas, and especially to evaluate quantitatively the relationship between cropping systems and the hydrological cycle and water balance in farmland and its environs.

ii. To develop a methodology or model for integrated assessment of impacts of climate change and adaptations on agricultural production systems, mainly on the aspect of land and water management.

iii. To assist the development and improvement of the Regional Climate Model (RCM) for more accurate prediction with higher resolution of future changes in regional climate.

iv. To assess the vulnerability of agricultural production systems to natural changes and to suggest possible and effective measures for enhancing the sustainability of agriculture through integrated impact and adaptive assessment of climate change.

1.3.2   Case Study Area: The Seyhan River Basin of Turkey

The research was implemented on the east coast of the Mediterranean Sea, mainly in the Seyhan River Basin in Turkey. Firstly, a comprehensive assessment of the basic and present structure of the agricultural production system was conducted with special reference to regional climate, land and water use, cropping pattern, and irrigation systems.

The Seyhan River Basin (Fig. 1.2) is dominated by a Mediterranean climate with winter precipitation. Rain-fed wheat is widespread in the upper hilly areas of the basin. Large-scale irrigated agriculture extends throughout the lower delta, where maize, citrus fruit, cotton, wheat, and vegetables are cultivated. These crops depend on water supplied by reservoirs that store the runoff of winter rain and snow in the upper mountainous areas.

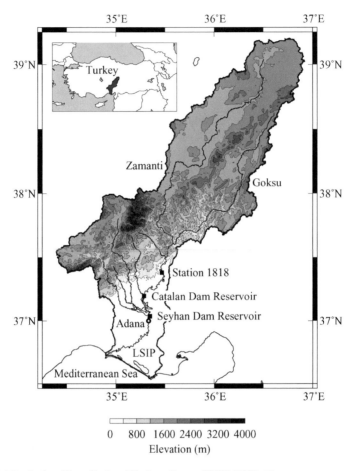

Fig. 1.2 The Seyhan River Basin of Turkey. *Source* RIHN (2007: 3)

1.3.3 Climate Change Scenarios: Step A

The future climate change scenarios of the basin in the 2070s were generated by the two most advanced GCMs and one RCM with downscaling methods based on the A2 scenarios of the Special Reports on Emissions Scenarios by IPCC. Here, the A2 scenarios were used because the future considerable changes to the climate are assumed to predict many substantial impacts for identifying future problems. The outputs of the GCMs were downscaled for a ten-year climate scenario during the 2070s by 25 km grid intervals across the whole of Turkey and by 8.3 km grid intervals in the area covering the Seyhan River Basin. Two independent GCM projections were downscaled by only one RCM. Another GCM with a very high horizontal resolution was used as a reference to assess the reliability of the downscaling done in this research.

In the ICCAP, downscaling the outputs of the GCMs by the RCM was applied according to the state-of-the-art technology at that time for generating the local climate change scenarios. If finer resolution of local climate change is available, it is desirable to use it.

In the ICCAP project, a unique and innovative technique was developed for generation of the future local climate scenario to reduce the bias of the GCM outputs. It is called "pseudo global warming method" and its details are provided in the papers by Dr. Fujio Kimura and others, who developed and applied this practical solution (Sato et al. 2006; Kimura et al. 2007).

According to the generated scenarios, the surface temperature in Turkey may increase by 2.0 °C (projected using the GCM developed by the Meteorological Research Institute of Japan: MRI-GCM) and 3.5 °C (projected using the GCM developed by the Center for Climate System Research of University of Tokyo and the National Institute of Environmental Studies of Japan: CCSR/NIES-GCM). The total precipitation in Turkey may decrease by about 20% except in the summer. The projected trend of changes in temperature and precipitation in the Seyhan River Basin is similar to the changes in the whole of Turkey, while precipitation may decrease by about 25% (Fig. 1.3).

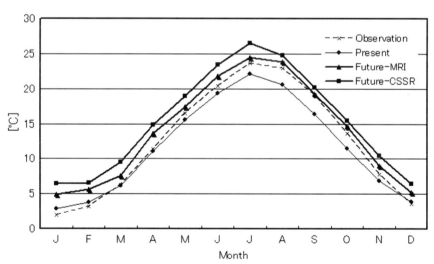

MRI: Meteorological Research Institute of Japan, CSSR: Center for Climate System Research, University of Tokyo

Fig. 1.3 Changes in monthly average temperatures in the Seyhan River Basin in the 2070s. *Source* Kimura et al. (2007: 27)

1.3.4 Hydrological Regime and Water Availability: Steps B, D and H

Precipitation in the basin is projected to decrease by about 170 mm, while evapotranspiration and runoff will decrease by about 40 mm and 110 mm respectively. Because of snowfall decreases and temperature rises, the snow amount will considerably decrease.

Compared to the present conditions, the decreased precipitation may cause a considerable decrease of inflow to the Çatalan and Seyhan reservoirs, in which the peak of monthly inflow might occur earlier than at present. Fewer flood events will occur under the warmer conditions.

The direct impact of future sea-level rise on groundwater salinity will not be serious, while increased evaporation and decreased precipitation with sea-level rise would cause a significant increase in salinity of the lagoon. Therefore, further groundwater withdrawal may result in saltwater intrusion. Build-up of a higher saline zone in the aquifer beneath the lagoon could cause water-logging on the land surface. Water-logging and increased salinity in shallow groundwater may cause salt accumulation on the land surface (Fujinawa et al. 2007).

In the project, environmental issues are not assessed. In the present, however, the river flow downstream of the diversion dam for irrigation is extremely low and would be increased according to the rising demand for conserving the ecosystem of the river in the future. Then, when the future impacts decrease the availability of water stored in reservoirs, the regulation of that so-called e-flow would be secured with the introduction of a guaranteed moderate discharge.

1.3.5 Vegetation: Step C

The actual and potential vegetation of the present were estimated using satellite images and field data. Areas of maquis and woodland with broadleaved evergreen trees of potential present vegetation were in practice occupied by field crops and Pinus brutia as secondary forest respectively. Areas of steppe and maquis will increase in the 2070s, while those of coniferous evergreen forests will decrease. The biomass of maquis and deciduous broadleaved woodlands will increase in the future and coniferous evergreen forests will markedly decrease, while the total biomass in the area will be only 45% of the present one.

1.3.6 Field Crop Production: Step E

Two crop growth simulation models were developed. The models projected that wheat and maize yields in Adana areas may increase at most by 15% from the current yield in the 2070s, whereas the simplified process model (SimWinc) (one of the models used) projected that wheat yields would decrease by 10% if the CO_2 concentrations were not incorporated in the estimate.

The yield estimated by two models suggests that the effect of elevated CO_2 almost offsets the impact of elevated temperature and reduced rainfall on wheat and maize grain yield. The global warming effects on wheat yield in Adana projected by both the wheat growth model and the economic model are around 13%, whereas other wheat growth model projections range between 25 and 37% (Nakagawa et al. 2007).

1.3.7 Economy: Step F and E

The econometric analysis estimated the climate change impacts on the production of wheat and barley and the farmers' economy and behaviour. Changes in crop yield were also predicted, together with the effects of prices, drought, high temperatures, and CO_2 concentrations. Changes in the sown area were predicted, together with the price and soil moisture effects. According to the predictions, the wheat and barley yields will decrease by 18% and 24% respectively in the 2070s with climate change. The area sown with wheat and barley will decrease slightly. Consequently, the total production of wheat will decrease by 3% and the production of barley will decrease by 13%. In addition, estimates in the case that Turkey becomes a member state of the EU show that the yield decrease of both of wheat and barley will be lower than the decrease if Turkey is not be admitted to the EU.

1.3.8 Water Sufficiency and Productivity – Irrigation and Drainage, Hydrological Regime and Water Resources Availability: Step G

The *Irrigation Management Performance Assessment Model* (IMPAM) was developed and validated in this research and was also applied to a small-scale monitored area in the Lower Seyhan Irrigation Project (LSIP). Using the IMPAM, the crop growth and water budget of the whole delta was simulated, and the results have revealed that irrigation demand in the future will increase due to an extended irrigation period. However, the change seems to be within the range of its adaptive capacity.

The water table was more sensitive to the degree of management than to climate change. In general, the risk of a higher water table seems less possible due to a projected decrease in precipitation and water supply. Waterlogging is predicted to partially occur along the coast (Fig. 1.4).

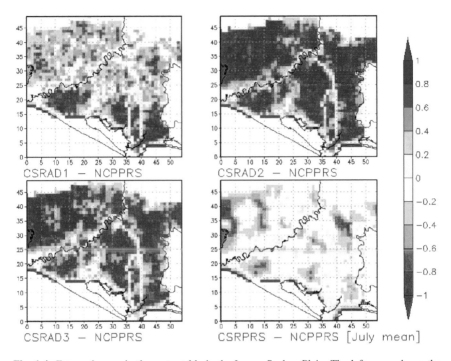

Fig. 1.4 Future changes in the water table in the Lower Seyhan Plain. The left-upper shows the changes with Adaptation 1, the left-lower with Adaptation 2, right-upper with Adaptation 3, and right-lower with the present situation, where the Adaptations are introduced in Step I (*Notes* These cases use CCSR/NIES-GCM). *Source* Hoshikawa et al. (2007: 228–229)

1.3.9 Adaptation in Agriculture and Water Resources Management: Step I

Future crops with future climate and reduced water availability are simulated first, using the expected value-variance (E-V) model. As land use would shift to more cash-generating crops than at present, the citrus area would remain constant at around 20% and, in the case of scarce water supply, watermelons would be grown in the 2070s. Watermelons are usually cultivated only once every five years to avoid replant failure (Umetsu et al. 2007)

Based on the results of impact prediction obtained in the study, the main features of the future conditions include less precipitation and water availability, increased irrigation demand, and expansion of vegetable production. The following three Adaptation Scenarios – No. 1 to 3 – are generated, covering the cropping patterns and the management of land and water use. In these scenarios the future basin conditions are assumed to be continuations of the present land use and water management system as well as the cropping pattern. Moreover, all four cases are simulated with future climate scenarios.

- Adaptation No. 1: Passive and lower investment, with reduction of rain-fed wheat in upstream.
- Adaptation No. 2: Active and higher investment, with expansion of irrigation in mid-stream and lower delta, dam construction, and enhanced irrigation efficiency.
- Adaptation No. 3: Higher investment, with expansion of irrigation in mid-stream and lower delta, and increased groundwater use in the delta.

With these adaptation scenarios, the future situation is projected and the impacts are assessed. The expansion of irrigated land in the middle basin with increased water demand and decreased river flow could lead to water scarcity for the lower plain of the basin, as shown in Fig. 1.5. Here, 'reliability' is defined as "water supply/water demand", that is an indicator to show how much the demand is

Fig. 1.5 Changes in water resources reliability (Top: by MRI-GCM, Bottom: by CCSR/NIES-GCM. Adapt 1 and 2: Adaptation Scenario No. 1 and No. 2). *Source* Fujihara et al. (2007: 94)

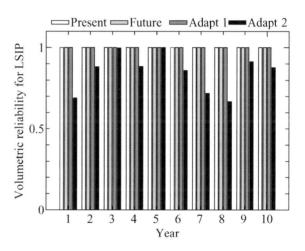

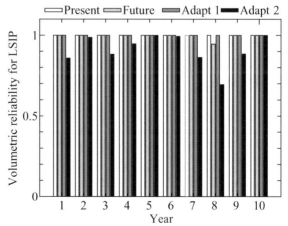

satisfied by supply from the reservoirs. This figure shows that irrigation in the delta region might face water shortages when water use upstream increases in the case of Adaptation Scenarios No. 2 or No. 3.

1.4 Epilogue

Although the ICCAP project, overviewed in this chapter within the context of the step-wise integrated approach, has accomplished model-wise reliable preliminary conclusions, the projection of future changes and prediction of future agriculture and its water use in a very limited area caused by global warming is still difficult to undertake due to the future uncertainties. In this context, prediction for a specific period and area could be considered almost 'impossible', although the accuracy and reliability of such forecasts is rapidly improving. The response of crops to climate change is also still in the basic study stage, even for a major staple crop like wheat, and this applies to water balance as well.

If the phenomena or factors associated with climate change and its apparent impacts are difficult to be projected and evaluated, one of the more effective and feasible measures for adapting to the impacts is to take actions incrementally, as in a trial-and-error manner, utilising the best available current knowledge and past experience, and collecting additional information as needed. In pursuing such an adaptive approach, the step-wise integrated assessment is effective and reliable. It is anticipated that the concept and method explained with these perspectives in this chapter could be applied to different cases and improved with experience and the introduction of advanced techniques.

Acknowledgements Some parts of this paper are contributions from the ICCAP Project promoted by the Research Institute for Humanity and Nature (RIHN) and the Scientific and Technical Research Council of Turkey (TÜBİTAK). This research was financially supported in part by the Japan Society for the Promotion of Science (JSPS), Grant-in-Aid No. 16380164, No. 19208022, No. 24248041 and No. 16H02763.

References

Fujihara Y, Tanaka K, Watanabe T, Kojiri T (2007) Assessing the Impact of Climate Change on the Water Resources of the Seyhan River Basin, Turkey. In the *Final Report of ICCAP*, Research Institute for Humanity and Nature and the Scientific and Technological Research Council of Turkey, 89–94.

Fujinawa K, Tanaka K, Fujihara Y, Kojiri T (2007) The Impacts of Climate Change on the Hydrology and Water Resources of the Seyhan River Basin, Turkey. In the *Final Report of ICCAP*, Research Institute for Humanity and Nature and the Scientific and Technological Research Council of Turkey, 73–79.

Hoshikawa K, Nagano T, Kume T, Watanabe T (2007) Evaluation of Impact of Climate Changes on the Lower Seyhan Irrigation Project, Turkey. In the *Final Report of ICCAP*, Research

Institute for Humanity and Nature and the Scientific and Technological Research Council of Turkey, 221–230.

IPCC (2014) Summary for Policymakers. In: *Climate Change 2014: Impacts, Adaptation, and Vulnerability. Contribution of Working Group II to the Fifth Assessment Report of the Intergovernmental Panel on Climate Change*. Cambridge: Cambridge University Press.

Kimura F, Kitoh A (2007) Downscaling by Pseudo Global Warming Method. In the *Final Report of ICCAP*, Research Institute for Humanity and Nature and the Scientific and Technological Research Council of Turkey, 43–46.

Kimura F, Kitoh A, Sumi A, Asanuma J, Yatagai A (2007) Downscaling of the global warming projections to Turkey. In the *Final Report of ICCAP*, Research Institute for Humanity and Nature and the Scientific and Technological Research Council of Turkey, 21–32.

Nakagawa H, Kobata T, Yano T, Barutçular C, Koç M, Tanaka K, Nagano T, Fujihara Y, Hoshikawa K, Kume T, Watanabe T (2007) Predicting the Impact of Global Warming on Wheat Production in Adana. In the *Final Report of ICCAP*, Research Institute for Humanity and Nature and the Scientific and Technological Research Council of Turkey, 163–168.

RIHN (2007) Final Report of ICCAP (The Research Project "Impact of Climate Change on Agricultural Production System in Arid Areas"), Research Institute for Humanity and Nature (RIHN) Japan; at: http://www.chikyu.ac.jp/iccap/finalreport.htm.

Sato T, Kimura F, Kitoh A (2006) Projection of Global Warming onto Regional Precipitation over Mongolia Using a Regional Climate Model. *Journal of Hydrology* 333:144–154.

Umetsu C, Palanisami K, Coşkun Z, Donma S, Nagano T, Fujihara Y, Tanaka K (2007) Climate Change and Alternative Cropping Patterns in Lower Seyhan Irrigation Project: A Regional Simulation Analysis with MRI-GCM and CSSR-GCM. In the *Final Report of* ICCAP, Research Institute for Humanity and Nature and the Scientific and Technological Research Council of Turkey, 231–243.

Watanabe T (2012) Integrated Approach to Climate Change Impact Assessment on Agricultural Production Systems. In: Anbumozhi V, Breiling M et al (eds) *Climate change in Asia and the Pacific*. Asian Development Bank Institute and SAGE, 138–155.

Watanabe T, Nagano T (2014) Integrated Assessment of Impacts of Climate Change on Basin Hydrology and Water Use in Agriculture. *Proceedings of the 22nd Congress of ICID*, CD-ROM.

Part I
Projections of Future Climate Change

Chapter 2
Climate Change Projection over Turkey with a High-Resolution Atmospheric General Circulation Model

Akio Kitoh

Abstract Future rainfall projections in Turkey and the surrounding regions simulated by two versions of the Meteorological Research Institute atmospheric general circulation model (MRI-AGCM3.2) are presented in this study. Time-slice experiments using a 20 km mesh AGCM were performed both for the present-day and the end of the 21st century. To assess the uncertainty of projections, twelve ensemble projections were also conducted using a 60 km mesh AGCM. High-resolution models reproduce regional details of rainfall in their present-day experiments, showing rainfall maxima along the coastal regions of the East Mediterranean and the Black Sea, corresponding well to the observations. Large reductions in precipitation are projected across Turkey towards the end of the 21st century. Projected precipitation decrease is higher in the western part of Turkey than the interior of the country. Precipitation decrease is high during spring and summer in the western part of Turkey and during winter and spring in the Adana region. Soil moisture will decrease in every month, with the highest reduction in April and May. Variations in the overall decreasing trend in average rainfall mean that rainfall intensity is projected to decrease for the Mediterranean coastal regions, while it is projected to increase over other land regions in Turkey.

Keywords Climate change · High-resolution · MRI-AGCM3 · Precipitation Turkey

2.1 Introduction

According to the multiple atmosphere-ocean general circulation models (AOGCMs) projections under the World Climate Research Program Coupled Model Intercomparison Project phase 3 (CMIP3), annual precipitation is very likely to decrease in most of the Mediterranean (IPCC 2007). Under the Special Report on Emission Scenario (SRES) A1B scenario, the model median annual mean

A. Kitoh, Office of Climate and Environmental Research Promotion, Japan Meteorological Business Support Center, Tsukuba, 305-0052, Japan; email: kitoh@jmbsc.or.jp.

© Springer Nature Switzerland AG 2019
T. Watanabe et al. (eds.), *Climate Change Impacts on Basin Agro-ecosystems*,
The Anthropocene: Politik—Economics—Society—Science 18,
https://doi.org/10.1007/978-3-030-01036-2_2

precipitation over the Mediterranean is projected to decrease by 12%, more (-24%) in the dry summer season and less (-6%) in the wet winter season (IPCC 2007). On the other hand, CMIP5 models project a smaller reduction of 5% in the annual mean precipitation under the Representative Concentration Pathways (RCP) 4.5 scenario over Southern Europe/the Mediterranean region (IPCC 2013). Over the Mediterranean Sea, the precipitation reduction and warming-enhanced evaporation is expected to cause a high loss of fresh water (precipitation minus evaporation). Over land, on the other hand, the evaporation reduction in the dry season counteracts the reduction in precipitation, resulting in a modest decrease in fresh water availability (Mariotti et al. 2008).

The horizontal resolution of the climate models used for these projections (between 400 and 120 km), however, is not sufficient to handle the land-sea coastline and topography adequately. It is anticipated that high-resolution models simulate orographic precipitation much better than lower-resolution models. The increasing ability of computing resources enables us to use climate models with much higher horizontal resolutions. However, running a high-resolution AOGCM for a long period is still difficult in practice. A time-slice experiment with a high-resolution atmospheric general circulation model (AGCM) is a plausible way to obtain future climate change projections with high resolution under reasonable amounts of computing resources (Kitoh et al. 2009).

Kitoh et al. (2008) used the 20 km mesh AGCM (MRI-AGCM3.1) to project the future stream flow in the Fertile Crescent region. They projected that the Fertile Crescent will lose its current shape and may disappear altogether by the end of this century under a high-warming scenario. They also showed that the annual discharge at the Seyhan River region in Turkey will decrease by 39% in one scenario and by 88% in another scenario from the present (1979–2003) to the future (2075–2099). Using the same output, over the Mediterranean region, Jin et al. (2010) projected a decrease of about 10% in the annual precipitation over both land and sea in the future. Evaporation, however, was projected to decrease over land and increase over sea in the Mediterranean area. Thus both the land and sea areas of the Mediterranean will become more arid in the future, and the sea areas will experience even greater decreases in precipitation than the land areas (Jin et al. 2010).

The MRI-AGCM3.1 has been updated and the new version (MRI-AGCM3.2) shows the general improvements in simulating heavy monthly-mean precipitation around the tropical Western Pacific, the global distribution of tropical cyclones, the seasonal march of the East Asian summer monsoon, and the blockings in the Pacific (Mizuta et al. 2012). Kitoh/Arakawa (2011) compared the precipitation reproducibility of MRI-AGCM3.1 and MRI-AGCM3.2 against high-resolution gridded precipitation data sets over the Middle Eastern region as well as the resolution dependency. In this paper, we investigate climate change in terms of precipitation over Turkey and the surrounding regions. The ensemble experiments with the 60 km mesh model version are also performed, with which statistical significance of climate change projections can be discussed.

2.2   GCM and Experiments

The GCM used in this study is MRI-AGCM3.2 (Mizuta et al. 2012). The model has a horizontal spectral triangular truncation of spherical function at wave number 959 with a linear grid for wave-to-grid transformation corresponding to about 20 km horizontal grid spacing, and has 64 layers in the vertical with a top at 0.01 hPa. The model has 1920 grids in longitude and 960 grids in latitude over the global domain. A new cumulus parameterisation scheme called the 'Yoshimura cumulus scheme' is introduced, where both concepts for the Arakawa-Schubert-type schemes and for Tiedtke-type schemes are incorporated (Yoshimura et al. 2015).

The present-day realisation is a so-called 'AMIP'-style simulation, where the observed monthly sea surface temperature (SST) and sea ice concentration (SIC) during 1979–2003 (HadISST1; Rayner et al. 2003) are prescribed as the lower boundary conditions. Mizuta et al. (2012) made verifications of the present-day climate projections, showing improvements in heavy monthly-mean precipitation around the tropical Western Pacific, the seasonal march of the East Asian summer monsoon, and the blockings in the Pacific.

For the future simulations, the targeted projection period is the near future (2015–2039) and the last quarter of the 21st century (2075–2099). For the lower boundary conditions in the near-future and future simulations, changes in SST and SIC were estimated from a CMIP3 multi-model ensemble mean under the SRES A1B scenario, to which detrended interannual variations in the observed SST were added (Mizuta et al. 2008), assuming that interannual variations in SST and SIC in the future simulations are similar to those of the present day. This assumption was made because there is great uncertainty about future changes in SST variability (IPCC 2007, 2013). The design retains observed year-to-year variability and El Niño and Southern Oscillation (ENSO) events in the future climate, but with a higher mean and a clearly increasing trend in SST.

To assess the uncertainty of climate change projections at the end of the 21st century, we performed ensemble simulations with the 60-km resolution model: multi-physics and multi-SST ensemble simulations (Table 2.1). The 60 km mesh model is equipped with multiple cumulus convection schemes. In this study, three cumulus convection schemes were used to develop the multi-physics ensemble simulations. These were the Yoshimura (YS) scheme, the Arakawa-Schubert (AS) scheme, and the Kain–Fritsch (KF) convection scheme. For the multi-SST simulations, in addition to the CMIP3 multi-model ensemble mean SST pattern, three different SST patterns were used. They are created using a cluster analysis, in which normalised tropical SST anomalies derived from the 18 CMIP3 models were grouped to avoid the subjective selection of a single model (Murakami et al. 2012;

Table 2.1 Experimental design. *Source* The author.

(a) Present-day climate simulations for 1979–2003, 25 years

Run name	Cumulus convection scheme	Sea surface temperature	Grid size (km)
SP_YS	Yoshimura (YS)	Observation HadISST	20
HP_YS	Yoshimura (YS)	Observation HadISST	60
HP_AS	Arakawa-Schubert (AS)	Observation HadISST	60
HP_KF	Kain-Fritsch (KF)	Observation HadISST	60

(b) Future climate simulations for 2075–2099, 25 years

Run name	Cumulus convection scheme	Sea surface temperature	Grid size (km)
SF_YS	Yoshimura (YS)	CMIP3[c] MME[d]	20
HF_YS	Yoshimura (YS)	CMIP3 MME	60
HF_YSc1	Yoshimura (YS)	Cluster 1	60
HF_YSc2	Yoshimura (YS)	Cluster 2	60
HF_YSc3	Yoshimura (YS)	Cluster 3	60
HF_AS	Arakawa-Schubert (AS)	CMIP3 MME	60
HF_ASc1	Arakawa-Schubert (AS)	Cluster 1	60
HF_ASc2	Arakawa-Schubert (AS)	Cluster 2	60
HF_ASc3	Arakawa-Schubert (AS)	Cluster 3	60
HF_KF	Kain-Fritsch (KF)	CMIP3 MME	60
HF_KFc1	Kain-Fritsch (KF)	Cluster 1	60
HF_KFc2	Kain-Fritsch (KF)	Cluster 2	60
HF_KFc3	Kain-Fritsch (KF)	Cluster 3	60

[a]First character of run name denotes horizontal grid size: S (Super high, 20 km), H (High, 60 km)
Second character denotes target period: P (Present-day), F (Future)
[b]Observational data by the Hadley Centre of the Met Office, United Kingdom (Rayner et al. 2003)
[c]Coupled Model Inter-Comparison Project phase 3
[d]Multi-Model Ensemble

Endo et al. 2012). Therefore, ultimately, we developed three realisations for the present-day (1979–2003) simulations, and 12 realisations for the future (2075–2099) simulations (Table 2.1).

Figure 2.1 shows the model topography with 161×80 grids and 53×27 grids in the target domain of this paper (20°E–50°E, 30°N–45°N) concerning the 20 km mesh and 60 km mesh models respectively.

2.3 Present-Day Simulation of Precipitation

Figure 2.2a, b, c show the long-term average annual precipitation for Turkey and the surrounding region by three different observation data. For the model validation, spatially high-resolution observed precipitation data is needed. The data used are:

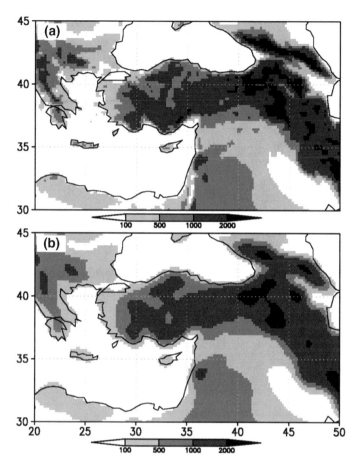

Fig. 2.1 Model topography in metres for **a** 20 km model and **b** 60 km model. *Source* The author

APHRO_ME1101 (APHRO hereafter), which is a daily gridded precipitation for the Middle East, developed by the project of Asian Precipitation – Highly-Resolved Observational Data Integration Towards Evaluation of water resources (APHRODITE) (Yatagai et al. 2009). This covered the study domains of 20°E–65°E to 15°N–45°N with 0.25° resolution during 1951–2007. The other two are CRU_TS2.1 with 0.5° × 0.5° grid from Mitchell/Jones (2005), developed and distributed by the Climatic Research Unit (CRU), University of East Anglia and GPCP_1DD_v1.1 by the Global Precipitation Climatology Project (GPCP) (Huffman/Bolvin 2009) with 1.0° × 1.0° grid. The APHRO_ME1101 and CRU are based on rain gauges, and GPCP is on satellites. In order to create compatibility and comparability with the restricted model data, which was available for 1979–2003, we used APHRO covering the 25-year period of 1979–2003, while CRU and GPCP covered 1979–1998 and 1997–2008 respectively. As our focus in

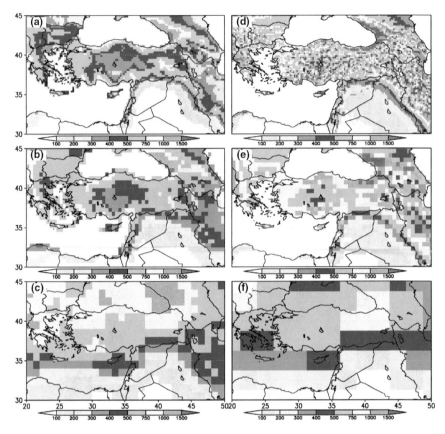

Fig. 2.2 **a–c** Observed and **d–f** simulated annual precipitation in mm for the present-day climatology. **a** APHRO (1979–2003 average), **b** CRU (1979–1998 average), **c** GPCP (1979–2008 average), **d** MRI-AGCM3.2S (20 km), **e** MRI-AGCM3.2H (60 km), **f** CMIP5 model ensemble mean. *Source* The author

model validation was the long-term average precipitation, the different data periods used did not affect the results.

The APHRO (Fig. 2.2a) clearly shows a crescent-shaped precipitation belt with more than 500 mm from Israel and the Adana region in Turkey through the region along the Zagros Mountains. It also shows the precipitation maxima along the Mediterranean and the Black Sea coast, as well as over the Caucasus Mountains. Precipitation in central Turkey (Anatolia) is around 1 mm/day (400 mm in annual total) and is less than that in the coastal area. There are also precipitation minima over the northwestern part of Iran and Azerbaijan. A dry area with precipitation less than 100 mm occupies most of the Arabian Peninsula, northern Africa and the eastern part of Iran.

The CRU (Fig. 2.2b) annual precipitation is overall similar to the APHRO, and shows the precipitation maxima along the Mediterranean and the Black Sea coast and over the Caucasus Mountains, along with a minimum over Anatolia. The CRU data also shows a crescent-shaped precipitation, which however is not so sharp compared to the APHRO data. When compared more precisely with the APHRO, the amount of the CRU precipitation is less over the crescent area, the Black Sea coast and the southern coast of the Caspian Sea. On the other hand, the CRU precipitation is larger than the APHRO over the Arabian Peninsula. The area-averaged annual precipitation for the land region (20°E–50°E, 30°N–45°N) is 447 mm in CRU, which is about 18% higher than APHRO (379 mm). In short, the CRU has more precipitation than the APHRO, while the APHRO has more precipitation over the narrow region of precipitation local maxima. Smaller area-averaged precipitation in APHRO than in CRU may be associated with that of APHRO which uses larger numbers of rain gauge observations compared to CRU mainly over Iran (Kitoh/Arakawa 2011).

The horizontal resolution affects the spatial characteristics of precipitation. Therefore, the GPCP data (Fig. 2.2c) do not show detailed structures compared with the previous two data sets. The GPCP has a hint of a local maximum along the Zagros Mountains, but fails to show a local minimum over Anatolia.

Figure 2.2d, e show the annual precipitation simulated by the model in the present-day conditions and the results obtained by the MRI-AGCM3.2 with two different horizontal resolutions. Consequently, the simulated annual precipitation distribution has yielded consistent results to the observed ones obtained by APHRO and CRU. The precipitation maxima, such as the crescent-type feature along the Zagros Mountains and coastal precipitation along the Mediterranean and the Black Sea coast together with a minimum over Anatolia, are all simulated by both the 20 km and 60 km mesh models. These models, however, tend to show the excessive amounts of precipitation. The CMIP5 model ensemble mean (Fig. 2.2f), however, fails to show the above-mentioned characteristics, except for a broad maximum over the Caucasus Mountains. Kitoh/Arakawa (2011) investigated the simulated precipitation climatology over the Middle East with this model more thoroughly.

2.4 Future Projections

Figure 2.3 shows the projected changes in the annual total precipitation of the near future (2015–2039 mean) and at the end of the 21st century (2075–2099 mean) compared to the present-day climate simulation (1979–2003 mean) by the 20 and 60 km mesh models. The 20 km mesh model projects a similar spatial pattern of changes in the near future (Fig. 2.3a) and the future (Fig. 2.3c), but with wider areas with statistically significant changes in the future than in the near future. The annual mean precipitation is projected to decrease in the Eastern Mediterranean Sea, Greece, western Turkey, Lebanon, Israel, and along the Zagros Mountains. This crescent-shaped belt corresponds to the ancient Fertile Crescent (Kitoh et al. 2008). The annual precipitation is projected to increase in the future in a small area over the Caucasus Mountains and the north-eastern part of Turkey. These projections are

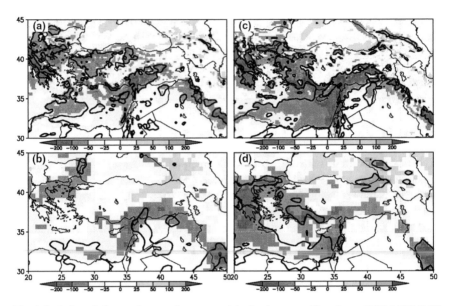

Fig. 2.3 Projected changes in annual mean precipitation in mm. **a** Near future MRI-AGCM3.2S (20 km), **b** near future MRI-AGCM3.2H (60 km), **c** future MRI-AGCM3.2S (20 km), **d** future MRI-AGCM3.2H (60 km). The contour indicates statistically significant at 90% level. *Source* The author

generally consistent with the results obtained by an earlier version of the model (Kitoh et al. 2008).

The 60 km mesh model results (Fig. 2.3b, d) are similar to the corresponding 20 km mesh model results, both in their spatial pattern and the magnitude of the changes. However, differences are found over western Turkey in the near future projections, where the 20 km mesh model projects large decreases in precipitation. Decrease in precipitation in the Fertile Crescent region is projected in both versions, but not significant in the 60 km model. It is also noted that areas with a significant decrease in precipitation are more widely distributed in the higher resolution model.

Figure 2.4 shows the projected changes in seasonal mean precipitation at the end of the 21st century by the 20 km mesh model. The largest decrease in precipitation is projected in the spring (March–April–May; MAM), when a significant precipitation decrease is projected over Greece, western and central Turkey and the Fertile Crescent area from Israel and the Adana region in Turkey through the region along the Zagros Mountains. In the dry summer season (June–July–August; JJA), a general tendency in precipitation decrease is foreseen over the whole region, with higher significance in the western part of Turkey. However, in the autumn (September–October–November; SON), there are almost no significant changes in precipitation. In the winter (December–January–February; DJF), decreased precipitation is projected over the Eastern Mediterranean Sea and the Adana region in Turkey. By contrast, increased precipitation is projected over the Caucasus

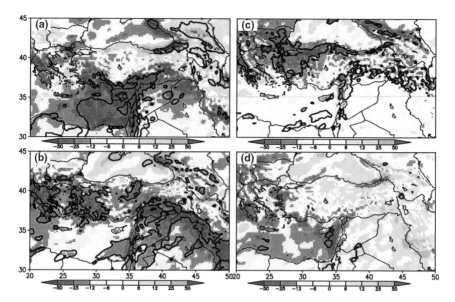

Fig. 2.4 Projected changes in seasonal mean precipitation in mm by MRI-AGCM3.2S (20 km). **a** DJF, **b** MAM, **c** JJA, **d** SON. *Source* The author

Mountains and mountain regions of north-eastern Turkey. In summary, the decreasing precipitation is projected during spring and summer for the western part of Turkey and during the winter and spring for the Adana region.

Warmed sea surface temperature in the future results in increased evaporation from the ocean (Fig. 2.5a). Over land, changes in evaporation are not significant except over narrow mountain areas from eastern Turkey through the Zagros Mountains, and over the Caucasus Mountains, where precipitation significantly decreases. Moreover, there are hints of a decrease in evaporation over the desert areas.

Figure 2.5b shows the projected changes in precipitation minus evaporation, which is derived from Figs. 2.3c and 2.5a. Over the Fertile Crescent area, including the south-eastern part of Turkey, this value is significantly negative, resulting in less surface runoff (Fig. 2.5c) and less soil moisture (Fig. 2.5d) in the future. The same applies to western Turkey and Greece, where the decreasing precipitation is responsible for less surface runoff and less soil moisture. Over the Caucasus Mountains, the increasing evaporation nullifies the increasing precipitation, and subsequently changes in runoff and soil moisture become smaller.

Figure 2.6 shows the seasonal cycle of monthly mean precipitation, evaporation, runoff and soil moisture over land area for 25°E–45°E, 35°N–42.5°N. In this region, there is a large seasonal cycle in precipitation, with its maxima during November–April and its minima during July–September. Evaporation has a different seasonal cycle to that of precipitation, with its peak in May–June, when soil is wet and solar radiation becomes greater. After July, the decreasing soil moisture

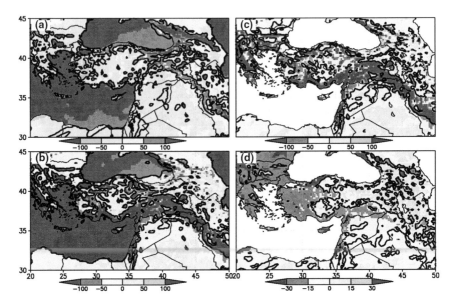

Fig. 2.5 Projected changes in annual means by MRI-AGCM3.2S (20 km). **a** Evaporation in mm, **b** precipitation minus evaporation in mm, **c** surface runoff in mm, **d** total soil moisture in kg m^{-2}. Contours denote statistically significant grid points (at the 5% level). *Source* The author

limits the actual evaporation from the surface. The magnitude of the seasonal cycle of evaporation is less than that of precipitation. Therefore, precipitation minus evaporation is positive during October–April and negative during May–September in the present-day climate. Runoff has a maximum in April, when the surface is the wettest. Runoff becomes negligible in the late summer season. The soil moisture has its maximum in April and its minimum in September, reflecting the seasonal cycle in precipitation minus evaporation.

In the future climate, a reduction in precipitation is projected from December to August. Evaporation will increase almost all the year. Soil moisture decreases in the future climate, but warmer surface temperature results in an increase in surface evaporation. Evaporation is projected to decrease in June–July. This is probably related to a larger decrease in soil moisture just before the month, i.e., May–June. Change in precipitation minus evaporation, thus water availability, becomes negative throughout the year. The season when precipitation minus evaporation is positive becomes shorter from seven months (October–April) in the present to five months (November–March) in the future. The runoff will decrease in the future for every month. It is noted that accumulated changes in precipitation, evaporation and runoff will cause shifts in soil moisture. A large reduction in runoff is simulated during April and May, followed by a soil moisture decrease in later months. There is a notable shift in seasonality of runoff and soil moisture; their peak season arrives earlier in the future climate.

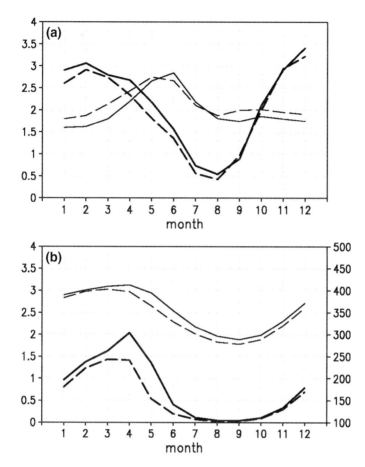

Fig. 2.6 a Seasonal cycle of monthly mean precipitation (thick lines) and evaporation (thin lines) in mm d^{-1} averaged over land area for 25°E–45°E, 35°N–42.5°N. Solid line: present. Dashed lines: future. **b** As in (**a**) but for runoff (thick lines) in mm d^{-1} (left axis) and total soil moisture (thin lines) in kg m^{-2} (right axis). *Source* The author

The next task of this paper was to investigate the changes in precipitation extremes indices. We used the 60-km model ensemble results to study the extremes, due to the insufficient number of years in the 20-km model. Figure 2.7 shows the projected changes in precipitation extremes indices at the end of the 21st century by the 60 km mesh MRI-AGCM3.2H ensemble experiments. Here we show changes in the average precipitation (Pav), the simple precipitation daily intensity index (SDII), the maximum five-day precipitation total (R5d), and the maximum consecutive dry days (CDD). Here SDII is defined as the total precipitation divided by the number of days with precipitation greater than or equal to 1 mm. R5d is defined as the maximum precipitation total in five consecutive days. CDD is defined as the maximum number of consecutive dry days with precipitation less than 1 mm.

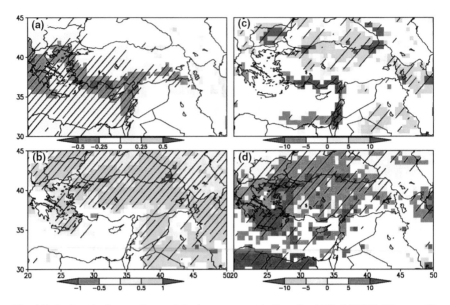

Fig. 2.7 Projected changes in precipitation extremes indices by MRI-AGCM3.2H ensemble experiments. **a** Average precipitation (Pav), **b** simple precipitation daily intensity index (SDII), **c** maximum five-day precipitation total (R5d), and **d** maximum consecutive dry days (CDD). Statistically significant grid points (at the 5% level) are shaded in colour, and grid points where all 12 (≥ 10) experiments show the same sign of changes are closely (widely) hatched. *Source* The author

Changes in the average precipitation by the 60-km model ensemble (Fig. 2.7a) are similar to those by the single 60-km model run (Fig. 2.3d). However, ensemble experiments allow us to show larger areas with significantly decreased precipitation over the Mediterranean Sea, Greece, western Turkey, Israel, Lebanon, and along the Zagros Mountains. The areal extent of the statistically significant regions corresponds to that in a single 20-km model simulation (Fig. 2.3c). Being different with an overall decreasing trend in average precipitation over these regions, the SDII is projected to increase over almost all land regions except Israel and some Mediterranean coastal regions. The heavy rainfall (R5d, Fig. 2.7c) is projected to increase over the Black Sea coastal regions and the mountain region of north-western Iran. R5d will decrease over Israel, Lebanon and southern coastal Turkey. Over western Turkey, the average precipitation will decrease (Fig. 2.7a) but the SDII will increase (Fig. 2.7b), implying a decrease in the number of wet days. Figure 2.7d shows that there will be a large increase in CDD in western Turkey, where the future climate would be characterised by a longer dry period but with a more intense rainfall once it occurs. Alpert et al. (2002) showed an increase of observed extreme daily rainfall over the Mediterranean region in spite of the decrease in the total rainfall during 1951–1995. Model results are inconsistent with this observed trend of the late 20th century, implying for need for further studies to examine this trend in the future.

2.5 Concluding Remarks

The 20 km and 60 km mesh global AGCM simulations are performed for the present (1979–2003) and future (2075–2099) conditions respectively. Because the horizontal resolution of our model is much higher than that of the ordinary AOGCMs, the synoptic scale atmospheric circulations are very well simulated, together with orographic precipitation.

Over Turkey, a large reduction in precipitation is projected from December to August. We found some regional and seasonal differences in precipitation changes across Turkey where the projected precipitation decrease is much larger in the western part of the country than the interior. Moreover, the decreasing precipitation is projected during spring and summer in the western part of Turkey and during winter and spring in the Adana region. The soil moisture is projected to decrease for every month, with the largest reductions in May and June. Variations in an overall decreasing trend in the average rainfall mean that the rainfall intensity is projected to increase over almost all land regions in Turkey except the Mediterranean coastal regions.

The combination of the 20 km mesh model single realisation and the 60 km mesh model ensemble experiment allows us to estimate regional level climate change projections in regional details and also the uncertainties of the projections. These high-resolution AGCM results would be very useful for regional climate change assessments, because the global model does not have the artificial lateral boundaries that the regional climate models must use. Also, the time-slice experiment method employed here reduces the AOGCM climate biases in their present-day climate simulations. As the horizontal resolution of 20 km and/or 60 km is comparable to that in the ordinary regional climate model (RCM), the model output from our high-resolution AGCM can be used for further investigation of the climate impact assessment study (Kitoh et al. 2016). Also dynamical downscaling with a few km mesh RCM is possible by using these AGCM data as lateral boundary conditions in regions of interest. Consequently, the methodology and approach used in this paper should be encouraged for use in relevant projections.

Acknowledgements The author acknowledges the Research Project on the Impact of Climate Changes on Agricultural Production System in Arid Areas (ICCAP) conducted by the Research Institute for Humanity and Nature (RIHN) for prompting his interest in climate changes in arid areas. The current work was supported by the SOUSEI Program of the Ministry of Education, Culture, Sports, Science and Technology (MEXT) of Japan.

References

Alpert P, Ben-Gai T, Baharad A, Benjamini Y, Yekutieli D, Colacino M, Diodato L, Rais C, Homar V, Romero R, Michaelides S, Manes A (2002) The paradoxical increase of Mediterranean extreme daily rainfall in spite of decrease in total values. *Geophysical Research Letters* 29(10):1536. https://doi.org/10.1029/2001gl013554.

Endo H, Kitoh A, Ose T, Mizuta R, Kusunoki S (2012) Future changes and uncertainties in Asian precipitation simulated by multiphysics and multi-sea surface temperature ensemble experiments with high-resolution Meteorological Research Institute atmospheric general circulation models (MRI-AGCMs). *Journal of Geophysical Research Atmospheres.* https://doi.org/10.1029/2012jd017874.

Huffman GJ, Bolvin DT (2009) GPCP one-degree daily precipitation data set documentation.

IPCC (2007) Climate Change 2007: *The Physical Science Basis. Contribution of Working Group I to the Fourth Assessment Report of the Intergovernmental Panel on Climate Change*, Cambridge: Cambridge University Press.

IPCC (2013) Climate Change 2013: *The Physical Science Basis. Contribution of Working Group I to the Fifth Assessment Report of the Intergovernmental Panel on Climate Change*, Cambridge: Cambridge University Press.

Jin F, Kitoh K, Alpert P (2010) Water cycle changes over the Mediterranean: a comparison study of a super-high-resolution global model with CMIP3. *Philosophical Transactions of the Royal Society A* 368:5137–5149.

Kitoh A, Arakawa O (2011) Precipitation climatology over the Middle East simulated by the high-resolution MRI-AGCM3. *Global Environmental Research* 15(2):139–146

Kitoh A, Yatagai A, Alpert P (2008) First super-high-resolution model projection that the ancient "Fertile Crescent" will disappear in this century. *Hydrological Research Letters* 2:1–4.

Kitoh A, Ose T, Kurihara K, Kusunoki S, Sugi M, KAKUSHIN Team-3 Modeling Group (2009) Projection of changes in future weather extremes using super-high-resolution global and regional atmospheric models in the KAKUSHIN Program: Results of preliminary experiments. *Hydrological Research Letters* 3:49–53.

Kitoh A, Ose T, Takayabu I (2016) Dynamical downscaling for climate projection with high-resolution MRI AGCM-RCM. *Journal of the Meteorological Society of Japan* 94A:1–16.

Mariotti A, Zeng N, Yoon JH, Artale V, Navarra A, Alpert P, Li LZX (2008) Mediterranean water cycle changes: Transition to drier 21st century conditions in observations and CMIP3 simulations. *Environmental Research Letters* 3:044001. https://doi.org/10.1088/1748-9326/3/4/044001.

Mitchell TD, Jones PD (2005) An improved method of constructing a database of monthly climate observations and associated high-resolution grids. *International Journal of Climatology* 25:693–712.

Mizuta R, Adachi Y, Yukimoto S, Kusunoki S (2008) Estimation of the future distribution of sea surface temperature and sea ice using the CMIP3 multi-model ensemble mean. *Technical Reports of the Meteorological Research Institute* 56:28.

Mizuta R, Yoshimura H, Murakami H, Matsueda M, Endo H, Ose T, Kamiguchi K, Hosaka M, Sugi M, Yukimoto S, Kusunoki S, Kitoh A (2012) Climate simulations using MRI-AGCM3.2 with 20-km grid. *Journal of the Meteorological Society of Japan* 90A:233–258.

Murakami H, Mizuta R, Shindo E (2012) Future changes in tropical cyclone activity projected by multi-physics and multi-SST ensemble experiments using the 60-km-mesh MRI-AGCM. *Climate Dynamics* 39:2569–2584.

Rayner NA, Parker DE, Horton EB, Folland CK, Alexander LV, Rowell DP, Kent EC, Kaplan A (2003) Global analyses of sea surface temperature, sea ice, and night marine air temperature since the late nineteenth century. *Journal of Geophysical Research.* https://doi.org/10.1029/2002jd002670.

Yatagai A, Arakawa O, Kamiguchi K, Kawamoto H, Nodzu MI, Hamada A (2009) A 44-year daily gridded precipitation data set for Asia based on a dense network of rain gauges. *Scientific Online Letters on the Atmosphere* 5:137–140.

Yoshimura H, Mizuta R, Murakami H (2015) A spectral cumulus parameterisation scheme interpolating between two convective updrafts with Semi-Lagrangian calculation of transport by compensatory subsidence. *Monthly Weather Review* 143:597–621.

Chapter 3
Development of Precise Precipitation Data for Assessing the Potential Impacts of Climate Change

Akiyo Yatagai, Vinay Kumar and Tiruvalam N. Krishnamurti

Abstract In this chapter we introduce the rain-gauge-based grid precipitation data APHRODITE, and show an experimental result of applying the synthetic super-ensemble (SSE) method to winter precipitation over the Middle East. As the change in precipitation according to climate variation is essential, in this study we used the precise observational precipitation as well as the outputs of numerical simulations. The APHRODITE precipitation data is widely used for understanding monsoon variability, various downscaling for impact assessment studies of global warming and validating precipitation estimates from satellites and models. Since the rain-gauge products are more accurate than those of satellites and used as 'teacher' data in various situations, APHRODITE is used for the SSE method developed at Florida State University. It is a unique method to combine several model outputs and precise observation data to make the best forecast. We first show the application of SSE to the Middle East area. We used the simulated precipitation of the five coupled general circulation model (CGCM) outputs, which are part of the CMIP5 project. The five models were chosen due to the availability of the APHRODITE model data up to 2007, along with the 10 years of (1997/1998–2006/2007) monthly precipitation (December, January and February) over the Middle East region (20°E–65°E, 15°N–45°N).

For the seasonal climate forecasts, a SSE technique was used and a cross-validation technique was adopted, in which the year to be forecasted was excluded from the calculations for obtaining the regression coefficients. As a result, seasonal forecasts of the Middle East precipitation were considerably improved by the use of APHRODITE rain-gauge-based data. These forecasts are much superior to those from the best model of our suite and ensemble mean. The use of statistical downscaling and SSE for multi-model forecasts of seasonal climate significantly improved precipitation prediction at higher resolution.

A. Yatagai, Professor, Hirosaki University, 3 Bunkyocho, Hirosaki, Aomori 036-8561, Japan; e-mail: yatagai@hirosaki-u.ac.jp.

V. Kumar, Researcher, Florida State University, Department of Earth, Ocean and Atmospheric Science, Tallahassee, FL, United States.

T. N. Krishnamurti, Professor Emeritus, Florida State University, Department of Earth, Ocean and Atmospheric Science, Tallahassee, FL, United States.

© Springer Nature Switzerland AG 2019
T. Watanabe et al. (eds.), *Climate Change Impacts on Basin Agro-ecosystems*,
The Anthropocene: Politik—Economics—Society—Science 18,
https://doi.org/10.1007/978-3-030-01036-2_3

These results demonstrate that high-resolution precipitation data from a dense network of rain gauges is essential for improving seasonal rainfall estimation over the Middle Eastern region. However, unfortunately, SSE does not represent the large-scale decreasing trend pattern, except in the eastern part of Turkey and part of Israel.

Keywords APHRODITE · CMIP5 · Fertile Crescent · Synthetic Super Ensemble

3.1　Introduction

Considering the impact of climate change on human life, the most important meteorological parameter is precipitation. Water resources maintain human life and agricultural production, but on the other hand cause severe disasters. Hence, it is important to understand the change in precipitation according to the variation in climate and its impact on global warming. This leads us to the inevitable use of the outputs of numerical simulations.

However, precipitation forecasting in mountainous regions is a challenging task and not merely a matter of horizontal/vertical resolution, which needs the improvement of model schemes, including parameterisation of the cumulus convection. Further adjustment of simulated precipitation with the observed value is necessary and is generally applied to the daily regional numerical forecast. Despite the recent trend of obtaining higher resolutions of the numerical models (general circulation models (GCMs) and regional climate models (RCMs)), there have not been sufficient precipitation grid data to validate and adjust their forecast precipitation.

Hence, the first author of this chapter started to create precise daily grid precipitation data over the Middle East for the ICCAP project to validate the results of the RCM and the super-high-resolution GCM and to develop a statistical downscaling method (Yatagai et al. 2008). This was the start of the Asian Precipitation – Highly Resolved Observational Data Integration Towards the Evaluation (APHRODITE) of the Water Resources project.

APHRODITE products (Yatagai et al. 2009, 2012) are currently widely used for understanding the Asian monsoon variability and for the various downscaling of the impact assessment studies of global warming. They are also used for improving numerical forecast models, and the monsoon Asia product (APHRO_MA) was used for improving the Asian monsoon seasonal forecast with a multimodal super-ensemble (SE) method (Yatagai et al. 2014) developed by Florida State University.

It is a unique method to combine several model outputs and precise observation data to make the best forecast. However, the SE method has not previously been applied to the Middle East. Hence, we will show the performance of SE with regard to the winter rainfall of the Middle East by using APHRODITE (APHRO_ME) and several climate model outputs that are used for climate simulation, including global warming studies (based on CMIP5 data sets).

We describe APHRODITE in Sect. 3.2, and SE in Sect. 3.3. We then show the climate super-ensemble example over the Middle East in Sect. 3.4. The discussion and summary comments are described in Sect. 3.5.

3.2 APHRODITE Product over the Middle East

Before starting APHRODITE, the East Asia daily precipitation analysis (Xie et al. 2007) was designed to drive hydrological models over the Yellow River basin in China. For such hydrological purposes, the use of mountainous precipitation is crucial. Hence, the orographic effect was considered by employing the Parameter Regression Independent Slope Model (PRISM, Daly et al. 1994) for climatology and applying interpolation of the ratio value to climatology. The early stage of the APHRODITE product for ME, APHRO_ME_V0804 (Yatagai et al. 2008) algorithm was simpler than that of Xie et al. (2007), where accordingly the monthly as well as the daily precipitation data was collected from Turkey, Israel and Iran. Hence, the APHRO_V0804 reproduced orographic rainfall patterns were obtained

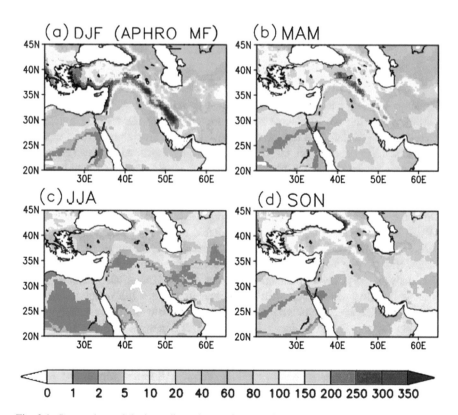

Fig. 3.1 Seasonal precipitation climatology of APHRO_ME_V1101 (unit: mm/3 months). **a** Winter (December–January–February, DJF), **b** Spring (March–April–May, MAM), **c** Summer (June–July–August, JJA), and **d** Autumn (September–October–November, SON). *Source* The authors

along transects from these areas. Ultimately, we have succeeded in showing the arch pattern along the so-called Fertile Crescent (Yatagai et al. 2008), and the APHRO_ME used for validating high-resolution models over the Middle East to diagnose the future hydrological change over the Fertile Crescent region. This area, including Mesopotamia (the lands between the Tigris and the Euphrates rivers) possesses historical merit for the origins of agriculture (Kitoh et al. 2008).

Later, the APHRODITE project created the daily grid precipitation data over the entire Asian domain in relation to world climatology (Hijmans et al. 2005). The following APHRODITE ME products involved the continuously collected rain-gauge precipitation data, the upgraded quality control scheme and the analysis scheme (Hamada et al. 2011; Yatagai et al. 2009, 2012).

Figure 3.1 shows the seasonal precipitation climatology (averaged for 1997/1998–2006/2007 that is used in the later section) of APHRO_ME_V1101. The figure also reveals that the precipitation along the coastal areas and in the mountains (e.g. Zagros) became sharper compared to APHRO_ME_V0804 (Yatagai et al. 2008). The changing water resources over the mountains, which are of utmost importance to the arid and semi-arid regions, can also be analysed by this data. The station data used for creating APHRODITE over the Middle East is given in Yatagai et al. (2008).

3.3 Improvement of the Weather/Climate Models with Observation Data

The super-ensemble technique is a smart way to reach a consistent forecast. In its training phase the model forecasts are regressed during the training period with reliable observed data sets for the same period to obtain weights. The forecast of super-ensemble depends on the quality of weights obtained during the training period and somewhat on the member models' forecast. The spatial and temporal resolution of the APHRODITE data is a unique data set for modellers and fore-casters. The APHRODITE rainfall data captures all the regional, orographic regions and large-scale details of rainfall very precisely. That is the reason why the APHRODITE rainfall data is the most suitable to utilise for extreme rainfalls as well. The APHRODITE data set is used as a benchmark for multimodal in the case of numerical weather prediction (NWP) and also for climate prediction validation and analysis purposes (Krishnamurti/Kumar 2012; Kumar/Krishnamurti 2012). This study calculates various skill scores specifically for the monsoon Asia and Indian region. These results are robust and comparable with other rainfall products available for the Indian region. The APHRODITE rainfall is being used as the benchmark data for NWP models in this study. Here six high-resolution meso-scale models (from WRF-ARW) at 25 km resolution were used to predict rainfall for a duration of 1–6 days. In another study, Tropical Rainfall Measuring Mission (TRMM) rainfall was calibrated against APHRODITE rainfall data over monsoon Asia and significantly improved the seasonal prediction of Asian monsoon

(Krishnamurti et al. 2009; Yatagai et al. 2014). In this study, each model was downscaled against APHRODITE data sets and then those downscaled member models were used for a super-ensemble forecast at the higher resolution of 25 km. This approach was followed to improve the NWP and climate predictions.

3.4 Climate Super-Ensemble Technique for the Middle East

Here we show the application of the climate super-ensemble technique on seasonal precipitation over the Middle East.

3.4.1 Data and Preprocessing

3.4.1.1 APHRODITE Precipitation Data

As a reference for rain-gauge-based precipitation, we used APHRO_ME_V1101 (Yatagai et al. 2012), with 0.25° resolution over the Middle East (APHRO hereafter). Since winter is a rainy season (Fig. 3.1), we only used the 10-year data (1997/1998–2006/2007) of December, January and February (DJF). As a first step, the 30 winter months starting from December 1997 till February 2007 were selected and the monthly mean precipitation was computed. The regions subjected to the super-ensemble computation are the same as those of APHRO_ME_V1101 (20°E–65°E, 15°N–45°N).

3.4.1.2 Model Data

We used the simulated precipitation of the five coupled general circulation model (CGCM) outputs, which are part of the CMIP5 project http://cmip-pcmdi.llnl.gov/cmip5/availability.html/. The five models were chosen because of the availability of the model data till 2007 for use with the APHRODITE available data. Since the CMIP5 'historical' run ends in 2005, we had to use the data of the 'extension' run till 2007. Two of the five models are from the NASA Goddard Institute for Space Studies (GISS) and the others from the Centre National de Recherches Meteorologiques (CNRM), the Meteorological Research Institute of Japan (MRI) and the Norwegian Climate Centre (NorESM). Details of the models and references are summarised in Table 3.1.

The forecast precipitation data from the five GCMs were converted to monthly (D, J, F) precipitation for the 10 years. After attaining the monthly precipitation values, each model result was inserted into the 0.25° grid by bilinear interpolation.

Table 3.1 Characteristics of the five coupled General Circulation Models (GCMs) from CMIP5. *Source* The authors

Institute	AGCM	OGCM	Reference
CNRM	ARPEGE-Climat v5.2	NEMO v3.2	Voldoire et al. (2012)
GISS-H	ModelE2	HYCOM	Schmidt et al. (2013)
GISS-R	ModelE2	Russell ocean model	Schmidt et al. (2013)
MRICGCM	MRI-AGCM3	MRI.COM3	Yukimoto et al. (2012)
NorESM	CAM4	MICOM	Bentsen et al. (2012)

3.4.2   Method of the Synthetic Super-Ensemble Analysis

3.4.2.1   Synthetic Super-Ensemble

As described above, super-ensemble consists of two phases, namely, 'training' and 'forecast.' In the former, a parameter matrix is defined so that the error between multiple models and observation is minimised. The parameter is defined grid by grid and month by month. The slightly modified super-ensemble method (synthetic super-ensemble – SSE) was constructed (Yun et al. 2003; Chakraborty/ Krishnamurti 2009) for the climate because a CGCM does not represent the circulation field in a year. In this method, the expansion of the forecast and observation fields over time was accomplished via the principal component (PC) and spatial empirical orthogonal functions (EOFs). From the 10-year DJF data set, only one year of data was separated for use in forecasting and validation (for example, when Jan 2003 is predicted, Dec 2002, Jan 2003 and Feb 2003 are not used. That means that 27 months of data are used for defining the parameters of the one month).

We computed the correlation, bias, and equitable threat scores (ETS) between the forecast precipitation of the SSE and the independent data set of the training phase.

3.4.2.2   Study Flow

The steps undertaken for the present work are as follows:

1. Coarse resolution precipitation data from four coupled climate models (resolution 2.5°) were bi-linearly interpolated to the 0.25° grid.
2. A cross-validation technique was adopted, in which the year to be forecast was excluded from calculations of the regression coefficients. Coefficients varied spatially and monthly during the study years.
3. These regression coefficients were applied to the forecast year to obtain downscaled model forecasts for that year.

4. The above steps (1–3) were repeated for each year of 1998–2007 to obtain downscaled forecasts of individual models.
5. The final outcome (SSE) was monthly precipitation forecasts on 0.25° grids across the Middle East region, from the four coupled models.
6. To assess the performance of SSE, we also computed ensemble means (EM) of seasonal forecasts from our suite of multiple models. The seasonal precipitation forecasts were validated against the APHRO_ME. Metrics for forecast validation included the standard correlation coefficients, root mean square (RMS) errors, equitable threat scores (ETS) and bias against APHRO_ME.

3.4.3 Results

3.4.3.1 Precipitation Pattern and Time Series

Figure 3.2 shows precipitation patterns of APHRO along with the results of forecasts for winter 2002/2003 (DJF). The super-ensemble and other forecast experiments were executed monthly, and the figure shows the three-month sum. The spatial correlation between APHRO and the 5 models were 0.503, 0.175, 0.094, 0.667 and 0.648, whereas the spatial correlation with the ensemble mean (EM) of the year was 0.546 and with the SSE was 0.924. In this case, the SSE shows an extremely high correlation, and that of the EM (0.546) is high but lower than the best two models (0.667 and 0.648).

Figure 3.3a shows the time series of these correlation coefficients. The same characteristics and the extremely high correlation were obtained throughout the period of 1998–2007. However, the correlation obtained by the EM is less than the best model but in between these successful three models. Similarly, Fig. 3.3b shows the time series of the root mean square error of the models and simulation results against APHRO. The SSE shows a much lower RMS error throughout the period. The RMS of the EM is between those of the top three models. The averaged statistics (Correlation and RMS) for the ten years are summarised in Table 3.2.

Figure 3.4 shows the seasonal (DJF) mean and 10-year averaged results. As summarised in Table 3.2, the correlation coefficients of SSE (h) are very high (0.912), and those of the RMS are very low (44.9). Moreover, the SSE pattern (Fig. 3.4h) resembles that of the APHRO_ME (Fig. 3.4a). In Fig. 3.2, the example of the 2002/2003 forecast shows high correlation, with the overall pattern of the SSE (Fig. 3.2h) also resembling that of the APHRO (Fig. 3.2a). However, the high rainfall intensities observed in the westernmost part of Turkey (28E, 36N) and the westernmost part of Iran (45E, 35N) are not that intensified in the SSE, and such extremes are not outstanding in the average of several years (Fig. 3.4h and Table 3.2).

To evaluate the model forecast according to the precipitation intensity, we calculated the equitable threat score (Krishnamurti/Kumar 2012) and bias according to

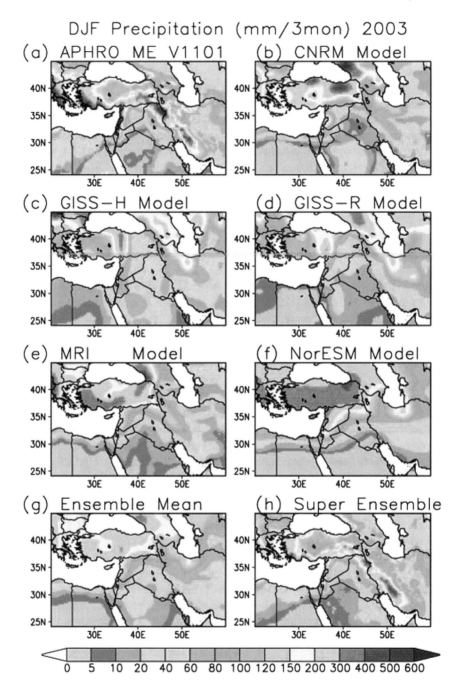

Fig. 3.2 Precipitation patterns in winter 2002/2003 (December 2002, January and February 2003) (unit: mm/3 months). **a** APHRO_ME (observation); **b** model CNRM; **c** model GISS-H; **d** model GISS-R; **e** model MRI; **f** model NorESM; **g** ensemble mean of the five models, and **h** synthetic super-ensemble. *Source* The authors

Fig. 3.3 Time series of the **a** correlation coefficients and **b** the RMS errors (lower) of each model, ensemble mean (EM) and synthetic super-ensemble (SSE) against APHRODITE. *Source* The authors

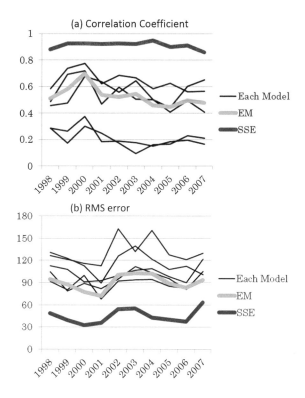

Table 3.2 10-year averaged statistics (correlation coefficients and root mean square errors) of each model and the ensemble mean (EM) and synthetic super-ensemble (SSE) against benchmark APHRODITE data. *Source* The authors

Models	CNRM	GISS-H	GISS-R	MRI	NorESM	EM	SSE
Correlation coefficient	0.566	0.216	0.204	0.641	0.526	0.527	**0.912**
RMS error	94.6	131.1	116.3	88.1	103.0	90.2	**44.9**

the precipitation intensity (threshold). These are computed based on the seasonal precipitation (i.e. 10-year samples) pairs. The ETS of SSE were extremely high compared to those of each model and EM at any threshold values in the amount of precipitation, whereas the bias of the SSE was very low in the low precipitation intensity threshold ($\sim$2.5 mm/day), where the EM demonstrated the lowest bias at a high threshold ($\sim$3 mm/day). For the heavy rainfall case ($\sim$4 mm/day), the EM performs the best and the SSE follows EM and the best two models. The lower ETS of SSE in the heavy seasonal precipitation case was consistent with the above-mentioned example in 2002/2003, which qualitatively (spatially) shows the SSE that represents the heavy precipitation areas but the intensity at a low magnitude. In this context, the heavy precipitation case should be improved for further

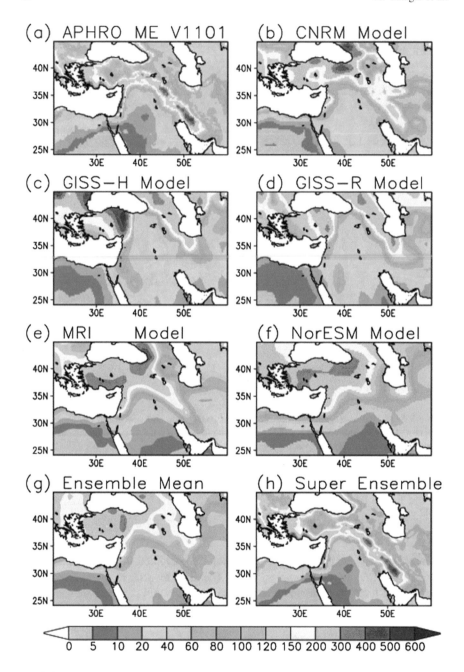

Fig. 3.4 Climatological precipitation pattern in winter (December, January and February) (unit: mm/3 months). **a** APHRODITE; **b** model CNRM; **c** model GISS-H; **d** model GISS-R; **e** model MRI; **f** model NorESM; **g** ensemble mean of the 5 models, and **h** synthetic super-ensemble. *Source* The authors

Fig. 3.5 a Equitable Threat
Score (ETS) and **b** bias of
each model, EM, SSE against
APHRODITE. Unit of
threshold is mm/day. *Source*
The authors

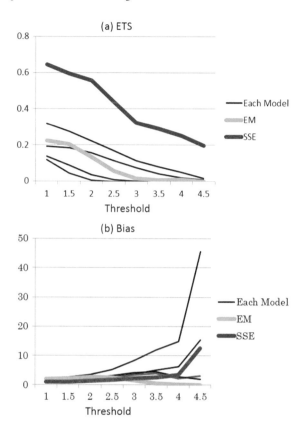

studies. These results demonstrate that high-resolution precipitation data from a
dense network of rain gauges is essential for improving seasonal rainfall estimation
over the Middle Eastern region (Fig. 3.5).

3.4.3.2   Trends

In the last section, we showed an experiment of Synthetic Super-Ensemble (SSE),
using the APHRODITE precipitation data over the Middle East
(APHRO_ME_V1101) and five CMIP5 GCMs. Regarding CMIP5, we used a
historical run, namely the experimental outputs of the observed sea surface temperature (SST) for 1998–2007.

   The linear trend pattern of winter (DJF) precipitation is shown in Fig. 3.6.
APHRODITE shows the decreasing trend over Turkey, east of the Black Sea and
Iraq. Strictly speaking, APHRODITE is not suitable for trend analyses, since input
data differ from year to year and the number of input data is significantly reduced
after 2004 (Yatagai et al. 2008). However, generally speaking, trends of the total
precipitation are not largely affected by the difference in the number of stations

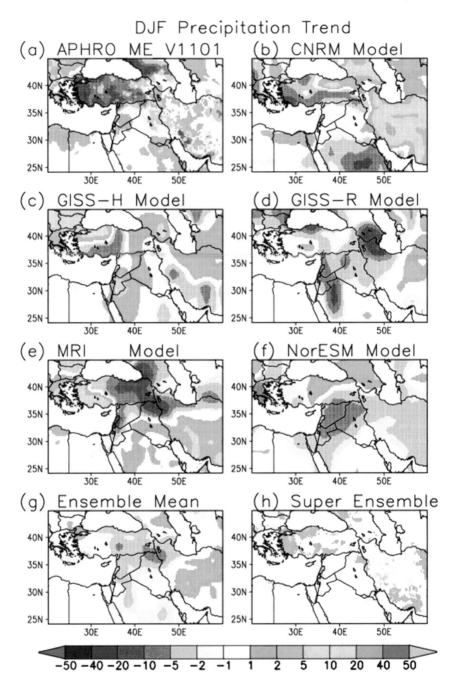

Fig. 3.6 Linear trend of winter (DJF) precipitation. **a** APHRODITE; **b** model CNRM; **c** model GISS-H; **d** model GISS-R; **e** model MRI; **f** model NorESM; **g** ensemble mean of the 5 models, and **h** synthetic super-ensemble. Unit: mm/10-year. *Source* The authors

(Kamiguchi et al. 2010). Here we show the results for comparison with that of the CMIP5. The significant decrease over Turkey is coincident with that of the longer time-scale (1971–2004) (Yatagai 2011).

First, unfortunately, the SSE (Fig. 3.6h) does not represent the large-scale decreasing trend pattern, except for the eastern part of Turkey and part of Israel. The reason for unrepresentativeness is under investigation. The SSE method does not consider the year-to-year variability but simply treats the 10-year DJF precipitation data as 10 samples. We define parameters with 9 samples, and make super-ensemble results for the remaining 1 year.

Conversely, although CMIP5 models have large model-to-model differences in trend representation, the EM (Fig. 3.6g) shows generally similar patterns compared to that of the SSE. The EM shows a decreasing trend pattern over Turkey, though the minimum at Eastern Turkey is not represented. In terms of trend patterns, the EM of historical run performs better than the SSE.

3.5 Discussion

There are many methods to project regional future changes. No method requires the observed and quantitative precipitation facts based on the present 10 to several decades' data. Hence, we introduced the rain-gauge-based grid precipitation data APHRODITE, and showed an experimental result of employing the synthetic super-ensemble method to winter precipitation over the Middle East. The models used here are from CMIP5, which are used for global warming studies. We only used 'historical' runs, namely simulation with observed SST. As a result, winter precipitation over the Middle East is significantly improved. So, we may get a better local-scale projection by applying super-ensemble for future runs of CMIP5 data with the same parameters between each model and APHRODITE.

However, the two remaining current issues to be solved when we apply SE to future runs are that the SSE that showed no meaningful results when the rainfall intensity is strong, and that the SSE does not represent the observed linear trend pattern.

Generally speaking, an ensemble method is not suitable for representing extreme large precipitation. Hence, it is also not ideal to apply the ensemble methods to precipitation over arid/semi-arid regions, since the atmospheric circulation with such events is different from that of the mean state in such a region. Further, a linear trend is affected by the 'first' and 'end' of the year high value. Consequently, the 10 sample data may have not been ideal where the magnitude of the SSE trend (Fig. 3.6h) is generally weak.

The super-ensemble (SE) and the synthetic super-ensemble (SSE) have been developed to minimise forecast errors without accounting for the trend representativeness. With this experimental study, we found that the EM with CMIP5 historical shows better trend patterns, at least over Turkey. Therefore, there is room for improvement in the SSE for localised trend analysis.

Finally, we offer some comments on APHRODITE, having shown the results of the possible use of winter (seasonal total) APHRODITE precipitation over the Middle East. As described in Sect. 3.2, the APHRODITE is a daily product and the algorithm was constructed to obtain a quantitative daily precipitation value to drive hydrological models. Hence, it is suitable to use APHRODITE for land surface models and hydrological models. However, the end-of-a-day (24-h accumulation time) data is different from region to region and caution is necessary when using APHRODITE for daily forecasts and adjustment of satellite precipitation estimates. As the latest IPCC report (IPCC 2013) emphasises the change in extreme events according to global warming, i.e. human-induced climate changes, APHRODITE could be used for the evaluation of extremely large values, but caution is necessary on the following points. Users should be vigilant about (1) whether the rain-gauge is used at the grid box or not, (2) the algorithm, especially for re-gridding, (3) the fact that different end-of-a-day data are sometimes used in a domain (grid), and (4) identifying "no rain" areas.

References

Bentsen M, Bethke I, Debernard JB, Iversen T, Kirkevåg A, Seland Ø, Drange H, Roelandt C, Seierstad IA, Hoose C, Kristjánsson JE (2012) The Norwegian Earth System Model, NorESM1-M – Part 1: Description and basic evaluation. *Geoscientific Model Develevelopment Discussions* 5:2843–2931. https://doi.org/10.5194/gmdd-5-2843-2012.

Chakraborty A, Krishnamurti TN (2009) Improving global model precipitation forecasts over India from downscaling and FSU super-ensemble. Part II: Seasonal climate. *Monthly Weather Review* 137:2736–2757.

Daly C, Neilson RP, Phillips DL (1994) A statistical-topographic model for mapping climatological precipitation over mountainous terrain. *Journal of Applied Meteorology* 33:140–158.

Hamada A, Arakawa O, Yatagai A (2011) An automated quality control method for daily rain gauge data. *Global Environmental Research* 15:165–172.

Hijmans RJ, Cameron SE, Parra JL, Jones PG, Jarvis A (2005) Very high resolution interpolated climate surfaces for global land areas. *International Journal of Climatology* 25:1965–1978.

IPCC (2013) Summary for Policymakers. In: Climate Change 2013: *The Physical Science Basis. Contribution of Working Group I to the Fifth Assessment Report of the Intergovernmental Panel on Climate Change* [Stocker, T.F., D. Qin, G.-K. Plattner, M. Tignor, S.K. Allen, J. Boschung, A. Nauels, Y. Xia, V. Bex and P.M. Midgley (eds.)]. Cambridge: Cambridge University Press.

Kamiguchi K, Arakawa O, Kitoh A, Yatagai A, Hamada A, Yasutomi N (2010) Development of APHRO_JP, the first Japanese high-resolution daily precipitation product for more than 100 years. *Hydrological Research Letters* 4:60–64.

Kitoh A, Yatagai A, Alpert P (2008) First super-high-resolution model projection that the ancient "Fertile Crescent" will disappear in this century. *Hydrological Research Letters* 2:1–4.

Krishnamurti TN, Mishra AK, Simon A, Yatagai A (2009) Use of a Dense Rain-gauge Network over India for Improving Blended TRMM Products and Downscaled Weather Models. *Journal of the Meteorological Society of Japan* 87A:393–412. https://doi.org/10.2151/jmsj.87a.393.

Krishnamurti TN, Kumar V (2012) Improved seasonal precipitation forecasts for the Asian Monsoon using a large suite of atmosphere ocean coupled models. *Journal of Climate* 25:65–88.

Kumar V, Krishnamurti TN (2012) Improved seasonal precipitation forecasts for the Asian Monsoon using a large suite of atmosphere ocean coupled models. *Journal of Climate* 25:39–64.

Schmidt GA, Kelley M, Nazarenko L, Ruedy R, Russell GL, Aleinov I, Bauer M, Bauer SE, Bhat MK, Bleck R, Canuto V, Chen YH, Cheng Y, Clune TL, Del Genio A, De Fainchtein R, Faluvegi G, Hansen JE, Healy RJ, Kiang NY, Koch D, Lacis AA, LeGrande AN, Lerner J, Lo KK, Matthews EE, Menon S, Miller RL, Oinas V, Oloso AO, Perlwitz JP, Puma MJ, Putman WM, Rind D, Romanou A, Sato M, Shindell DT, Sun S, Syed RA, Tausnev N, Tsigaridis K, Unger N, Voulgarakis A, Yao MS, Zhang J (2013) Configuration and assessment of the GISS ModelE2 contributions to the CMIP5 archive. *Journal of Advances in Modeling Earth Systems* 6(1):141–184.

Voldoire A, Sanchez-Gomez E, Salas, Mélia D, Decharme B, Cassou C, Sénési S, Valcke S, Beau I, Alias A, Chevallier M, Déqué M, Deshayes J, Douville H, Fernandez E, Madec G, Maisonnave E, Moine MP, Planton S, Saint-Martin D, Szopa S, Tyteca S, Alkama R, Belamari S, Braun A, Coquart L, Chauvin F (2012) The CNRM-CM5.1 global climate model: description and basic evaluation. *Climate Dynamics* 40(9):2091–2121. https://doi.org/10.1007/s00382-011-1259-y.

Xie P, Yatagai A, Chen M, Hayasaka T, Fukushima Y, Liu C, Yang S (2007) A gauge-based analysis of daily precipitation over East Asia. *Journal of Hydrometeorology* 8:607–627.

Yatagai A, Xie P, Alpert P (2008) Development of a daily gridded precipitation data set for the Middle East. *Advances in Geosciences* 12:165–170.

Yatagai A (2011) Trends in orographic rainfall over the Fertile Crescent, Middle East. *Global Environmental Research* 15:147–156.

Yatagai A, Kamiguchi K, Arakawa O, Hamada A, Yasutomi N, Kitoh A (2012) APHRODITE: Constructing a Long-term Daily Gridded Precipitation Dataset for Asia based on a Dense Network of Rain Gauges. *American Meteorological Society* 93:1401–1415. http://dx.doi.org/10.1175/BAMS-D-11-00122.1.

Yatagai A, Krishnamurti TN, Kumar V, Mishra AK, Simon A (2014) Use of APHRODITE rain-gauge-based precipitation and TRMM3B43 products for improving Asian monsoon seasonal precipitation forecasts. *Journal of Climate* 27:1062–1069.

Yukimoto S, Adachi Y, Hosaka M, Sakami T, Yoshimura H, Hirabara M, Tanaka TY, Shindo E, Tsujino H, Deushi M, Mizuta R, Yabu S, Obata A, Nakano H, Koshiro T, Ose T, Kitoh A (2012) A New Global Climate Model of the Meteorological Research Institute: MRI-CGCM3 – Model Description and Basic Performance. *Journal of the Meteorological Society of Japan* 90A:23–64.

Yun WT, Stefanova L, Krishnamurti TN (2003) Improvement of the super-ensemble technique for seasonal forecasts. *Journal of Climate* 16:3834–3840.

Chapter 4
The Atmospheric Moisture Budget over the Eastern Mediterranean Based on the Super-High-Resolution Global Model – Effects of Global Warming at the End of the 21st Century

Pinhas Alpert and Fengjun Jin

Abstract Several reanalysis and model data sets, i.e., ERA-40, CRU and 20-km MRI-GCM, are employed to study the current and future changes in the wet season moisture fields over the Eastern Mediterranean (EM) including Turkey. The changes in moisture fields at the end of the present century, i.e. 2075–2099, are compared to the present period and discussed. It is shown that the very high-resolution 20 km GCM much better represents the current EM precipitation regime. Future projection of moisture fields suggests an increasing evaporation of about 12% and decreasing precipitation of about 7% over the EM at the end of this century. A significant decrease in precipitation was noticed over west Turkey, west Syria, Israel and Lebanon, with values of over 200 mm/wet season. In particular, the famous Fertile Crescent precipitation strip located over the ME also becomes much drier. The total moisture budget, usually expressed by the precipitation minus evaporation (P-E), confirms that a drier scenario is expected for the water body area and most of the coastline countries including southern Turkey. Analysis of the potential mechanism that controls the drying scenario shows that the precipitation recycling does not change between the present and the future. However, the moisture transport patterns over the EM explain the drying as follows. The sub-tropical mean flow of the low troposphere moves the moisture out of these regions, and there are not enough extra moisture sources to compensate in spite of the enhanced evaporation. One major conclusion is that the EM / ME topographic forcing including the physiographical changes effects are dominant. Therefore, high-resolution modelling plays a critical role in the atmospheric processes for this region.

P. Alpert, Professor, senior lecturer, Tel-Aviv University, Department of Geophysics, Tel-Aviv, Israel; e-mail: pinhas@post.tau.ac.il.

F. Jin, Researcher, Xiamen Meteorological Administration, Xiamen, China; e-mail: jfj9999@hotmail.com.

© Springer Nature Switzerland AG 2019
T. Watanabe et al. (eds.), *Climate Change Impacts on Basin Agro-ecosystems*,
The Anthropocene: Politik—Economics—Society—Science 18,
https://doi.org/10.1007/978-3-030-01036-2_4

Keywords East Mediterranean · Evaporation · Global warming
Moisture budget · Precipitation · Super-high-resolution model

4.1 Introduction

The Middle-East (ME), located on the border between the mid-latitudes and sub-tropics, is unique in both its meteorological and climatological aspects, being predominantly a semi-arid to arid region with steep climate gradients. Lack of water is one of the greatest problems in the ME, and it has been and will continue to be a key element that affects social development, people's livelihoods, and even political stability. This problem may become worse under global warming and make the ME extremely vulnerable to any reductions (natural or anthropogenic) in its available surface water, rendering it highly sensitive to changes in climate. The Intergovernmental Panel on Climate Change (IPCC) Fourth Assessment Report (AR4 as well as AR5) suggested that the Eastern Mediterranean (EM) region would become significantly drier under a future climate scenario, with potentially devastating impacts on the population (IPCC 2007). Therefore, it is vital to gain a better understanding of the distribution of the atmospheric moisture budget of this region and the projected changes, especially for the main two components of atmospheric moisture budget, i.e. precipitation (P) and evaporation (E).

The exact mechanism controlling the precipitation in the ME is complex, and precipitation amounts and distributions are largely affected by the topography and land-sea distribution (Özsoy 1981). However, the precipitation regime of the ME has been studied during the past few decades by using different kinds of data sets, such as observation data, reanalysis data and satellite data, as well as climate model data (Alpert et al. 2002, 2008; Mariotti et al. 2002a, b). Some of the studies focus on the synoptic situation for the generation of precipitation, such as that of Zangvil/Druian (1990), who investigated the relationship between the upper air trough and the location of precipitation in Israel. Krichak/Alpert (2005) studied the relationship between EM precipitation and the indices of the East Atlantic West Russia pattern. Other studies deal with the climatology of precipitation; for instance, Alpert et al. (2002) analysed observational databases over several areas of the Mediterranean basin during the 20th century, and concluded that there exists a dominant increase in extreme daily rainfall events together with a slight decrease in total values. Price et al. (1998) investigated the relationship between El Niño and precipitation in Israel by using historical observations. Seager et al. (2007) studied the climate change of south-western North America by using an ensemble regional climate model; their results also suggested that the Mediterranean region will be drier at the end of this century. Some studies focus on the hydrological cycle or the moisture budget components over the Mediterranean region, for example, Mariotti et al. (2002a, b), Jin et al. (2010, 2011), and Jin/Zangvil (2010).

Since its introduction, the climate model has been widely used for both global and regional climate studies, particularly with some high temporal and spatial resolution models. However, the global climate model (GCM) is usually uses a coarse spatial resolution of about 100–300 km; therefore, it cannot capture well the small-scale factors which are crucial for the local climate, particularly over the Mediterranean. On the other hand, the regional climate model (RCM) has relatively fine spatial and temporal resolution compared to the GCM. But, in addition to RCMs being computationally expensive, they also need lateral boundary condition data, which usually come from the GCM in order to drive the RCM. The very high spatial resolution GCM model employed here addresses the disadvantages which exist in both the GCM and RCM. It avoids the problems of the unfit-in-scale of the lateral boundary condition, but can also incorporate interactions between the global scale and the regional scale explicitly. Here, in addition to several traditional data sets, a high-resolution 20 km grid GCM that was developed at the Meteorological Research Institute (MRI) of the Japan Meteorological Agency (JMA) is also employed in order to investigate the current and future precipitation regime in the ME.

4.2   Methods

4.2.1   Data

To study the current precipitation regime of the EM, several data sets have been used here. These include, first, the global time series data set based on rain gauge measurements (land only) from the climate research unit (CRU, for short; Mitchell/Jones 2005). The grid horizontal resolution is 0.5×0.5 degrees, and the time period is available from 1901 to 2002. Second is the European Centre for Medium-range Weather Forecast (ECMWF) reanalysis data set (ERA-40, for short; Kallberg et al. 2004). This data covers the time period from mid-1957 to 2002. Originally, ERA-40 has a spectral representation based on a triangular truncation at wave number 156 or 1.125 degree horizontal resolution using a Gaussian grid (Gibson et al. 1997). However, the spatial resolution of ERA-40 data used in this study is 2.5×2.5 degrees. The third database is from the Israel Meteorological Service (IMS) and based on daily precipitation for several selected observation stations in Israel at different time periods. The fourth database is the MRI's super-high spatial resolution (about 20 km) grid GCM (MRI/JMA GCM), which is a climate-model version of the operational numerical weather prediction model used in the JMA. A detailed description of the model is given in Mizuta et al. (2006). Two runs of the 20 km GCM cover the time periods 1979–2007 for the current/control run and 2075–2099 for the future run. The control run used the observed monthly sea surface temperature (SST) and sea-ice distribution, while the future run used the SST anomalies of the multi-model ensemble projected by the third phase of the Coupled Model Intercomparison Project (CMIP3) under the Special Report on Emission Scenario (SRES) A1B emission scenario. Details of the method are found in Mizuta et al. (2008). The monthly mean precipitation taken from the data sets 1, 2 and 4 is used here, while the daily mean precipitation is also available for data set 4. Since the current 20 km run covers the time period 1979–2007, while the ERA-40 and CRU data are available only until 2002, the time period selected for the current atmospheric moisture budget research is 1979–2002 in order to make all main three data sets overlap the same period.

4.2.2   Research Area and Study Time Period

The study area here covers the main part of the EM and a good part of the ME, and was chosen to be 27°–41° N and 22°–50° E with a total area of about 3.96×10^6 km^2. Also, in order to study the moisture field over the ME, a sub-region within this area was defined by the latitude 30°–37° N and longitude 30°–40° E, with Israel located approximately in the centre of this area. Also, in order to study the orographic precipitation over the mountain area of Turkey (the second part of the current study), a rectangular area of 37°–38° N and 36°–45° E was selected.

The research period for the current climate simulation is 1979–2002, while for the future it is 2075–2099. Since the main rainy season in the EM region is October–April, only these seven months were identified as the wet season, while the rest of the months are referred to as the dry season.

4.2.3   Water Vapour Budget (WVB) Equations and Recycling Ratio

Following Rasmusson (1968, 1971) and Yanai et al. (1973), by ignoring the cloud liquid water, the traditional equation of WVB per unit mass of air can be written as:

$$\frac{\partial q}{\partial t} + \vec{V} \cdot \nabla q + \omega \frac{\partial q}{\partial p} = e - c \qquad (4.1)$$

where q is the specific humidity, p is atmospheric pressure, $\vec{V}$ is the horizontal wind vector, ω is the vertical p-velocity, and e and c are the cloud evaporation and condensation rates per unit mass. For discussion of the importance of cloud liquid water in some specific regions, see e.g. Shay-El et al. (2000). By using the mass continuity equation and vertical integration of Eq. (4.1), the traditional atmospheric moisture budget equation takes the form:

$$\frac{1}{g}\frac{\partial}{\partial t}\int_s^T qdp + \frac{1}{g}\int_s^T \vec{V} \cdot \nabla qdp + \frac{1}{g}q \cdot \nabla\vec{V}dp = E - P \qquad (4.2)$$

where, g is the acceleration of gravity, S and T indicate the earth Surface and an upper (i.e. Top) integration limit respectively, E and P are the surface evaporation and precipitation rates respectively. On the left side of Eq. (4.2), the first term is the time change of atmospheric perceptible water (dPW), and the second and third terms are for the horizontal water vapour advection and the horizontal velocity divergence in the presence of moisture respectively. The second and the third term together form the moisture flux divergence (MFD) term. Using Green's Theorem and ignoring the first storage term due to climatological concern, MFD can be expressed as:

$$MFD = \frac{1}{g}\int_S^T \nabla \cdot q\vec{V}dp = \frac{1}{Ag}\int_S^T \int qv_n \, dl \, dp = \frac{OF}{A} - \frac{IF}{A} \qquad (4.3)$$

where A is the area of the region, v_n is the normal wind component on the region's boundary, dl is a length increment along that boundary, and OF/A and IF/A are the total water vapour outflow from and inflow into the region respectively. The advantage of Eq. (4.3) is that it clearly identifies the externally advected (i.e. IF = InFlow) water vapour, IF/A.

In order to quantify the relative contribution of water vapour originating in local evaporation, E, and the moisture originating from outside the region, IF/A, we have used the Recycling Ratio (R) formula introduced by Zangvil et al. (1992):

$$R = \frac{E}{E + \frac{IF}{A}} \qquad (4.4)$$

The physical meaning of R is that it refers to the process by which a portion of the *precipitated water* that *evapotranspired* from a given area contributes to the precipitation over the same area.

4.3 Results of the Moisture Balance Components and Discussions

4.3.1 Seasonal Precipitation

The average total precipitation for the wet season (Oct–Apr) of the ME and zoomed in on the EM from 1979 to 2002 is given in Fig. 4.1. The zoom over the EM coast is performed to examine the capability of different climatic rainfall fields to capture the very steep and very crucial north-to-south gradient. In general, the under 50 mm precipitation contour line can be clearly defined from these three charts with more or less the same locations. The latitudinal and longitudinal gradient is the most predominant EM precipitation feature. A clear precipitation strip with one peak zone of precipitation approximately located at 37° N forms the famous Fertile Crescent strip due to the topographic rain effects generated by the mountains of Taurus, Elburz and Zagros in this area. However, the peak of the total precipitation of the crescent strip from ERA-40, CRU and 20 km GCM are quite different, with the corresponding values of 500–700, 700–900 and 900–1100 mm respectively. Another maximum of the average total precipitation can also be identified along the eastern and northern coastlines of EM (later referred to as a north-to-south 'tongue' of precipitation), with the amount of precipitation of 350–500, 500–700 and over 1100 for the ERA-40, CRU and 20 km GCM respectively. The zoomed-in ME in Fig. 4.1 (right panels) shows the more detailed distribution of the precipitation in this region, with a sharp eastward decreasing gradient of precipitation that starts from the eastern coastline of the Mediterranean. This gradient can be explicitly defined only in the CRU and 20 km GCM (i.e., lower two panels), and the 20 km GCM shows even sharper patterns, to be discussed later in comparison with rain gauges observations. The 20 km GCM further shows its two centres of peak precipitation in the east and north-coast line of EM, with the value over 1100 mm/season, but the results from CRU and ERA-40 are significantly lower compared to that in the 20 km GCM. Does the over-evaluated amount of precipitation from 20 km GCM reflect the real precipitation regime in this area?

To verify how the average total seasonal precipitation from the three different data sets fits with the observations, six station points were selected from south to north, which make an approximate south-to-north cross-section along the EM coast, and cover the countries of Egypt, Israel, Lebanon and Turkey. The detailed information about these six station points is shown in Table 4.1. The reason why the six selected stations are all located near the coastline is that both the land-sea interaction and the significant change of topography from sea to land have a strong influence on the precipitation regime in the coastal area. It should be noted that we do not expect one-to-one correspondence between observational data and global gridded data because of different spatial representativeness, but we can still get useful information about the crucial meridional gradient of precipitation. Figure 4.2 shows the seasonal averaged total precipitation for the six selected stations. The results from ERA-40 are, as expected due to coarser resolution, significantly underestimated compared to the observation data, except for the stations of Cairo and Beer-Sheva. The ERA-40 rainfall catches less than half of the total precipitation in Tel-Aviv and Beirut. This finding is consistent with Mariotti et al. (2002a), in which several data sets were used to study the hydrological cycle of the Mediterranean. The CRU and the 20 km GCM results show a better estimation, and the 20 km GCM's results are surprisingly close to the observational data. The bias and the standard deviation of errors for each model are shown in Table 4.1 and they further confirm this fact. However, the CRU is unable to reproduce the peak

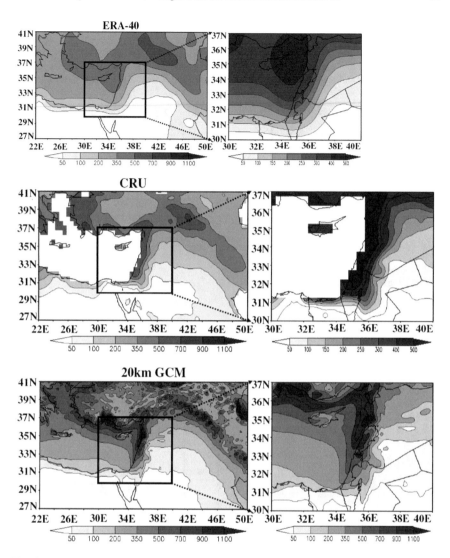

Fig. 4.1 Total seasonal (Oct–Apr) precipitation for the Eastern Mediterranean (EM)+Middle-East (left panel) and zoomed in over the EM (right panel). Averaging time period is 1979–2002. Unit: mm/season. *Source* The authors

precipitation in Beirut, except with an error of over 150 mm. It can be concluded that the 20 km GCM better captures the total amounts of precipitation for the selected six stations.

It should be pointed out that the IPCC AR4 GCM runs (e.g. Evans 2009) cannot simulate the narrow 'tongue' of relatively high precipitation parallel to the south EM coast (Syria, Lebanon and Israel). This is probably due to the insufficient

Table 4.1 Geographic location of the stations used for the models evaluation. Observed total seasonal (Oct–Apr) precipitation is based on sources listed under the table. The two right-most columns, i.e. Bias and SD (E), show the mean bias and the standard deviation of the errors for each model. The accurate values of precipitation were obtained by the same interpolation method (GrADS). For comparison, the ERA-40, CRU and 20 km GCM run are listed in units of mm/season

Items	Cairo	Beer-Sheva	Tel-Aviv	Haifa	Beirut	Adana	Bias	SD (E)
Longitude	31.37°E	34.90°E	34.77°E	34.98°E	35.51°E	35.32°E	–	–
Latitude	30.05°N	31.25°N	32.02°N	32.82°N	33.98°N	37°N	–	–
Observed	26[a]	201[b]	527[b]	534[b]	840[c]	550[c]	0	0
ERA-40	8	165	245	320	380	425	189	167
CRU	21	220	560	520	680	670	1	91
20 km GCM	39	210	480	590	780	550	5	43

[a]Weather Underground
[b]Israel Meteorological Service
[c]Weatherbase.com

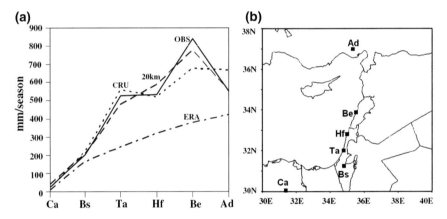

Fig. 4.2 **a** Comparison of average total observed seasonal precipitation with three model data for the selected six stations, which make an approximate south-to-north cross-section along the EM coast. The six stations are, from south-to-north, Egypt—Cairo (Ca); Israel—Beer-Sheva (Bs), Tel-Aviv (Ta), Haifa (Hf); Lebanon—Beirut (Be) and Turkey—Adana (Ad). The three models are the European reanalysis (ERA), the Climate Research Unit (CRU) and the Meteorological Research Institute's 20 km GCM run (20 km). Unit: mm/season. **b** Eastern Mediterranean map indicating the location of the six stations. *Source* The authors

horizontal resolution. Similarly, the rainfall detail along the Fertile Crescent lacks the important topographical features which are so important over this sensitive and water-scarce region, as pointed out by Kitoh et al. (2008).

4.3.2   Monthly Distribution of Precipitation

It is also interesting to know how the super-high-resolution climate model with 20 km captures the monthly mean precipitation over the study area. For this analysis, the state of Israel was selected. The reason for this specific selection was not only because it is a sharp transition zone between hyper-arid and relative humid regions, but also because it is a complicated topographic zone for this small country. Therefore, the region of Israel was arbitrarily divided into three parts, which are the northern, the centre and the southern part (see Fig. 4.3). For each part, two stations were chosen to calculate the monthly mean precipitation based on rain gauge data during the time period of 1979–2002. Figure 4.4 shows that, in general, there is a good agreement between the precipitation from the rain gauge and the 20 km GCM, with correlation coefficients for monthly precipitation varying from 0.97, 0.93 and 0.96 for north, centre and south parts respectively – a statistical significance of 99%. Also, the model credibly describes the dry period from May to August, where only very little precipitation amounts are observed. However, Fig. 4.4 shows that the model underestimated the autumn precipitation, and a larger error can be seen in Jerusalem at the altitude of 750 m, in addition to problems of spatial representativeness, probably due to the fact that the spatial resolution of the model is still not fine enough to accurately describe the Jerusalem orographic rainfall. Shafir/Alpert (1990) highlighted the importance of the very high-resolution orography in Jerusalem. Another model deviation is its overestimation of the precipitation for most of the wet seasons in Elat. However, the absolute quantities of precipitation in Elat are very small (less than 0.2 mm/d). Hence, it can be concluded that the 20 km GCM performs quite well in simulating the current monthly rainfall distribution in the research region.

4.3.3   Orographic Precipitation over the Mountain Area of Turkey

It is important to note that there is a problem in observing rainfall over high mountains due to the lack of rain gauges (Alpert et al. 2011). Moreover, in areas like the high mountains of Turkey perhaps a super-high-resolution model may better represent the rainfall climatology. This is demonstrated in the present section over the Anatolian highlands by gradually increasing the average altitude and comparing the super-high-resolution model rainfall with observations.

Figure 4.5b shows the differences between the 20 km GCM and the CRU data in simulating the orographic rainfall over the high mountainous section of Turkey in both the wet season (Oct–April) and the dry season (May–September); the four domains are indicated in Fig. 4.5a. Figure 4.5b indicates that, over the same domain, the area mean precipitation from the 20 km GCM is higher than that of CRU in both the wet and the dry seasons, with the mean value higher by about 25%

Fig. 4.3 Geographic map
indicating the selected six
stations focused on Israel. The
empty square (□) denotes
northern stations, the cross (+)
the centre and the solid square
(■) the southern part of Israel.
Contour lines show the
topography (m) at 200 m
intervals. *Source* The authors

and 39% respectively. This becomes especially pronounced for the highest altitude
domain of 1,856 m average altitude, where these differences reach the highest
values of about 48 and 46% for the wet and dry seasons. Notice that the absolute
average differences are of about 1.3 and 0.25 mm/day over that inner domain,
which is as large as 3×0.5 degrees (an approximate area of 15,000 km^2).

One major model result is that the wet season area mean precipitation is gradually increasing with the increase in altitude, as can clearly be seen in the 20 km
GCM, but it is not clear at all in the CRU data. This seems to suggest that the
high-resolution model has indeed an essential role in capturing the orographic

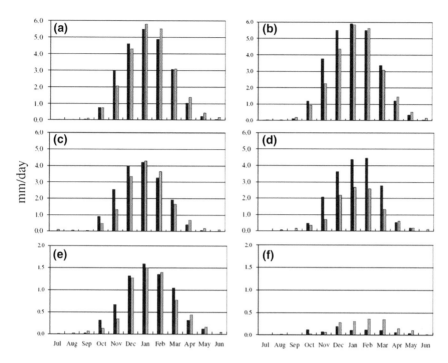

Fig. 4.4 Comparison of monthly mean precipitation from observed rain gauge (black column) and 20 km GCM (grey column) for the selected six stations in Israel, based on their locations. Unit: mm/day. The selected stations are: Northern Israel—**a** Har-Knaan, **b** Eilon; Center Israel—**c** Tel-Aviv, **d** Jerusalem; Southern Israel—**e** Beer-Sheva, **f** Elat. *Source* The authors

rainfall effect which is missed in the observations due to the severe lack of rain-gauges at high altitudes (for further discussion see Alpert et al. 2011).

4.3.4   Evaporation

The evaporation (E) based on the ERA-40 and the 20 km GCM indicates that, as expected, the water body shows larger evaporation values than the land area. The ERA-40 does not show the sharp land-sea boundary as in the 20 km. Three maxima centres of evaporation over the ME are noticed. The Red Sea and the Persian Gulf peaks can be seen both in the ERA-40 and the 20 km GCM. The maximum evaporation over the EM is about 900–1100 mm/season in the ERA-40 compared to over 1100 mm/season in the 20 km GCM. This suggests that the evaporation, E, is underestimated in the reanalysis data over the Mediterranean region, as also suggested by Mariotti et al. (2002a).

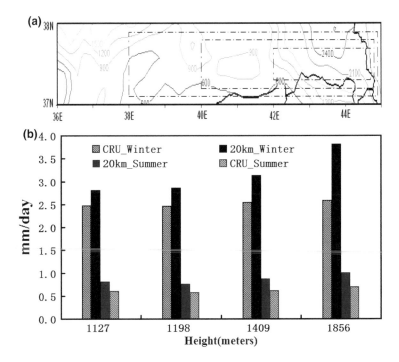

Fig. 4.5 a The altitude map for the selected four domains over Turkey. Unit: metres. **b** Area mean precipitation changes with the mean altitude for the selected four domains in both the wet season (Oct–Apr; denoted winter in the figure) and the dry season (May–September; denoted summer in the figure) for the 20 km control run as well as CRU (1979–2002). Unit: mm/day. The average altitudes of the four domains vary from 1,127 to 1,198, 1,409 and 1,856 m, as indicated by the X-axis. *Source* The authors

4.3.5 End of the 21st Century Predicted Changes in the Moisture Budget Components

Based on Eqs. (4.3) and (4.4), the moisture budget components for both the current and future runs are shown in Table 4.2. The table shows that the seasonal area mean P decreases from 1.34 to 1.24 mm/day, while the E increases from 1.63 to 1.82 mm/day in the future. The changes of E and P lead to the E-P being *doubled* in the future, with the value of 0.58 mm/day, indicating that the research area will become significantly drier. For the moisture flux, both total inflow and outflow are increasing by about 14 and 12% in the future, indicating an enhancement of the hydrological cycle. However, total outflow is larger than total inflow, indicating that the research region is a moisture divergence zone. For the boundary moisture flux, an increasing inflow on both the northern (79%) and southern (43%) boundaries can be seen in Table 4.2. Hence, the region to the north and south will transport more moisture to the research area. The increasing inflow from the abovementioned

Table 4.2 Comparison of the moisture budget components between the present and the future. The changes are calculated by deducting the current amount from the future amount. Except for the Recycling Ratio (%) and changes (%), all other components units are mm/day. *Source* The authors

Components		Current	Future	Changes (%)
Inflow	West	2.22	2.19	−1
	East	0.00	0.00	0
	North	0.19	0.34	79
	South	0.70	1.00	43
Outflow	West	0.00	0.01	0
	East	1.92	2.31	20
	North	0.82	0.66	−20
	South	0.62	0.77	24
Total_INF		3.11	3.53	14
Total_OUTF		3.36	3.75	12
OUT-INF		0.25	0.22	−12
P		1.34	1.24	−7
E		1.63	1.82	12
E-P		0.29	0.58	100
Recycling ratio (%)		34	34	0

boundaries, as well as the increasing E over the Eastern Mediterranean, results in the higher outflow in the east boundary, as the research area is dominated by the easterly flow pattern of the atmospheric circulation. The changes of other boundaries outflow are also consistent with the projected wetter regions in the study area, such as the decreasing north outflow and increasing south outflow.

Figure 4.6 shows the spatial distribution of the E (panel a), P (b) and P-E (c) changes between the future (2075–2099) and the current climate (1979–2007). The E increase is clearly noticed over the water body, with a maximum value of 150–200, 200–250 and over 250 mm/season over the EM, Red Sea and the Persian Gulf respectively (Fig. 4.6a). The maximum E-increases in the EM is located along the northern boundary, reaching the magnitude of 150–200 mm/season. A slight increasing of E over the Fertile Crescent can also be seen. A dramatic E-decrease is found over the Sinai Peninsula, Israel and Jordan, with the maximum value exceeding 100 mm/season.

The precipitation differences show (Fig. 4.6b) that the P of the entire EM is decreasing, with an average value of over 100 mm/season, with maximum P decreases over the northern and eastern coastline areas of the EM with a magnitude of above 250 mm/season. The western part of Turkey and most of the Fertile Crescent are also projected to be drier, as reported also by Kitoh et al. (2008). Figure 4.6b also suggests that the eastern coastline countries, i.e. Israel, Lebanon and the western part of Syria, will become much drier in the future by about 200 mm/season. On the other hand, a precipitation increase belt is found at the most easterly part of research region, including the eastern part of Iraq, East Turkey and the western part of Iran. A potential explanation for the P increases is perhaps that the increasing evaporation over the surrounding region makes more moisture

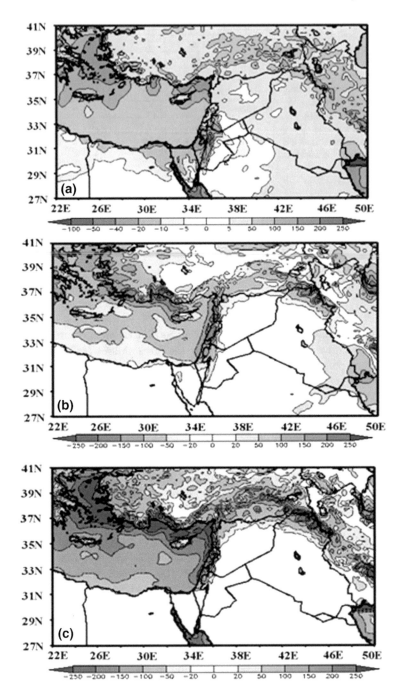

Fig. 4.6 Differences in seasonal totals E (**a**), P (**b**) and P-E (**c**) between the future (2075–2099) and the present (1979–2002). Unit: mm/season. *Source* The authors

available over this area; at the same time, the mountain region provides strong orographic forcing.

The P-E field is an important indicator in the study of long-term climate changes in the moisture fields. The advantage of using the P-E term is that it shows the total moisture sinks or sources as determined by the sign of P-E. The moisture budget equation shows that P-E exactly equals the vertically integrated moisture convergence term when ignoring the storage term. The P-E changes between the future and current climate are shown in Fig. 4.6c. Negative P-E changes indicate that the area will lose moisture and vice-versa. In general, it has a similar pattern to the precipitation change, i.e. Fig. 4.6b. However, when examining carefully, it can be found that the region with the precipitation increase in Fig. 4.6b shrank dramatically. The Red Sea and the Persian Gulf regions both show negative changes which cannot be seen in Fig. 4.6b, suggesting that these two water bodies will have a moisture deficit due to the increasing evaporation in the future, in spite of no changes in precipitation over these areas. A completed Fertile Crescent strip, clearer than that of the P difference chart, further emphasises the future drying tendency in this region.

4.3.6 Discussion on the Potential Drying Mechanism in the Future

Assuming that the changes in the moisture fields predicted by the 20 km GCM realistically represent the future, the following question may be asked: What might be the main mechanisms for the drying scenario over the research area? As mentioned before, the mechanisms that control the precipitation over the research area are complicated. They include the changes in the atmospheric general circulation pattern, local synoptic systems, and also some tele-connections. However, a relatively simple and intuitive way to test this drying mechanism could be related to the following question: Is this phenomenon due to the precipitation recycling which becomes less efficient or is it due to the change in the moisture transports in the future? The first hypothesis can easily be eliminated, as Table 4.2 shows that there is no significant change in the Recycling Ratio between future and current simulations, due to the projected increases in both local evaporation and the total inflow in the future. However, the moisture transport does show some very interesting changes.

Figure 4.7a shows the mean vertically integrated moisture flux of the current run over the study area. In general, the moisture fluxes transport moisture from west to east smoothly, and the moisture flux over the Eastern Mediterranean is obviously larger than in the surrounding areas since more moisture is available over the sea. The differences in moisture flux fields between the future and the present are shown in Fig. 4.7b. Over the Eastern Mediterranean and Egypt, the southward latitudinal component of the moisture flux is enhanced with an increasing gradient from the

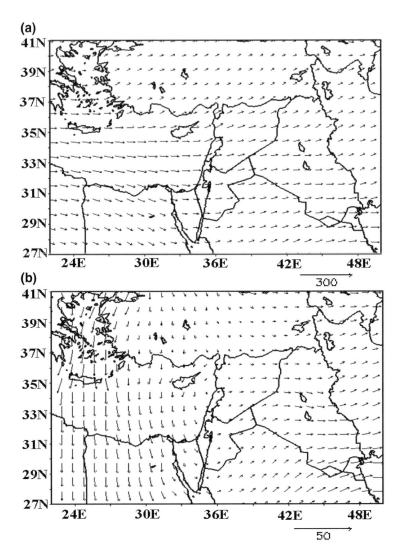

Fig. 4.7 Area mean vertical integrated moisture flux (**a**) of the present, and the difference between the future (2075–2099) and the present (1979–2007) by calculation of future minus present (**b**). Unit: kg m^{-1} s^{-1}. Notice the change in the scale (factor 1:6) between (**a**) and (**b**), which is for the purpose of highlighting the moisture flux changes. *Source* The authors

east to the west Mediterranean in the future. This means that more moisture over the east Mediterranean will be transported towards the south (Egypt) in the future. In addition, the eastward component of the moisture flux over the east Mediterranean does not change, even with a slight decreasing indicator over the east Mediterranean coastline. At the same time, the increasing eastward moisture flux over the Middle Eastern region suggests a stronger moisture divergence zone over the east

Mediterranean coastline area in the future. This causes the EM and southern Turkey areas to become drier in the future.

The boundary moisture fluxes in Fig. 4.7b also fit the separately calculated moisture budget components shown in Table 4.2 quite well. For instance, the inflow at the north lateral boundary (mostly over northern Turkey) increases along with the increasing outflow at the south and east boundaries. Therefore, the changes in the moisture transport patterns correspond well to the drying pattern as follows. The increased moisture evaporated from the Eastern Mediterranean Sea does not provide extra moisture to the drier land area, but instead additional flux to more remote regions e.g., the tropics. For the water bodies, the combination of strong evaporation, enhanced atmospheric moisture divergence and low precipitation also make these areas drier in the future.

The large-scale drying mechanism was related to some adjustment of the atmospheric general circulation, such as the poleward expansion of the Hadley Cell as proposed by Lu et al. (2007). The descending air of Hadley Cell suppresses precipitation by drying the lower troposphere and therefore expands the subtropical dry zones. At the same time, and related to this, the rain-bearing mid-latitude storm tracks also shift poleward. Held/Soden (2006) suggested that extratropical drying is mainly the consequence of the increase in lower-tropospheric water vapour. This leads to the decrease in convective mass fluxes, increases in horizontal moisture transport, and the increase in the evaporation minus precipitation.

4.4 Summary

The MRI's 20 km GCM is shown to simulate quite well the present period two main moisture budget components, which are the precipitation and evaporation over the Middle East. As expected, the precipitation and evaporation are badly under-estimated by the coarser resolution of the ERA-40 data, particularly emphasised by the large errors in the estimation of the observed precipitation distribution in a south-to-north cross section from Egypt to Israel, Lebanon and Turkey. The simulation of orographic precipitation over the high mountainous area of Turkey also suggests that the 20 km GCM is better at capturing the altitudinal changes of the orographic rainfall. The main three water bodies – EM, the Red Sea and the Persian Gulf – are projected to be moisture deficit at the end of this century, i.e. reduced P-E. Also, the famous Fertile Crescent is projected to become dramatically drier at the end of this century. Most of the EM adjacent Middle East countries, such as west Turkey, west Syria, the whole of Israel and Lebanon, are projected to be drier. However, east Iraq and part of Iran are projected to become wetter by the end of this century. One of the findings of this research is that the EM and the ME topographic forcing (including the physiographical changes, like the land-sea or land-use) effects on rainfall are quite dominant. Therefore, high-resolution modelling plays a critical role in the atmospheric processes for this region and is demonstrated nicely by the narrow 'tongue' of rainfall along the Syrian, Lebanese and Israeli coastline.

This high sensitivity to the horizontal resolution seems to apply to the whole Mediterranean region, as stated by Lionello et al. (2006). The study of the mechanism control of the drying scenario shows that the precipitation recycling does not change between the present and the future. However, analysis of the changes in the moisture transport patterns over the Eastern Mediterranean explains the drier tendency in the future as follows. The subtropical mean flow of the low troposphere moves the moisture out of these regions, and there are not enough extra moisture sources to compensate for this drying in the future, even with the enhanced evaporation.

Acknowledgements For this research, we acknowledge GLOWA-JR support by the Federal Ministry of Education and Research (BMBF) and Israel's Ministry of Science and Technology. Also, partial support was given by DESERVE (Dead Sea Research Venue) and the Israel Water Authority. Thanks to A. Kitoh for providing the super-high-resolution global runs performed by the MRI/JMA.

References

Alpert P, Ben-Gai T, Baharad A, Benjamini Y, Yekutieli D, Colacino M, Diodato L, Ramis C, Homar V, Romero R, Michaelides S, Manes A (2002) The paradoxical increase of Mediterranean extreme daily rainfall in spite of decrease in total values. *Geophysical Research Letters* 29:31.1–31.4.

Alpert P, Krichak SO, Osetinsky I, Dayan M, Haim D, Shafir H (2008) Climatic trends to extremes employing regional modeling and statistical interpretation over the E. Mediterranean. *Global Planet Change* 63:163–170.

Alpert P, Jin F, Shafir H (2011) Orographic precipitation simulated by a super-high resolution global climate model over the Middle East. In Fernando HJS, Klaić Z, McCulley JL (eds) *National Security and Human Health Implications of Climate Change*, Springer Publication in cooperation with NATO, pp 301–306.

Evans JP (2009) 21st century climate change in the Middle East. *Climate Change* 92:417–432. https://doi.org/10.1007/s10584-008-9438-5.

Gibson JK, Kallberg P, Uppala S, Nomura A, Hernandez A, Serrano A (1997) *ERA Description*. ECMWF Reanalysis Project Report Series No 1:77.

Held IM, Soden BJ (2006) Robust responses of the hydrological cycle to global warming. *Journal of Climate* 19:5686–5699.

IPCC (2007) *Fourth Assessment Report: Working Group II Report "Impacts, Adaptation and Vulnerability"*; at: http://www.ipcc.ch/ipccreports/ar4-wg2.htm.

Jin F, Zangvil A (2010) Relationship between moisture budget components over the eastern Mediterranean. *International Journal of Climatology* 30:733–742. https://doi.org/10.1002/joc.1911.

Jin F, Kitoh A, Alpert P (2010) Global warming projected water cycle changes over the Mediterranean, East and West: A comparison study of a super-high resolution global model with CMIP3. *Philosophical Transactions of the Royal Society A* 368:5137–5149. https://doi.org/10.1098/rsta.2010.0204.

Jin F, Kitoh A, Alpert P (2011) Climatological relationships among the moisture budget components and rainfall amounts over the Mediterranean based on a super-high resolution climate model. *Journal of Geophysical Research* 116(D9). https://doi.org/10.1029/2010JD014021.

Kallberg P, Simmons A, Uppala S, Fuentes M (2004) *The ERA-40 Archive*, ERA-40 Project Report Series No. 17, European Centre for Medium-range Weather Forecast, Reading: UK; 31.

Kitoh A, Yatagai A, Alpert P (2008) First super-high-resolution model projection that the ancient Fertile Crescent will disappear in this century. *Hydrological Research Letters* 2:1–4. https://doi.org/10.3178/HRL.2.1.

Krichak SO, Alpert P (2005) Decadal trends in the East-Atlantic West Russia pattern and the Mediterranean precipitation. *International Journal of Climatology* 25:183–192.

Lionello P, Malanotte-Rizzoli P, Boscolo R (2006) Mediterranean climate variability. *Developments in Earth & Environment Sciences*, 4. Elsevier BV.

Lu J, Vecchi G, Reichler T (2007) Expansion of the Hadley cell under global warming. *Geophysical Research Letters* 34:L06805. https://doi.org/10.1029/2006GL028443.

Mariotti A, Struglia MV, Zeng N, Lau KM (2002a) The hydrological cycle in the Mediterranean region and implications for the water budget of the Mediterranean Sea. *Journal of Climate* 15:1674–1690.

Mariotti A, Zeng N, Lau KM (2002b) Euro-Mediterranean rainfall and ENSO – a seasonally varying relationship. *Geophysical Research Letters* 29(12):59-1–59-4. https://doi.org/10.1029/2001GL014248.

Mitchell TD, Jones PD (2005) An improved method of constructing a database of monthly climate observations and associated high-resolution grids. *International Journal of Climatology* 25:693–712. https://doi.org/10.1002/joc.1181.

Mizuta R, Oouchi K, Yoshimura H, Noda A, Katayama K, Yukimoto S, Hosaka M, Kusunoki S, Kawai H, Nakagawa M (2006) 20-km-mesh global climate simulations using JMA-GSM model Mean climate states. *Journal of the Meteorological Society of Japan* 84:165–185.

Mizuta R, Adachi Y, Yukimoto S, Kusunoki S (2008) *Estimation of the Future Distribution of Sea Surface Temperature and Sea Ice Using the CMIP3 Multi-model Ensemble Mean*. Technical Report of the Meteorological Research Institute No 56.

Özsoy E (1981) *On the Atmospheric Factors Affecting the Levantine Sea, European Center for Medium Range Weather Forecasts*, Technical Report No 25, Shinfied Park, Reading, U.K.

Price C, Stone L, Huppert A, Rajagopalan B, Alpert P (1998) A possible link between El Niño and precipitation in Israel. *Geophysical Research Letters* 25:3963–3966.

Rasmusson EM (1968) Atmospheric water vapor transport and the water balance of North America. Part II: Large-scale water balance investigations. *Monthly Weather Review* 96:720–734.

Rasmusson EM (1971) A study of the hydrology of eastern North America using atmospheric vapor flux data. *Monthly Weather Review* 99:119–135.

Seager R, Ting MF, Held I, Kushnir Y, Lu J, Vecchi G, Huang HP, Harnik N, Leetmaa A, Lau NC, Li CH, Velez J, Naik N (2007) Model projections of an imminent transition to a more arid climate in southwestern North America. *Science* 316(5828):1181–1184.

Shafir H, Alpert P (1990) On the urban orographic rainfall anomaly in Jerusalem – A numerical study. *Atmospheric Environment* 24B(3):365–375.

Shay-El Y, Alpert P, DaSilva A (2000) Preliminary estimation of horizontal fluxes of cloud liquid water in relation to subtropical moisture budget studies employing ISCCP, SSMI and GEOS-1/DAS data sets. *Journal of Geophysical Research* 105(D14):18067–18089.

Yanai M, Esbensen S, Chu JH (1973) Determination of average bulk properties of tropical cloud clusters from large-scale heat and moisture budgets. *Journal of the Atmospheric Sciences* 30:611–627.

Zangvil A, Druian P (1990) Upper air trough axis orientation and the spatial distribution of rainfall over Israel. *International Journal of Climatology* 10:57–62.

Zangvil A, Portis DH, Lamb PJ (1992) *Interannual variations of the moisture budget over the Midwestern United States in relation to summer precipitation. Part II: Impact of local evaporation on precipitation. Extended Abstracts*. Yale Mintz Memorial Symposium on Climate and Climate Change. Jerusalem, Israel, Israel Meteorological Society and American Meteorological Society 101.

Part II
Climate Change Impacts on Basin Hydrology and Agricultural Water Management

Chapter 5
Impacts of Climate Change on Basin Hydrology and the Availability of Water Resources

Kenji Tanaka, Yoichi Fujihara, Fatih Topaloğlu, Slobodan P. Simonovic and Toshiharu Kojiri

Abstract Surface energy, water balance components and related hydrological variables of the Seyhan River basin Turkey were estimated through the off-line simulation of a land surface model forced by the output of a regional climate model for both present and future conditions. Future climate conditions were produced by two different general circulation models, and land-cover conditions were determined under three land-use scenarios. The two climate conditions and the three land-use scenarios were combined in six different simulations for the future. Vegetation parameters were adjusted for the future land-use scenarios by applying the average future seasonal cycle for each vegetation class. The maximum snow water equivalent for the study area was almost 0.4 Gt in the present climate and decreased to as little as 0.1 Gt in the future climate. In the present climate, annual evaporation in irrigated areas is about 800 mm, and about 500 mm of irrigation water must be supplied to maintain soil wetness during the growing season. Over the study area, annual average values were projected to decrease by 170 mm for precipitation, about 40 mm for evaporation and about 110 mm for runoff. The proportional impact on runoff is particularly significant.

Keywords Basin hydrology · Climate change · Land surface model
Seyhan River basin · Water resources availability

K. Tanaka, Associate Professor, Disaster Prevention Research Institute, Kyoto University Gokasho, Uji, Kyoto 611-0011, Japan; e-mail: tanaka.kenji.6u@kyoto-u.ac.jp.

Y. Fujihara, Associate Professor, Ishikawa Prefectural University 1-308, Suematsu, Nonoichi, Ishikawa 921-8836, Japan; e-mail: yfuji@ishikawa-pu.ac.jp.

F. Topaloğlu, Professor, Department of Agricultural Construction and Irrigation, Faculty of Agriculture, Çukurova University, Adana, Turkey; e-mail: topaloglu@cu.edu.tr.

S. P. Simonovic, Professor, Department of Civil and Environmental Engineering Program, The University of Western Ontario 1151 Richmond Street, London, Ontario, Canada, N6A 3K7; e-mail: ssimonovic@eng.uwo.ca.

T. Kojiri, researcher, Disaster Prevention Research Institute, Kyoto University Gokasho, Uji, Kyoto 611-0011, Japan.

© Springer Nature Switzerland AG 2019
T. Watanabe et al. (eds.), *Climate Change Impacts on Basin Agro-ecosystems*,
The Anthropocene: Politik—Economics—Society—Science 18,
https://doi.org/10.1007/978-3-030-01036-2_5

5.1   Introduction

As the world population grows and the demand for food increases, the productivity of agriculture in arid areas needs to improve, despite severe restrictions on water availability. Moreover, future global climate change will challenge agriculture in arid regions by bringing about substantial changes in temperature, rainfall and evapotranspiration. This study attempted to estimate the components of surface energy and water balance and related hydrological variables of the Seyhan River basin, Turkey, using a land surface model (LSM) forced by the product of a regional climate model (RCM) for both present and future conditions. The RCM provided the detailed meteorological field for the Seyhan River basin by physically based (or dynamic) downscaling, an appropriate technique for this basin, where surface meteorological stations are scarce.

It is usually difficult to directly use the RCM products in hydrological applications because these applications typically require greater accuracy than that achieved by the RCM. For that reason, this study investigated the model biases that the RCM imposes and developed a method to correct biases by making the best use of available meteorological data. To assess the impact of climate change on agriculture in the study area, including human factors (farm management, cropping patterns, etc.), the simulation considered three land-use scenarios and used two future climate scenarios to evaluate uncertainty in the assessment of surface energy and water balance.

5.2   Model and Input Parameters

5.2.1   Basin Characteristics

The surface hydrology group began by defining the physical boundary of the Seyhan River Basin. Once the basin boundary is defined, other basin characteristics can be extracted from global data resources and local information. Three different digital elevation models (DEM) were considered: the Turkish DEM (250 m mesh), Gtopo30 (30 s mesh), and SRTM (3 s mesh). The first of these was judged deficient in describing the lowland area (lower Seyhan delta). The other two data sets were considered equivalent. As the planned target resolution of the hydrological model was 1 km, Gtopo30 was used for basic topographic data. The locations of the boundary and river channels were extracted by using HYDRO1k (https://lta.cr.usgs.gov/HYDRO1K). By using these data sets and large-scale maps, the physical boundary of the Seyhan River Basin was carefully defined, and its catchment area was estimated as 24,625 km^2.

A data set of land use/land cover was produced by the *Vegetation sub-group* (see Chap. 9 in this book) from Landsat satellite images. Land cover in the Seyhan River basin was classified primarily as grasslands (31.74%), dry croplands (22.22%), evergreen needle leaf forests (19.37%), irrigated croplands (15.21%), water body (7.19%), and others (4.27%).

Soil physical parameters including porosity, field capacity, and root zone depth were extracted from the ECOCLIMAP database (Masson et al. 2003), a new global data set with a 1 km resolution intended for use in initialising model parameters in land surface models.

5.2.2 Land Surface and Irrigation Schemes

The Simple Biosphere including Urban Canopy (SiBUC) land surface scheme (Tanaka/Ikebuchi 1994) was designed to treat land-use conditions (natural vegetation, cropland, urban area, water body) in detail. In particular, SiBUC includes irrigation schemes that maintain the soil moisture within appropriate ranges defined for each growing stage of each crop type (Kozan et al. 2003). The irrigation rules for cropland are described by the planting date, harvesting date, duration of each growing stage, the lower limit of soil wetness during each growing stage, and amount of water supplied at one time. The soil wetness minima for each growing stage and each crop type were adopted from an agricultural manual issued in China.

In this study, maize and citrus were selected as representative irrigated crops. According to the *Irrigation and Drainage sub-group* (Chap. 7 in this book), the irrigation periods for maize and citrus are from 23 May to 6 August and from 14 May to 9 October respectively. For the future climate simulation, the growing period was shortened by 10 days, considering the effect of higher temperatures on crop growth (Table 5.1).

5.2.3 Model Domain

The product of the RCM (8.3 km resolution) was used to force the land surface model (LSM). Seven surface meteorological components (precipitation, downward short-wave and long-wave radiation, wind speed, air temperature, specific humidity, and air pressure) were provided by the *Climate sub-group* (see Chaps. 6 and 13 in this book) at hourly time intervals.

Table 5.1 Growing periods used for present and future simulations. *Source* Discussion with Irrigation and Drainage sub-group

		Start (DOY)	End (DOY)	Period (days)
Present	Maize	143	218	75
	Citrus	134	282	148
Future	Maize	143	208	65
	Citrus	134	272	138

The simulation period for the present climate was from 1994 to 2003, when precipitation was normal. The future climate condition (2070s) was produced by the "pseudo warm-up" method (Sato et al. 2006), which determines the boundary condition for the RCM by linear coupling of reanalysis data (observations) and the trend of global warming as estimated by general circulation models. In this way, the pseudo warm-up method incorporates the synoptic-scale variability of the current condition (observations). Given a simulation period of only 10 years, the projected future climate condition does not necessarily represent the average future condition, but the pseudo warm-up method, by considering the original present condition as normal, provides some assurance that future condition also may be regarded as normal.

For the future climate condition, data sets were produced from two different general circulation models (MRI and CCSR). For the land-cover condition, three land-use scenarios were prepared after discussions with project members (A0, no adaptation; A1, adaptation 1; A2, adaptation 2). Thus, simulations with six combinations of climate conditions and land-use scenarios were conducted for the future condition (Table 5.2).

Figure 5.1 shows the model domains of the RCM and LSM. The RCM domain covered the whole Seyhan River basin, and the LSM domain was the $2.75° \times 2.75°$ area bounded by 34.25°E, 37.0°E, 36.5°N, and 39.25°N. The LSM area was divided into a grid of 33×33 cells with sides measuring 5 min of arc (approximately 10 km). SiBUC uses a mosaic approach (Avissar/Pielke 1989) to incorporate all kinds of land use. Figure 5.2 shows the coverage of the four major land-cover conditions under each of the three land-use scenarios, and Table 5.3 lists their percentages of the total basin area. Future climate conditions at current dry cropland areas are outside the suitable zone for rainfed agriculture. So all dry croplands were converted into grassland in adaptation scenario 1. Also, irrigated areas in coastal parts of the Seyhan delta were abandoned because of the salinity issue. In the adaptation scenario 2, some of the forest was converted into grassland. Also, some of the dry croplands were converted into irrigated areas on the assumption that new irrigated areas would develop in the middle of the basin.

Table 5.2 Climate and land use condition for each simulation. *Source* The authors

Run name	Climate	Land use
P0	Present	Current
M0	Warm-up (MRI)	No adaptation
M1	Warm-up (MRI)	Adaptation 1
M2	Warm-up (MRI)	Adaptation 2
C0	Warm-up (CCSRI)	No adaptation
C1	Warm-up (CCSRI)	Adaptation 1
C2	Warm-up (CCSRI)	Adaptation 2

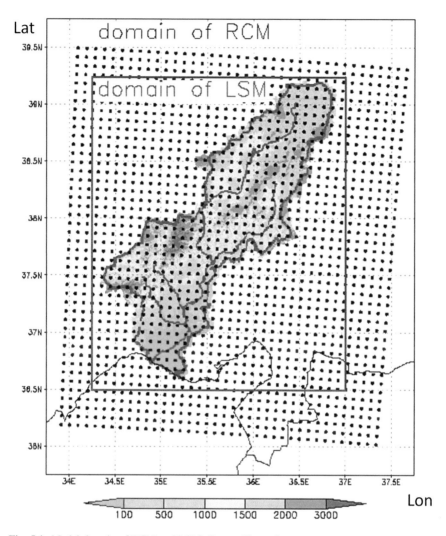

Fig. 5.1 Model domain of RCM and LSM. *Source* The authors

5.2.4 Vegetation Dynamics

The satellite-derived Normalised Difference Vegetation Index (NDVI), especially its time series, is widely used for describing the land-surface status (Kozan et al. 2004). Here, NDVI was used to express the activity of vegetation. The SPOT-VEGETATION Product (http://www.spot-vegetation.com/), used to extract vegetation dynamics, is a ten-day composite data set with a resolution of approximately 1 km. Cloud noise was removed by the BISE method (Viovy/Arino 1992). Also, a data set of average seasonal cycle was produced from the six-year data

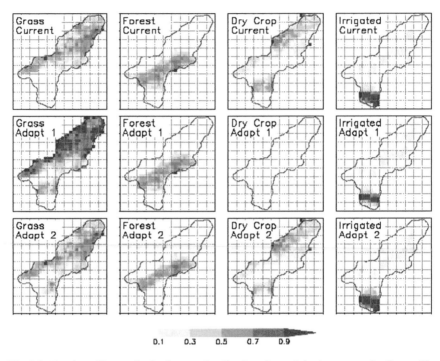

Fig. 5.2 Fraction of four major land-cover classifications for each land-use scenario. *Source* The authors

Table 5.3 Fraction of four major land-cover classifications for each land-use scenario (unit: percentage). *Source* Vegetation sub-group

	Forest	Grass	Irrigated	Dry crop
A0	19.37	31.74	15.21	22.22
A1	19.37	53.97	13.45	0.00
A2	16.44	34.67	16.51	20.92

Forest: Evergreen needle leaf forest. Grass: grassland, short vegetation. Irrigated: Irrigated farmland (total). Dry crop: rain-fed wheat

period (from 1999 to 2004). Leaf area index (LAI) was calculated from NDVI data and the vegetation class. Other time-varying vegetation parameters, such as the greenness fraction (Nc) and vegetation coverage (Vc), were extracted from the ECOCLIMAP database. Because SiBUC uses a mosaic scheme to account for sub-grid scale heterogeneity, these vegetation parameters were aggregated within each land-cover classification in each 10 km LSM grid cell. These data sets enabled adequate description of the spatial distribution and time evolution of vegetation parameters.

There is no reliable information about the vegetation status in the future climate condition. Three scenarios were prepared for the future land use. Vegetation

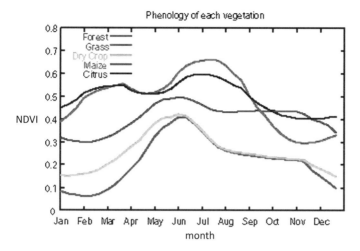

Fig. 5.3 Average seasonal cycle of NDVI for each vegetation type. *Source* The authors

parameters were adjusted whenever a pixel was changed from its current land cover. Average seasonal cycles of vegetation parameters (NDVI, Nc, Vc) were calculated for each vegetation type on the basis of current land-cover information, then assigned to the 1 km pixels that had a change in the land cover in the A1 and A2 scenarios. Figure 5.3 shows the average seasonal cycles of NDVI for each vegetation type. Current parameter values were assigned to the remaining pixels. Finally, these parameters were aggregated for each land-cover classification in each 10 km LSM grid cell.

5.3 Bias Correction of the RCM Output

5.3.1 Analysis of Meteorological Data

All of the available meteorological data from the Turkish State Meteorological Service (DMI) for the study area from 1971 to 2002 were checked and analysed. Table 5.4 summarises the available data within and near the Seyhan River Basin.

Figure 5.4a shows the distribution of stations with daily precipitation data. These data make it possible to create a gridded meteorological data set. Figure 5.4b, c are examples of such data sets showing the average annual precipitation (climatology) and the 1999 precipitation. Because only ten stations are located within the basin, these data sets do not capture the realistic distribution of precipitation in this large (over 20,000 km^2) and highly mountainous area (altitude range is more than 3,000 m). They were produced to be used as forcing data for SiBUC and validation data for the output of the RCM.

Table 5.4 List of available DMI stations. *Source* The authors

Meteorological element	Number
Precipitation (daily)	13
Air temperature (hourly)	16
Tmax, Tmin (daily)	13
RHmax, RHmin, Wind (daily)	13
SWdown (daily)	9

Domain: E34.0-E37.0, N36.5-N39.5, Tmax, Tmin: daily maximum and minimum air temperature, RHmax, RHmin: daily maximum and minimum relative humidity

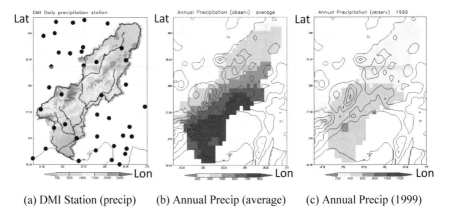

(a) DMI Station (precip)   (b) Annual Precip (average)   (c) Annual Precip (1999)

Fig. 5.4 Surface meteorological station and gridded precipitation data. *Source* The authors

A representative diurnal cycle for air temperature was produced from hourly data from eighteen stations. Typical diurnal cycles were prepared for every month at eighteen stations (216 patterns), and a best-fit cycle was selected from these on the basis of daily mean, maximum, and minimum temperatures. Hourly values were derived from this cycle at each of the daily stations. The station data (both observations and reproduced values) were extended into a gridded data set using spatial interpolation. Pressure and temperature values were adjusted to sea level values before the interpolation process, then adjusted back to the correct altitude.

5.3.2 Bias Correction of Version 2 Product (V2.0 to V2.5)

5.3.2.1 Precipitation

There were thirteen stations with records of precipitation for more than 90% (108) of the months during the ten-year period 1994–2003. Total precipitation was

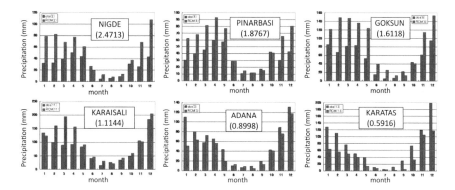

Fig. 5.5 Comparison of monthly precipitation at each station (Blue: Observation, Red: RCM). *Source* The authors

calculated to determine the mean bias, defined as the RCM precipitation divided by observed precipitation. This bias ranged from 0.59 to 2.47 and was generally small in flat areas and large in mountainous areas. Figure 5.5 compares the ten-year average seasonal cycle of modelled and observed monthly precipitation at six stations. It shows that model bias varied from station to station and from month to month, requiring the bias correction to vary in both space and time. Although the rain gauge network was rather sparse, a gridded precipitation data set was produced by spatial interpolation to calculate the monthly precipitation bias. Figure 5.6 shows the spatial distribution of the precipitation bias. Note that the observation-based gridded precipitation data set did not include the topographic dependency in its spatial interpolation process, but because the precipitation distribution was well captured by the RCM, limits were imposed on extreme values of the bias, restricting them between 0.667 and 1.500. The RCM precipitation for each grid cell was divided by the values shown in Fig. 5.6 to yield bias-corrected precipitation, as shown in Fig. 5.7 for each year and for the average annual cycle. Use of the limited bias range led to corrected values that consistently lay between the observations and the RCM values.

5.3.2.2 Downward Shortwave Radiation

Nine stations had records of downward short-wave radiation for more than 90% (108) of the months during 1994–2003. Average short-wave radiation was calculated to evaluate the mean bias, which ranged from 0.44 to 0.70 and was generally smaller in the inner part of the study area. Figure 5.8 shows the average monthly radiation at these stations before and after bias correction. Figure 5.9 shows the time series of daily mean radiation at station Adana. The modelled radiation from the RCM was much smaller than the observations, with great short-term variability. Correcting the RCM result using the monthly mean bias yielded unrealistic values

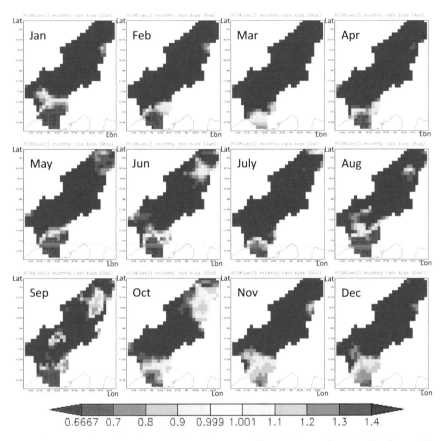

Fig. 5.6 Spatial distribution of precipitation bias (upper limit: 1.5, lower limit: 0.667). *Source* The authors

(sometimes exceeding the theoretical maximum value at the top of the atmosphere, S_{Otop}) when the uncorrected RCM value was close to observations. The radiation bias was found to have a dependency on the weather (sunny or cloudy) and season; thus, model biases were calculated for each of ten weather classifications (defined by the ratio of the RCM radiation to S_{Otop}) and for each month. After this disaggregation of all stations' data into 120 categories (10 classes × 12 months), the mean bias for each category (radiation bias matrix) was calculated and yielded radiation values that matched the observations.

5.3.2.3   Air Temperature

There were sixteen stations with records of air temperature for more than 90% (108) of the months during 1994–2003. Figure 5.10 shows the time series of diurnal

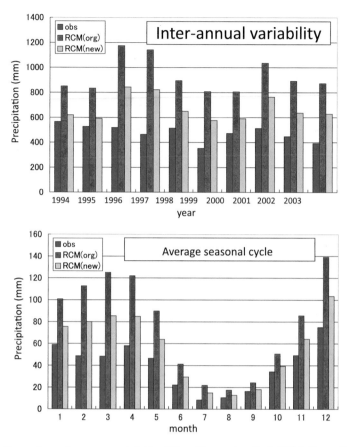

Fig. 5.7 Comparison of average precipitation (Blue: Observation, Red: RCM, Green: Corrected). *Source* The authors

Fig. 5.8 Comparison of monthly solar radiation (Blue: RCM, Red: Corrected). *Source* The authors

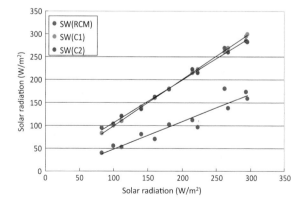

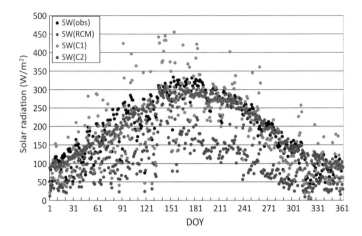

Fig. 5.9 Comparison of daily solar radiation (Black: Observation, Blue: RCM, Red: Corrected). Green dots are corrected by monthly mean bias (simple method). *Source* The authors

range and maximum and minimum temperatures at stations Pınarbaşı and Adana. Figure 5.11 shows the time series of hourly air temperature in January and July 2000 at Adana. Lower temperature bias was reported by the *Climate sub-group*, and a correction method was provided. Although this method improved the daily mean temperatures, it was not effective for the diurnal temperature range. Because the diurnal range of the RCM temperature was much smaller than the observed range, correcting the temperature bias required the diurnal range to be amplified. As was the case with downward short-wave radiation, amplification based on the monthly mean bias sometimes produced unrealistically large diurnal variations. Thus, model biases were calculated for each of the ten classes of the RCM temperature range and for each month. After this disaggregation into 120 categories, the mean bias for each category (temperature bias matrix) was calculated and yielded temperatures that matched the observations.

5.3.2.4 Humidity (Water Vapour Pressure)

The only available information on water vapour pressure was the daily maximum and minimum relative humidity, for which the times were not known. Relative humidity, being highly dependent on air temperature, is not well suited to determining the humidity bias. Instead, a new index of water vapour pressure called E_{ave} was defined, calculated from mean temperature and mean humidity as a measure of daily mean water vapour pressure. Figure 5.12 shows monthly average daily maximum and minimum temperatures, relative humidity, and E_{ave} at stations Gemerek and Karaisali. Biases for relative humidity were large at both stations, but biases for E_{ave} were very small at Gemerek and greatly reduced at Karaisali as a result of lower temperature biases. The humidity bias, defined by the ratio of E_{ave}

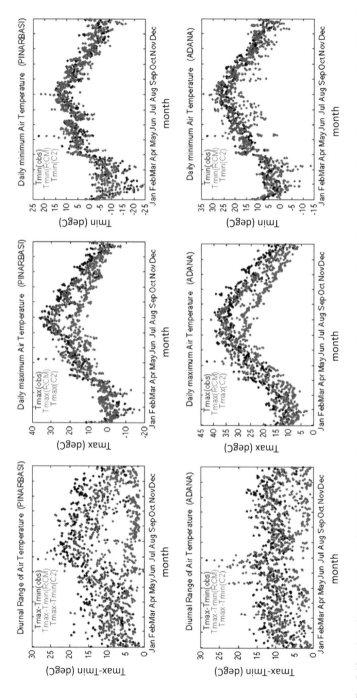

Fig. 5.10 Diurnal range, daily maximum, and daily minimum temperature at Pinarbasi (upper) and Adana (lower) (Black: Observation, Blue: RCM, Red: Corrected). *Source* The authors

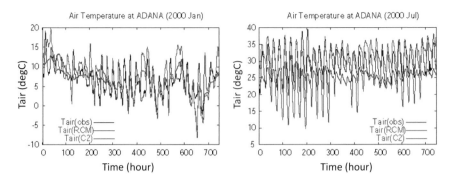

Fig. 5.11 Time series of hourly air temperature in January (left) and July (right) 2000 at Adana (Black: Observation, Blue: RCM, Red: Corrected). *Source* The authors

from RCM to E_{ave} derived from observations, was calculated at each station for each month and then spatially interpolated to produce a gridded data set of humidity bias.

5.3.3 Bias Correction of Version 3 Product (V3.0 to V3.5)

The same respective methods of bias correction used with V2.0, with some modifications and simplifications, were applied to the new and final RCM output.

Figure 5.13 shows the spatial distribution of precipitation bias, which was restricted to values between 2.25 and 0.5. These limits were set to minimise the error in the runoff simulated by the LSM (Fig. 5.14). The values of monthly mean bias in the short-wave radiation flux at each station (Table 5.5) were spatially interpolated to produce a gridded data set of monthly bias. The downward short-wave radiation required only a simple monthly mean bias correction (Fig. 5.15). Daily mean temperature was compared with observations, and regression coefficients, calculated at each station for each month, were spatially interpolated to correct this parameter at all grid points. Diurnal temperature range (Tmax – Tmin) was compared with observations, and an adjustment ratio was calculated at each station for five classes of temperature range (0–3, 3–6, 6–9, 9–12, and more than 12 °C) to yield the corrections shown in Fig. 5.16. As for humidity, the same bias correction method was applied to the new product. The wind bias was calculated at each station for five wind speed classes (0–2, 2–3, 3–4, 4–5, 5∼). These station biases were spatially interpolated to produce gridded monthly bias data sets for each wind-speed class.

This study featured an iterative procedure in which the model bias was analysed by the *Hydrology sub-group* (data user) and the bias information was provided to the *Climate sub-group* (modeller) several times to benefit from feedback between hydrologists and meteorologists (Fig. 5.17). As a result, the daily maximum air

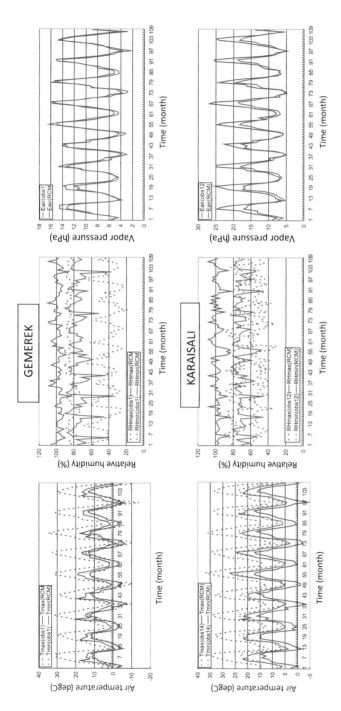

Fig. 5.12 Monthly average daily maximum and minimum temperature (left), relative humidity (centre), and daily mean vapour pressure at Gemerek (upper) and Karaisali (lower). Left, Centre (Broken line: Observation, Solid line: RCM, Red: maximum, Blue: minimum). Right (Blue: Observation, Red: RCM). *Source* The authors

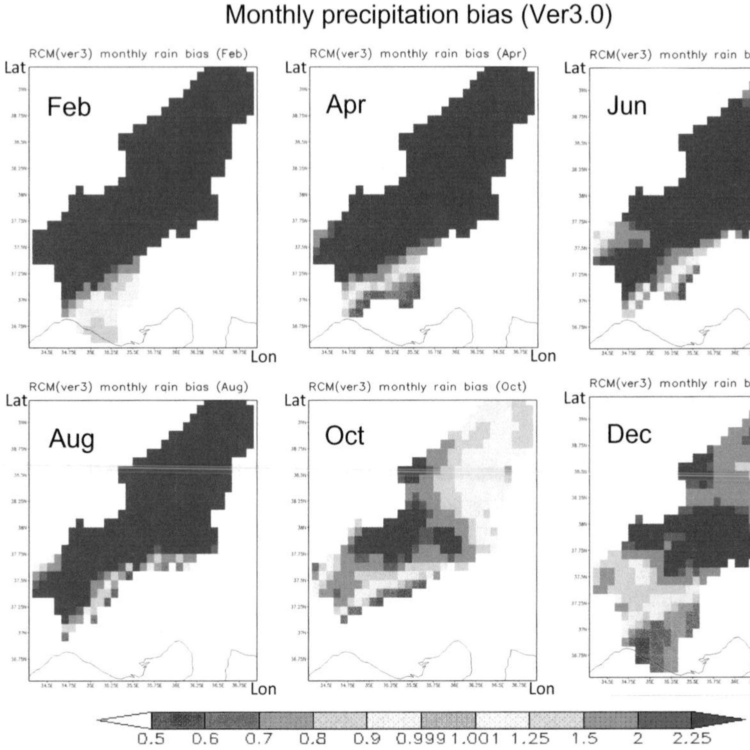

Monthly precipitation bias (Ver3.0)

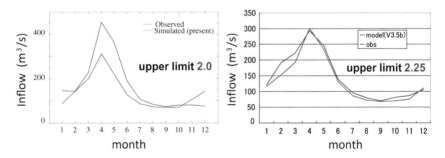

Fig. 5.13 Precipitation bias for each month. *Source* The authors

Fig. 5.14 Adjustment of upper limit of precipitation correction factor by hydrological analysis. *Source* The authors

temperature and solar radiation data (important parameters in crop growth modelling) were greatly improved from Version 2 to Version 3 (Fig. 5.18). This feedback process was a noteworthy aspect of the ICCAP.

Table 5.5 Monthly mean bias of downward shortwave radiation at eight stations. *Source* The authors

Month	GEMEREK	EREGLI	NIGDE	ADANA	PINARBASI	DEVELI	TOMARZA	GOKSUN
1	1.2091	1.2526	1.2697	1.005	1.3656	1.213	1.4017	1.3533
2	1.1603	1.1456	1.1621	0.962	1.2926	1.1293	1.2608	1.1597
3	1.1242	1.0789	1.2379	1.1371	1.2639	1.1372	1.214	1.078
4	1.0771	1.0208	1.1884	1.2901	1.125	1.0398	1.0924	1.042
5	1.1762	1.0802	1.191	1.0317	1.1388	1.1557	1.1478	1.0636
6	1.2178	1.1168	1.188	1.0727	1.1791	1.1775	1.1889	1.1177
7	1.169	1.0847	1.1233	1.095	1.1182	1.1135	1.0952	1.0727
8	1.0069	0.9538	1.1219	1.1005	1.0685	1.0821	1.0593	1.0642
9	1.0418	0.9774	1.1743	1.0886	1.0978	1.1186	1.0968	1.0887
10	1.0447	1.008	1.2055	1.1878	1.1517	1.1793	1.1655	1.1541
11	1.1188	1.1218	1.1874	1.132	1.1932	1.1361	1.2369	1.1707
12	1.2264	1.2653	1.2769	1.0084	1.424	1.2329	1.4212	1.4084

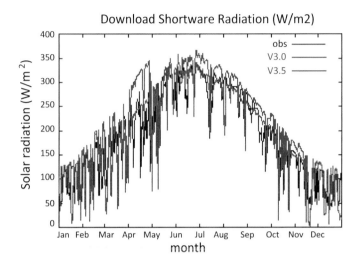

Fig. 5.15 Comparison of daily solar radiation (Black: Observation, Blue: RCM, Red: Corrected). *Source* The authors

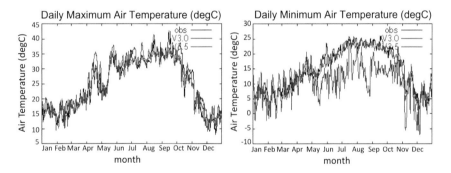

Fig. 5.16 Comparison of daily maximum and minimum temperature (Black: Observation, Blue: RCM, Red: Corrected). *Source* The authors

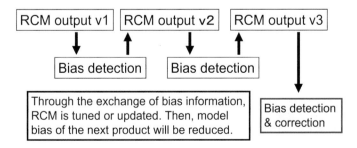

Fig. 5.17 Feedback process between data user and modeller. *Source* The authors

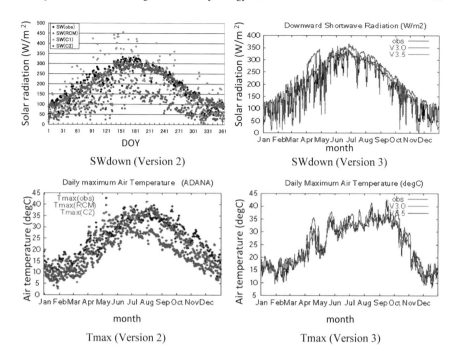

SWdown (Version 2)   SWdown (Version 3)

Tmax (Version 2)   Tmax (Version 3)

Fig. 5.18 Improvement of the RCM product (left: Version 2, right: Version 3). *Source* The authors

5.4 Results and Discussions

Figures 5.19, 5.20, 5.21 and 5.22 show the annual (ten-year average) water balance components (precipitation, runoff, snowfall, maximum snow water equivalent respectively) for the present climate and their change from the present in the *M0* and *C0* simulations, and the warm-up simulations with no adaptation (see Table 5.2). Figure 5.23 and Fig. 5.24 show the annual (ten-year average) evaporation and irrigation water requirement respectively, for the present climate, along with the differences due to climate change in the A1 and A2 land-use scenarios. The resulting impacts of climate change are shown for evaporation and the irrigation water requirement in Fig. 5.26 and Fig. 5.27 respectively.

Annual precipitation in the present climate is about 400 mm in the upstream region, above 1,000 mm in the middle region, and about 700 mm in the Seyhan delta (Fig. 5.19a). In both future simulations, precipitation decreased in the whole Seyhan basin, decreasing by more than 250 mm in the middle and delta regions. Runoff also decreased as a result of the reduced precipitation, especially in the mountainous region, and the increased evaporation (Fig. 5.23b, c).

Snowfall was reduced in the future simulations (Fig. 5.21), and the reduction was larger in the **C0** simulation, mainly because of warmer temperatures. As a

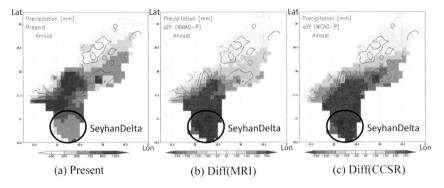

(a) Present   (b) Diff(MRI)   (c) Diff(CCSR)

Fig. 5.19 Annual precipitation of present climate and its difference in MRI and CCSR run. *Source* The authors

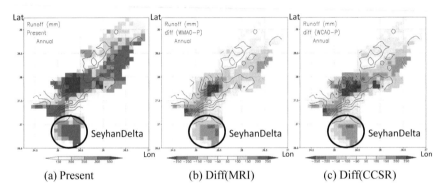

(a) Present   (b) Diff(MRI)   (c) Diff(CCSR)

Fig. 5.20 Annual runoff of present climate and its difference in MRI and CCSR run. *Source* The authors

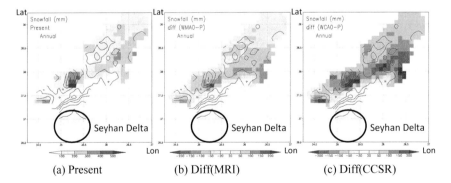

(a) Present   (b) Diff(MRI)   (c) Diff(CCSR)

Fig. 5.21 Annual snowfall of present climate and its difference in MRI and CCSR run. *Source* The authors

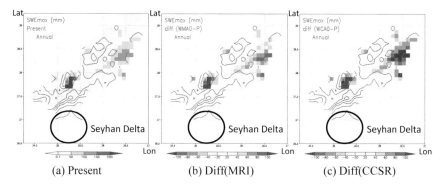

Fig. 5.22 Maximum SWE of present climate and its difference in MRI and CCSR run. *Source* The authors

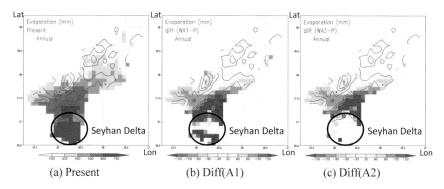

Fig. 5.23 Annual evapotranspiration of present climate and its impact in A1 and A2 scenarios. *Source* The authors

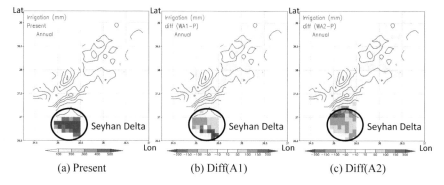

Fig. 5.24 Annual IWR of present climate and its impact in A1 and A2 scenarios. *Source* The authors

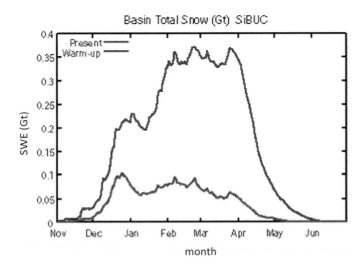

Fig. 5.25 Basin total storage of snow (Blue: present, Red: future). *Source* The authors

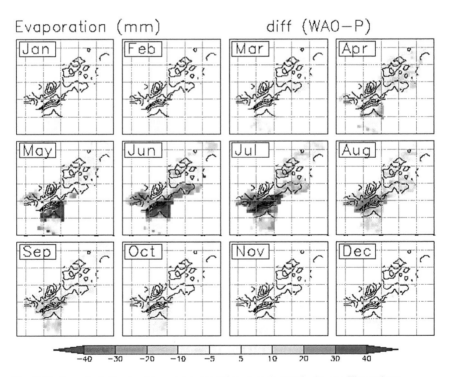

Fig. 5.26 Impact of climate change on evaporation at each month. *Source* The authors

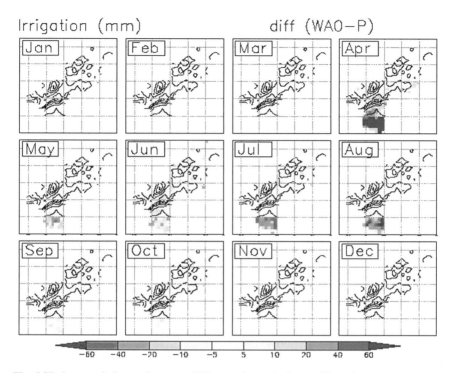

Fig. 5.27 Impact of climate change on IWR at each month. *Source* The authors

result, the maximum snow water equivalent decreased more in the **C0** simulation (Fig. 5.22). Figure 5.25 shows the seasonal evolution of the total amount of water stored as snow in the Seyhan River basin for the present climate and for the average future climate (mean of *M0* and *C0* simulations). Whereas snow water equivalent reached a maximum of almost 0.4 Gt in the present climate, it decreased to 0.1 Gt in the future climate.

In the Seyhan delta, present annual evaporation is about 800 mm, and about 500 mm of irrigation water must be supplied to maintain soil wetness during the growing season. Although simulated future precipitation decreased in the whole Seyhan River basin, evaporation increased in the area where snow water equivalent decreased, because the reduction in the period of snow cover means that those areas receive more short-wave radiation (see Fig. 5.26). The difference in evaporation between the two land-use scenarios (Fig. 5.23b, c) was large where irrigated areas were abandoned and where dry cropland was converted into grassland. Although the future growing period was shortened, the irrigation water requirements were projected to increase because of the higher evaporation demand in the growing season and the reduction in soil moisture at the beginning of the growing season (see Fig. 5.27).

Table 5.6 lists the basin-wide averages for the components of the annual water balance in present and future climate conditions under both land-use scenarios.

Table 5.6 Basin average annual water balance components (unit: mm). *Source* The authors

	Present	Future (A0)	Future (A1)	Future (A2)	Diff (A0)	Diff (A1)	Diff (A2)
Prec	634.0	464.3	464.3	464.3	−169.7	−169.7	−169.7
Evap	411.3	373.9	365.4	378.9	−37.4	−45.9	−32.4
Runoff	281.6	168.9	168.1	170.4	−112.7	−113.5	−111.2
Irrig	53.8	69.7	60.4	76.4	15.9	6.6	22.6
delS	−5.1	−8.8	−8.8	−8.6	−3.7	−3.7	−3.5

Projected precipitation decreased by about 170 mm, evapotranspiration decreased by about 40 mm, and runoff decreased by about 110 mm. Considering the amount of the current water balance component, the impact on runoff was especially large.

To estimate the impacts of climate change for each land cover type, model outputs were aggregated according to the dominant land cover (occupying more than 0.8 of a grid cell). Figure 5.28 shows the resulting time series of energy balance for the four land covers in the model. In this figure, thin lines are for present climate and thick lines are for future climate. In grasslands, net radiation and latent

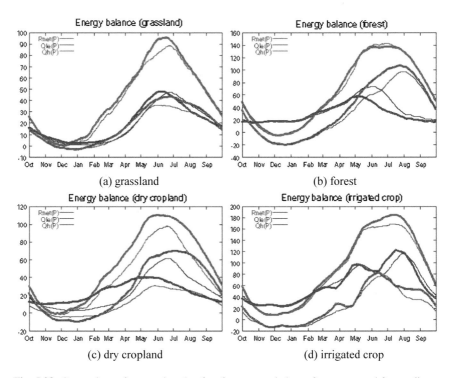

        (a) grassland   (b) forest

        (c) dry cropland   (d) irrigated crop

Fig. 5.28 Comparison of seasonal cycle of surface energy balance for present and future climate in four different land-cover conditions (Purple: net radiation, Blue: latent heat, Red: sensible heat). Thin lines are for present climate and thick lines are for future climate. *Source* The authors

heat became larger in May, and sensible heat became larger from May to August. In forests, latent heat became smaller and sensible heat became larger from May to August. In dry cropland, latent heat became larger in winter and spring and much smaller in the summer. The impact of climate change was greatest on dry cropland, because most of it was in the area where the reduction of precipitation was greatest. Irrigated cropland was similarly affected by the reduction of precipitation, but the presence of irrigation almost eliminated the impact on the surface energy balance.

5.5 Impact on Water Resources

The flow routing model HydroBEAM (Kojiri et al. 1997) and a dam operation model based on historical operations were used to simulate river discharge and assess the impact of climate change on the water supply system (Fujihara et al. 2008). Figure 5.29 shows the river routing network for the Seyhan River basin at 5-min resolution, which includes flow regulation by the Catalan and Seyhan Dams. Surface runoff and base flow for each grid cell, as well as changes in water demand due to climate change, were from SiBUC.

The estimated reliability of the future water supply is shown in Fig. 5.30. Reliability is defined as the volume of water supplied divided by the volume of the water demanded and is always 1 for the present climate condition, indicating that the dams can supply the entire demand. For the future case, reliability under the A0 and A1 land-use scenarios was 1 for normal years, but it was 0.8–0.9 for drought years. Under the A2 land-use scenario, reliability was less than 1 for most years and dropped to around 0.5 for drought years.

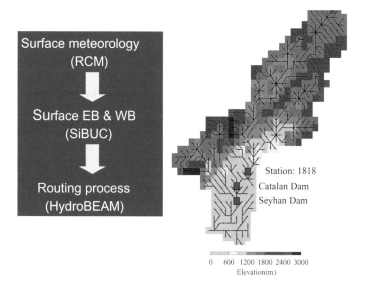

Fig. 5.29 Coupling of land surface model (SiBUC) and river routing model (HydroBEAM). *Source* The authors

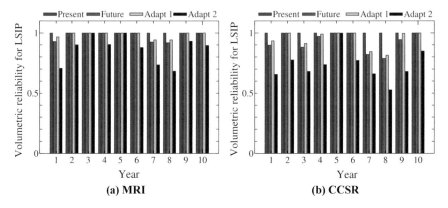

Fig. 5.30 Water supply reliability under **a** MRI future climate and **b** CCSR future climate. Future, Adapt1, and Adapt2 are future climate with A0, A1, and A2 land-use scenarios respectively. *Source* The authors

These results indicate that after several decades of climate change, it will be possible to supply adequate water during most years under the **A0** and **A1** land-use scenarios, despite decreased precipitation and increased water demand. However, it will almost always be difficult to satisfy water demand under the **A2** land-use scenario, and this scenario should be considered unsustainable.

Acknowledgements This research was financially supported by the Impact of Climate Changes on Agricultural Production System in the Arid Areas (ICCAP) project, administered by the Research Institute for Humanity and Nature (RIHN) and the Scientific and Technical Research Council of Turkey (TUBITAK). This work was also supported by the Global Environment Research Fund (S-5-3) of the Ministry of the Environment, Japan.

References

Avissar R, Pielke RA (1989) A parameterization of heterogeneous land surfaces for atmospheric numerical models and its impact on regional meteorology. *Monthly Weather Review* 117:2113–2136.

Fujihara Y, Tanaka K, Watanabe T, Nagano T, Kojiri T (2008) Assessing the impacts of climate change on the water resources of the Seyhan River Basin in Turkey: Use of dynamically downscaled data for hydrologic simulations. *Journal of Hydrology* 353(1–2):33–48.

Kojiri T, Kuroda Y, Tokai A (1997) Basin simulation and spatial assessment on water quality and quantity with multi-layer mesh-typed run-off model. *Proceedings of the International Conference on Large Scale Water Resources Development in Developing Countries – New Dimensions of Prospects and Problems WQ1–WQ8.*

Kozan O, Tanaka K, Ikebuchi S (2003) The estimation of water and heat budget in the Huaihe River Basin China – Detail representation of various cropland and irrigation. *Proceedings of the 1st International Conference on Hydrology and Water Resources in Asia Pacific Region* 2:763–768.

Kozan O, Tanaka K, Ikebuchi S, Qian M (2004) Land use and cropping pattern classification using satellite derived vegetation indices in the Huaihe river basin. *Proceedings of the 2nd International Conference on Hydrology and Water Resources in Asia Pacific Region* 2:732–740.

Masson V, Champeaux JL, Chauvin F, Meriguet C, Lacaze RA (2003) Global database of land surface parameters at 1-km resolution in meteorological and climate models. *Journal of Climate* 16(9):1261–1282.

Sato T, Kimura F, Kitoh A (2006) Projection of global warming onto regional precipitation over Mongolia using a regional climate model. *Journal of Hydrology* 333(1):144–154.

Tanaka K, Ikebuchi S (1994) Simple Biosphere Model Including Urban Canopy (SiBUC) for regional or basin-scale land surface processes. *Proceedings of the International Symposium on GEWEX Asian Monsoon Experiment*, pp 59–62.

Viovy N, Arino O (1992) The best index slope extraction (BISE): A method for reducing noise in NDVI time series. *International Journal of Remote Sensing* 13:1585–1590.

Chapter 6
Evaluation of Impact of Climate Changes in the Lower Seyhan Irrigation Project Area, Turkey

Keisuke Hoshikawa, Takanori Nagano, Takashi Kume and Tsugihiro Watanabe

Abstract This study quantitatively assesses the impacts of climate change on the irrigated agriculture of the Lower Seyhan Irrigation Project (LSIP) area in Turkey in the 2070s by factoring the projected future climate data into computational simulations of crop growth and the hydrological structure. According to simulation results by models registered to Coupled Model Intercomparison Project Phase 5 (CMIP5) with the representative concentration (RCP) 8.5 scenario, annual temperatures and precipitation around the Mediterranean region in 2081–2100 are, respectively, 4–5 °C higher and 10–20% smaller than those in 1986–2005. To assess the effects on these factors at the same time, a grid-based distributed hydrological model – IMPAM (Irrigation Management Performance Assessment Model) – was used. IMPAM, which includes modules for quasi three-dimensional soil-water dynamics, evapo-transpiration, crop-growth, irrigation and seepage, and drainage was developed by the authors for the simulation of the hydrology in irrigated agricultural areas. Three scenarios of adaptation to climate change were employed for the simulations: (a) adaptation without a large amount of investment for water management, (b) increasing the irrigation area, and (c) increasing the net amount of water applied to crops with decreasing water diversion from the river and the introduction of groundwater irrigation for 21,900 ha of orchards. Climate data for the 2070s that was used in this study was derived through RCM (Regional Climate Model) downscaling of the NCEP-reanalysis data and results of two GCMs (MRI-CGCM2 and CCSR/NIES-CGCM) with the "pseudo warming" method. The results revealed that the direct effect of global warming on the hydrology of the LSIP may not be large enough to affect agricultural production. The effect of the sea-level rise might be limited to the range of a few kilometres from the coastline

K. Hoshikawa, Assistant Professor, Toyama Prefectural University, Faculty of Engineering, 5180 Kurokawa, Imizu, Toyama 939-0398, Japan; e-mail: hoshi@pu-toyama.ac.jp.

T. Nagano, Associate Professor, Kobe University, Graduate School of Agricultural Science, 1-1 Rokkodai, Nada-ku, Kobe, Japan; e-mail: naganot@ruby.kobe-u.ac.jp.

T. Kume, Associate Professor, Ehime University, Faculty of Agriculture, Department of Rural Engineering, Tarumi, Matsuyama city, Japan; e-mail: kume@ehime-u.ac.jp.

T. Watanabe, Professor, Kyoto University, Regional Planning, Graduate School of Global Environmental Studies, Sakyo-ku, Kyoto 606-8501, Japan; e-mail: nabe@kais.kyoto-u.ac.jp.

© Springer Nature Switzerland AG 2019   99
T. Watanabe et al. (eds.), *Climate Change Impacts on Basin Agro-ecosystems*,
The Anthropocene: Politik—Economics—Society—Science 18,
https://doi.org/10.1007/978-3-030-01036-2_6

with scarce field crop areas. The reason why the effect of climate change on crop cultivation in the LSIP is not significantly larger than changes in water management is the existing water management, with plentiful water application, including losses. Even the scenarios with large decreases in water diversion from the Seyhan River brought positive effects to hydrological conditions and crop growth through the improvement of over-humidity. The average ratio of actual transpiration to potential transpiration (Ta/Tp ratio), which decreases as water stress on crops increases, varies from 0.81 to 0.89 with combinations of adaptation scenarios and projected climate, while all simulations based on the current management resulted in almost the same Ta/Tp ratio, 0.86. These facts indicate that the effects of changes in water management are much greater than the direct effects of climate change with regard to water balance and agricultural production in the LSIP.

Keywords Climate change · IMPAM · Irrigation management simulation model Performance assessment

6.1 Introduction

Agriculture that strongly depends on water resources and climate conditions is now facing the effects of climate change. The Mediterranean region is one of the areas where there is a distinct possibility that agriculture will experience the negative effects of climate change. According to simulation results by models registered to Coupled Model Intercomparison Project Phase 5 (CMIP5) with the representative concentration (RCP) 8.5 scenario, annual temperatures and precipitation around the Mediterranean region in 2081–2100 is 4–5 °C higher and 10–20% smaller than those in 1986–2005, respectively (IPCC WG1 2014). This means water demand for irrigation will increase in the future, while river runoff will decrease. Simulation results for rain-fed agriculture in Jordan indicated that a reduction in rainfall of 10–20% reduced the expected yield by 4–8% for barley and 10–20% for wheat (Al-Bakri et al. 2010). Crop production in irrigated areas may also decrease because of water shortages.

   Of course, farmers do not have to accept such decreases in production. There is increasing evidence that farmers in some regions are already adapting to observed climate changes. Although such adaptations are being carried out in limited areas now, there are a large number of potential adaptations for cropping systems and for the food systems of which they are part, many of them enhancements of existing climate risk management and all of which need to be embedded in the wider farm systems and community contexts (IPCC WG2 2014). Irrigated agriculture has various options for adaptation, such as modification of water supply systems, changes in irrigation schedules, etc., in addition to changes in crop types and cultivation schedules. Although possible options for adaptation of irrigated agriculture vary with climate, hydrological situations, existing infrastructure for irrigation, and the financial situation of each country, etc., the number of studies (N =

17) on irrigation optimisation for adaptation evaluated in the IPCC report is limited compared with those on the planting date and cultivar adjustment (N = 152).

This study quantitatively assessed how climate change will affect the irrigated agriculture of the Lower Seyhan Irrigation Project area in the 2070s by factoring the projected future climate data into computational simulations of crop growth and the hydrological structure in Turkey. Three scenarios of adaptation to climate change were employed for the simulations. This study did not seek to predict the actual situation for the irrigated area, but was aimed at providing information that should be considered regarding vulnerabilities in the present irrigation management and determining how irrigated agriculture could adapt to a changing climate.

6.2 Materials and Method

6.2.1 Framework of the Assessment

The effects of climate change were assessed based on two hydrological aspects, namely the water stress on the crops and the groundwater level. In the Lower Seyhan Irrigation District, waterlogging is one of the most important concerns. Simulations of groundwater level and crop growth were carried out employing a numerical model. The climate data of the 2070s, obtained through the downscaling of the global circulation model (GCM) simulations, was fed to the model. MRI-CGCM2 (Yukimoto et al. 2000) and CCSR/NIES-CGCM (Abe-Ouchi et al. 1996) were used. Adaptations to climate change were also assumed. Data related to crop and water management were prepared according to the assumed adaptation as well as the current management obtained through observations. Details of scenarios and parameter settings are described and listed in Sect. 6.2.4.

Water stress was accessed by the ratio of transpiration to the potential transpiration of the crops that decreased by drought or high humidity (Feddes et al. 1978). Although climate change can also affect crop production through the changes occurring in photosynthesis, flowering, and ripening processes, this study only included the effect of water stress to avoid the possible complications and uncertainties arising from such factors during the simulation results.

Simulations were carried out with different combinations of climate data sets (present and projected) and water/crop management data sets (current and adapted). Simulation with the combination of present climate and present management was carried out as the baseline.

6.2.2 Model

To assess the effects on these factors at the same time, a grid-based distributed hydrological model – Irrigation Management Performance Assessment Model

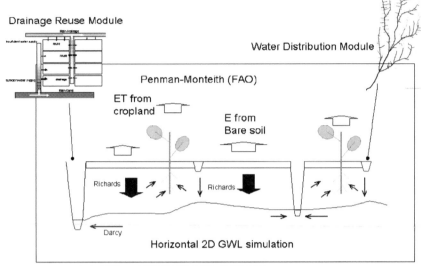

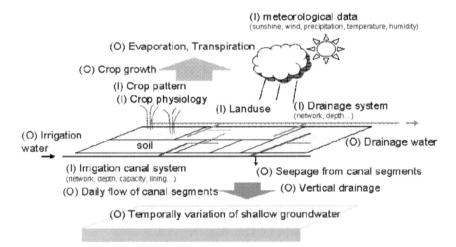

Fig. 6.1 Concept and I/O of IMPAM. *Source* The authors

(IMPAM) (Fig. 6.1) – was used in this study. IMPAM was developed by the authors for simulation of hydrology in irrigated agricultural areas (Hoshikawa et al. 2005). Its spatial scope is from the command area of a tertiary canal up to the whole irrigation project area bound by irrigation canal networks. It calculates the major water balance components in irrigated agricultural areas, namely the amount of irrigation water withdrawal to a subject area as well as the precipitation, seepage from irrigation canals and drainage, evaporation from the soil surface and transpiration from crops. The soil moisture dynamics in saturated and unsaturated zones are calculated separately by a two-dimensional horizontal model and a one-dimensional vertical model respectively.

Crop, irrigation, drainage, water delivery, well water withdrawal modules etc. are assembled on the quasi-three-dimensional soil water dynamics model that consists of the one-dimensional vertical and two-dimensional horizontal models. All major factors and components in hydrological processes, such as the crop calendar and its spatial distribution, irrigation and drainage facility arrangement, topography, etc., are included within the spatial scope of the model. Major inputs of IMPAM are indicated in Fig. 6.1. The hydrological processes that have to be described by this model are seepage from irrigation canals, groundwater flow to drainages, interaction between surface and groundwater (capillary rising and infiltration), soil surface evaporation, transpiration (soil moisture withdrawal by roots), and groundwater flow. Meteorology, irrigation schedule, land use-crop spatial distribution, and the irrigation-drainage channels' spatial distribution database are the main input items of this module.

Resolution of the horizontal grid can be set freely from 10 mup to about 1 km according to the purpose of the simulations.

The details of the methodologies of the model modules and elements are supplied below.

6.2.2.1   Soil Water Dynamics

Horizontal water movement in the saturated zone (temporal and spatial variation of groundwater levels) is expressed by the advection-dispersion equation (ADE) (Eq. 6.1).

$$S\frac{\partial h}{\partial t} = \frac{\partial}{\partial x}\left(K\frac{\partial h}{\partial x}\right) + \frac{\partial}{\partial y}\left(K\frac{\partial h}{\partial y}\right) + R \qquad (6.1)$$

where, h: head, K: transmissivity, S: coefficient of storage, R: recharge rate (Anderson et al. 2015). In irrigated areas, the recharge rate is given by balance of natural and artificial inflow and drainage. IMPAM considers bottom flux of the one-dimensional vertical soil-water dynamics model (upward positive) (q_{bot}), drainage (q_{drain}), seepage from canal segments ($q_{seepage}$), application loss at each farm-lot (q_{apl}), well withdrawal (q_{well}) as components of R, which is given by Eq. 6.2 (Hoshikawa et al. 2005).

$$q_{ssh} = q_{bot} + q_{drain} - q_{seepage} - q_{apl} + q_{well} \qquad (6.2)$$

Methodologies for calculating vertical one-dimensional water movement, including matrix potential flux and root water extraction from soil, are based on a theory used in the SWAP (Soil Water Atmosphere Plant) model (Van Dam et al.

1997), although some parts are simplified. Soil water movement is calculated with the partial differential equation of Richards (Eq. 6.3) for each horizontal node.

$$\frac{\partial \theta}{\partial t} = \frac{\partial [K(h)(\partial h/\partial z) + 1]}{\partial z} - q_{ssv} \qquad (6.3)$$

where, θ: volumetric coefficient of water content. $K(h)$: soil water conductivity [cm d^{-1}] given as a function of pressure head, q_{ssv}: sink/source term for the one-dimensional model that consists of root extraction and preferential flow.

6.2.2.2 Evaporation and Transpiration

Transpiration and evaporation are calculated by two steps. Firstly the transpiration without water stress, which is defined as the "potential transpiration" (T_p), and potential evaporation are calculated by the Penman-Monteith equation with climate data, minimum canopy resistance, leaf area index (LAI), and crop height. Then they are reduced by functions of soil moisture.

Soil surface evaporation is limited by unsaturated soil moisture conductivity. Transpiration is given as the sum of root water extraction S_a at each depth z (Eq. 6.4):

$$T_a = \int_{-D_{root}}^{0} S_a(z) \qquad (6.4)$$

where D_{root} is root depth – in other words, distance from the soil surface to the deepest point of root, which varies with crop types and increases with crop growth. In this study, it was derived from a table showing the relationship between the developing stage and root depth. The S_a is the product of potential root water extraction S_p and the coefficient α that is a function of Feddes et al. (1978) (Eq. 6.5).

$$S_a(z) = \alpha(z)S_p(z) \qquad (6.5)$$

The coefficient α that takes a value between 0 and 1 is determined by a function of soil moisture. It takes 1 when soil moisture is in upper and lower thresholds where crop roots intake water without stress. Out of the range of the two thresholds, the coefficient α decreases linearly toward 0. There are specific thresholds for each crop. The coefficient α less than 1 (water stress) occurs not only in drought conditions under the lower threshold, but also in humid condition higher than the upper threshold.

Although the potential root water extraction at each depth should be determined by potential transpiration (total root water extraction) and ratio of root length density at each depth to the total density, variation of the density with depth is often ignored (Van Dam et al. 1997). IMPAM simply calculates the Sp on the assumption that density distribution is uniform (Eq. 6.6).

$$S_p(z) = \frac{T_p}{D_{root}} \qquad (6.6)$$

6.2.2.3 Crop Growth

LAI, root depth and crop height that are used in the calculation of evaporation and transpiration are calculated as a function of accumulated temperature for each farm plot.

6.2.2.4 Irrigation and Seepage from Canals

The irrigation schedule for each plot is given by table (day-plot-depth) or calculated by the irrigation module that functions to keep soil water content no less than a threshold. Table 6.1 shows apart of a table of irrigation schedule employed in simulations for present water management carried out in this study. Irrigation depth (mm) for each day and plot is recorded in each cell.

Losses in conveyance and delivery often occupy large parts of water balance in irrigated agricultural areas. IMPAM calculates spatial and temporal distribution of

Table 6.1 Part of irrigation schedule for each plot (mm/day) employed in simulations for present water management. *Source* The authors

		Day of year							
		143	144	145	146	147	148	149	150
Plot number	159	0	0	30	30	0	0	0	0
	160	0	0	30	30	30	30	0	0
	161	0	0	30	30	30	30	0	0
	162	0	0	0	0	0	30	30	30
	163	30	30	0	0	0	0	0	0
	164	0	0	0	0	0	0	0	0
	165	0	0	0	30	30	30	30	0
	166	0	0	0	0	0	0	0	0
	167	0	0	0	0	30	30	30	30
	168	0	30	30	0	0	0	0	0
	169	0	0	0	0	30	30	30	30
	170	0	0	0	0	0	0	30	30
	171	0	30	30	0	0	0	0	0
	172	0	0	0	0	0	0	30	30
	173	0	30	30	0	0	0	0	0
	174	0	0	0	30	30	30	30	0

conveyance and delivery losses conceptually according to a time schedule of water delivery. Evaporation is ignored, and conveyance losses consist only of seepage loss in the model. The water delivery schedule is given as a table (canal segment-day) or is created by the Water Distribution Module of the model.

6.2.2.5 Drainage

Drainage contains three types of water: water directly discharged from irrigation canals (tail water), quick drainage of infiltrated water from the soil surface through cracks, and oozing water from the saturated zone. The oozing is calculated as a function of groundwater level, density and level of drain bottom.

6.2.3 Study Area

The Lower Seyhan Irrigation Project (LSIP) (Fig. 6.2a) is located on the Eastern Mediterranean coast of Turkey. Its construction began in the 1960s. The project area of 175,000 ha was divided into four areas and construction for each area was conducted in each project phase. Project Phases I–III (133,000 ha) were completed by 1985. The area Phase IV that remains incomplete is located in the lowest part of the project area. Although it has no water allocation, irrigated agriculture is also practised using the surplus water from the main canals of the completed areas.

The average annual precipitation from 1994 to 2003 in this area was 744 mm (observed at Adana), with most of the precipitation falling during the winter months. According to the projection results obtained by the major GCMs, precipitation and river runoff in the Mediterranean Region, including the Seyhan River Basin, will decrease under warmer climates in the future. In response to the Mediterranean climate, farmers on the upstream side of the basin have been

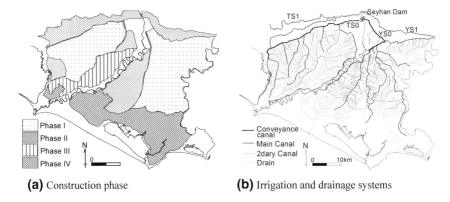

(a) Construction phase   **(b)** Irrigation and drainage systems

Fig. 6.2 Lower Seyhan Irrigation Project. *Source* The authors

cultivating rain-fed winter wheat. However, in the command area of the LSIP, the downstream agricultural production is active mainly during the dry season from spring to autumn and uses a water supply from two neighbouring reservoirs, the Seyhan and Catalan Reservoirs, which store runoff due to winter precipitation from the upstream. As of 2005, the winter wheat is cultivated in only 20% of the project area. Although cotton was a dominant summer crop in the LSIP before the 1980s, maize had replaced it by 2000 because of pest and disease problems and economic reasons. Cultivation of citrus has also been increasing gradually since the 1980s. In 2004, the cultivation areas of maize, cotton, citrus, vegetables, and watermelon comprised 45%, 9%, 13%, 4%, and 6% of the total area, respectively.

The irrigation canal system of the LSIP consists of two conveyance canals, main canals, secondary canals, and tertiary canals. The two conveyance canals (YS0 and TS0) are diverted at the Seyhan Regulator. All of the main canals branch off these two conveyance canals, except for YS1 and TS1, which are diverted directly from the Seyhan Reservoir (Fig. 6.2b). The annual amount of water diversion for the LSIP in 2000 was about 1.6×10^9 m^3. It is trending upwards since the 1990s because of the change in the cropping pattern.

The groundwater exists only between 1 and 2 m below the ground surface in most parts of the LSIP area. Although the Turkish government has been constructing the drainage systems in the LSIP area since the 1960s, some parts of the LSIP had severe problems of improper drainage and salt accumulation induced by the shallow water table (Çetin/Diker 2003). Salinity is stated to proceed due to the same discrepancies, which are valid even today (Şener 1986; Selek/Tunçok 2013). Average electrical conductivity (EC) of groundwater from 2003 to 2004 was 0.8 dS m^{-1} in area Phases I–III, while it was 12.4 0.8 dS m^{-1} in area in Phase IV, which is located at the lowest area near the sea (Kume et al. 2007; Nagano et al. 2009).

6.2.4 Scenarios

Simulations were carried out with three water and crop management scenarios as well as the current management. According to Fujihara et al. (2008), who carried out simulations for runoff of the Seyhan River with the projected climate data sets also used by this study, inflow to Catalan Reservoir at gauge station 1818 will markedly decrease in the future. Average runoff at station 1818 in April, when the annual peak flow occurs, will decrease from 300 m^3/s at the present to around 100 m^3/s in the 2070s. The ratio of water withdrawal to discharge at Seyhan Dam will increase from 0.4 to 0.6 if the current amount of water withdrawal does not change. Many studies have reported that a region is considered highly water stressed if this index exceeds 0.4 (Fujihara et al. 2008). In addition, a decrease in precipitation may induce expansion of irrigated areas in the upstream, which would also cause a decrease in inflow to the dams. Thus, climate change will bring restriction of water diversion for irrigation where changes in crop and irrigation management to adapt to situations with less water availability are assumed as one of the measures. In addition, the abandonment of the uncompleted project area of the LSIP (Phase IV area) and increase of groundwater use for irrigation will possibly occur. The three scenarios listed below took these changes into account.

6.2.4.1 Adaptation Scenario 1 (AD1)

The LSIP will adapt to climate changes while avoiding a large amount of investment in water management. Irrigation water losses (management water) will increase because of the deterioration of the facilities as well as the irrigation water demand at each field-lot due to the anticipated increase in dryness. Subsequently, the LSIP water users will increase the water withdrawal from the river to compensate for the increased water requirement coupled with the decreased availability of the water resources.

In addition, rain-fed winter wheat will not be cultivated because of the decrease in winter precipitation, in turn increasing the orchard cultivated areas. These trends of crop changes are common in the three scenarios.

The incomplete project area (Phase IV) that is irrigated with surplus water from the completed area is estimated to be abandoned in this scenario. In contrast to this prediction, the area is intensely used for cultivated crops today via an almost completed irrigation and drainage network. The irrigation canals of this system have been designed over land consolidation (Selek/Tunçok 2013).

6.2.4.2 Adaptation Scenario 2 (AD2)

Increase of irrigation area (completion of Phase IV area) and increase of net amount of water applied to crops will be attained, although water diversion from the Seyhan River will be decreased. Application losses (water waste at the end of the canal; tail water) and conveyance losses (leakage) will be decreased largely through maintenance of facilities and revision of the water management system. The present situation of the Phase IV irrigation area corresponds to this scenario.

6.2.4.3 Adaptation scenario 3 (AD3)

While all farms depend on surface water (rainwater and water supplied through canals) in scenarios 1 and 2, groundwater irrigation will be adopted for 21,900 ha of orchards (citrus and other fruit-tree crops). The amount of well irrigation is 780 mm in depth annually – as much as the surface water irrigation. In this scenario, 0.17×10^9 m^3 (about 150 mm for area) of total water is withdrawn from groundwater. In the present situation, part of the cultivated land (especially citrus orchards) is irrigated by groundwater wells, whereas most of the irrigation is undertaken via the appropriately planned irrigation canals by land consolidation. In the meantime irrigation via groundwater wells is prohibited and licences for new wells are not issued (Selek/Tunçok 2013).

6.2.5 Data and Parameters

6.2.5.1 Climate Data

Climate data for the 2070s that was used in this study was derived through RCM (Regional Climate Model) downscaling of NCEP-reanalysis data with the "pseudo warming" method (Kimura/Kitoh 2007). Results of two GCMs (MRI-CGCM2 and CCSR/NIES-CGCM) were used to obtain a differential climatology, consequently generating two data sets from the future to the present. In the following explanations, the two climate data sets for the 2070s will be named after the institutes that developed the GCMs: MRI and CCSR/NIES. The climate data set delivered through downscaling of the NCEP-reanalysis data (called NCEP) was used for control runs.

The evaporation of the MRI and CCSR/NIES is slightly larger than that of NCEP. The precipitation of the MRI and CCSR is much less than that of NCEP during winter to spring (Fig. 6.3).

6.2.5.2 Sea and Groundwater Level

The Dirichlet boundary conditions that give values directly on the boundary were used in this study to calculate groundwater levels for simulations of the northern and southern boundaries. The groundwater level at the northern boundary along the foot of the mountains was fixed about 5 m below the ground surface. The one at the southern boundary (the coastline) was 0 m in simulations with the present climate and 0.8 m in those with the projected climate based on a rising trend by the IPCC

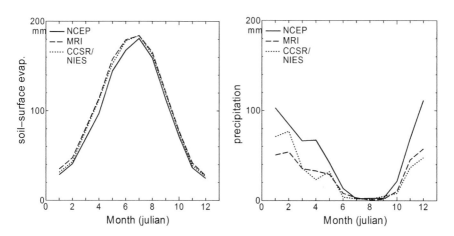

Fig.6.3 Monthly precipitation and potential soil-surface evaporation (ten-year average). Soil-surface evaporation was calculated by Penman–Monteith equation. *Source* The authors

report (Nicholls et al. 2007). The Neumann condition (zero flux) was assumed for the eastern and western boundaries.

As the initial groundwater level is unknown, the five-year spin-up was conducted before the calculation for 1994–2003 in each case.

6.2.5.3 Water and Crop Management

The water and crop management data were generated based on the basin change scenarios that determine the water requirement in the LSIP and the available water resources. Six water management and cropping patterns were assumed as two climate data sets and three basin-change scenarios were used in this study, Simulations were carried out with nine combinations of climate and management data sets, namely one for the control run (NCPPRS), two for the impact assessment on the present management (MRIPRS and CSRPRS) and six for the climate and management changes (MRIAD1, MRIAD2, MRIAD3, CSRAD1, CSRAD2 and CSRAD3) (Table 6.2).

The crop pattern map for the base case (current situation) was derived from a supervised classification of satellite images in 2003. The spatial cropping patterns (Fig. 6.4) in the assumed scenarios were created by a probability method based on the calculated crop ratios and spatial distribution pattern of the base case. The ratios of the crop patterns under adaptation scenarios 1, 2 and 3 under both of the two projected climate (MRI and CCSR/NIES) scenarios were based on the results of the economic analysis concerning the relationship between the available water resource and farmers' behaviour determined by Umetsu et al. (2007) (Table 6.3).

The amount of the current water diversion to the LSIP and available water for the LSIP under the basin change scenarios that were used in Umetsu et al. (2007) was based on the measurements conducted by the DSİ and the runoff analysis undertaken by Fujihara et al. (2008) respectively. The analysis of Fujihara et al. (2008) was based on the same projected climate data set as the one utilised in this study.

Nagano et al. (2005) based parameter settings for irrigation management at field level on the survey. Rotation irrigation was not practised in the LSIP area, and water was supplied at every canal throughout the irrigation seasons. The ratio of the

Table 6.2 Codes for combination of data sets for simulations. *Source* The authors

	Present	Adaptation 1		Adaptation 2		Adaptation 3	
	NCPPRS	MRIAD1	CSRAD1	MRIAD2	CSRAD2	MRIAD3	CSRAD3
NCP	NCPPRS						
MRI	MRIPRS	MRIAD1		MRIAD2		MRIAD3	
CSR	CSRPRS		CSRAD1		CSRAD2		CSRAD3

NCP, and CSR stand for NCEP and CCSR/NIES respectively

Crop and water management in 2003

ADxM/ADxC: Crop and water management for each scenario under climate MRI / CSR

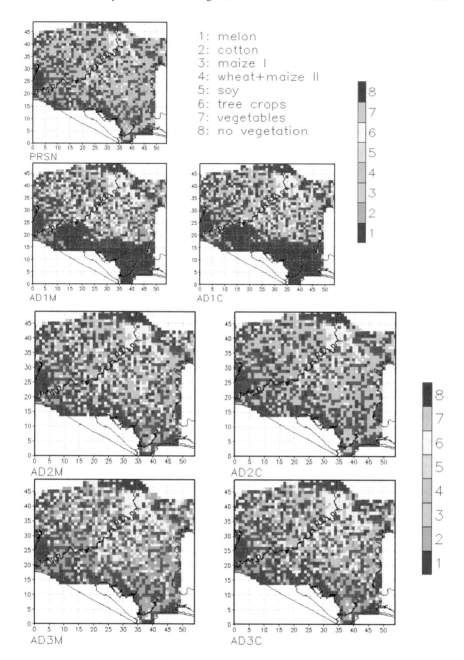

Fig. 6.4 Spatial distributions of crop patterns (Grid size: 1,000 m × 1,000 m). *Source* The authors

Table 6.3 Ratio of crops for each scenario (unit: %). *Source* The authors

	Present	Adaptation 1		Adaptation 2		Adaptation 3	
	NCPPRS	MRIAD1	CSRAD1	MRIAD2	CSRAD2	MRIAD3	CSRAD3
Maize	52	33	51	41	63	10	31
Wheat + Maize II	12	0	0	0	0	0	0
Cotton	9	24	4	15	0	48	26
Melon	7	8	13	10	16	3	8
Orchard	14	30	29	30	29	32	30
Soy	2	0	0	0	0	0	0
Vegetables	5	4	3	4	3	6	5

delivery water requirement to the total was quite large in the LSIP area. Seepage losses from the tertiary canals in the LSIP were relatively small as they were lined or constructed with flumes. However, two-thirds of the water supply directly flowed into the drainage channels as the tail water or infiltrate after the application loss around the inlet of each farm-lot. Consequently, the amount of the water supplied to a tertiary canal was three times the net water requirement in the LSIP area.

Based on Fujihara et al. (2007), Nagano et al. (2005) and the design conveyance efficiencies of the LSIP area, the parameters for water supply and delivery were set for the present and assumed situations (Table 6.4). The amount and schedule of the irrigation determined in the above processes were fed to IMPAM as a fixed schedule table.

Table 6.4 Annual water supply and delivery parameters. *Source* The authors

	Present[a] PRS	Ad.1 MRIAD1	CSRAD1	Ad.2 MRIAD2	CSRAD2	Ad.3 MRIAD3	CSRAD3
From the Seyhan River	1.56×10^9 m^3	1.82×10^9 m^3	1.82×10^9 m^3	1.32×10^9 m^3	1.31×10^9 m^3	0.98×10^9 m^3	0.97×10^9 m^3
Well water						0.28×10^9 m^3	0.28×10^9 m^3
Irrigated area	93,500 ha	93,500 ha		113,900 ha		113,900 ha	
Applied to crops (net)	0.50×10^9 m^3	0.59×10^9 m^3	0.59×10^9 m^3	0.70×10^9 m^3	0.69×10^9 m^3	0.66×10^9 m^3	0.61×10^9 m^3
Others							
Application losses	0.34×10^9 m^3	0.39×10^9 m^3	0.39×10^9 m^3	0.27×10^9 m^3	0.29×10^9 m^3	0.22×10^9 m^3	0.23×10^9 m^3
Conveyance losses	0.16×10^9 m^3	0.38×10^9 m^3		0.11×10^9 m^3		0.10×10^9 m^3	
Tail water	0.57×10^9 m^3	0.46×10^9 m^3		0.21×10^9 m^3		0.21×10^9 m^3	

[a]Net irrigation requirement (0.10×10^9 m^3), application losses (0.07×10^9 m^3) and conveyance losses (0.06×10^9 m^3) for the Phase IV area are not included for PRS in this table

The present command area in Table 6.4 does not contain the Phase IV area (20,400 ha was irrigated in this study). It was assumed that the unit amount of net water requirement and application losses for Phase IV are the same as in the other areas. These water requirements for the Phase IV area 0.23×10^9 m^3 were covered by the tail water (0.57×10^9 m^3) (Table 6.4), whereas the tail water in AD1 (0.46×10^9 m^3) was wasted by being diverted to the sea.

6.2.5.4 Other Settings and Parameters

For simulation of the vertical water movement in IMPAM, parameters of soli physics for silt with saturated water conductivity 0.26 ms^{-1} were given for the whole area based on field measurements. Horizontal transmissivity of the groundwater was set to 4,000 m^2 d^{-1} for the whole study area. The spatial resolution was 1,000 m $\times$ 1,000 m and the time step for crop growth and water management was one day. The hydrological dynamics in the one-dimensional vertical and two-dimensional horizontal models were calculated with a 0.5 day time step at most, which was divided into shorter periods as necessary for the convergence of the calculation.

6.3 Results

6.3.1 Groundwater Level

Figure 6.5 shows the temporal groundwater depth (depth from the ground surface) in the control run. The spatial distribution of the groundwater depth is largely in line with the actual one obtained through measurements at wells in the LSIP area.

Table 6.5 shows average groundwater depth from the ground surface (m) (ten-year mean) for the whole study area by each combination of climate data and scenario.

Figure 6.6 shows the differences of the groundwater levels between the control and runs with projected climate data sets and the present water/crop management (MRIPRS and CSRPRS). The groundwater levels in MRIPRS and CSRPRS were lower than that in the control run (NCPPRS), especially in the winter, but overall showed quite similar trends of changes. Although the 0.8 m higher seawater level was given to the simulations with projected climate data, the groundwater rise was observed within 3 km at most along the coastline in these two runs.

Figure 6.7 shows differences of groundwater levels between the control run and runs with one of the projected climate data sets (CSR) and assumed management to adapt climate changes (AD1–AD3). Significant differences were observed between

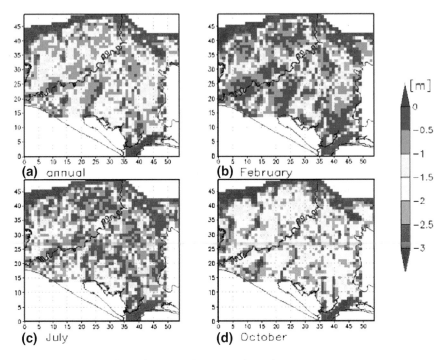

Fig. 6.5 Groundwater depths in NCPPRS. *Source* The authors

Table 6.5 Average groundwater depth from the ground surface (m) (ten-year mean) for the whole study area by each combination of climate data and scenario. *Source* The authors

	AD1	AD2	AD3	PRS
CSR	1.44	1.67	1.73	1.59
MRI	1.40	1.61	1.67	1.55
NCEP				1.31

the scenarios. The AD1 simulations resulted in lower groundwater levels in the project Phase IV area (the lowest part of the area) and higher groundwater level in Phase I–III areas (Fig. 6.7a). No significant difference in the groundwater level was observed between AD2 and AD3 (Fig. 6.7a, b), although the application loss and conveyance loss (leakage from canal, water waste from the end of canal, etc.) in AD3 were smaller than in AD2.

The inflow rates at the boundaries along the sea and mountainsides were 150 mm, 140 mm and 170 mm in MRIPRS, MRIAD1 and MRIAD2 respectively, while it was 100 mm in NCPPRS.

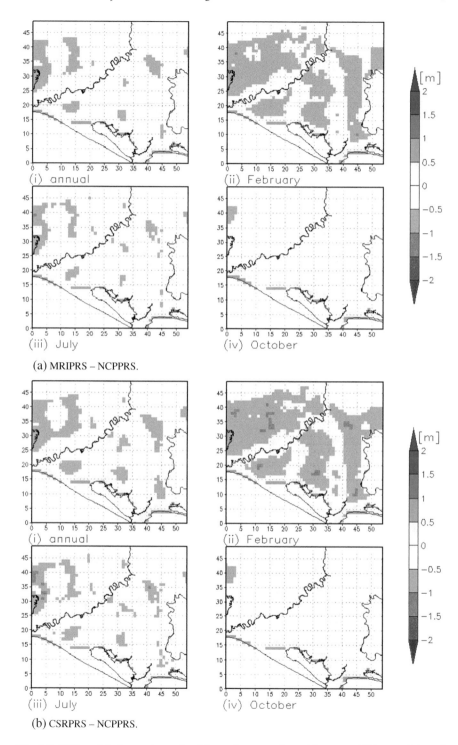

(a) MRIPRS – NCPPRS.

(b) CSRPRS – NCPPRS.

Fig. 6.6 Differential of groundwater level (annual mean and monthly mean of February, July and October) with the two climate data set. *Source* The authors

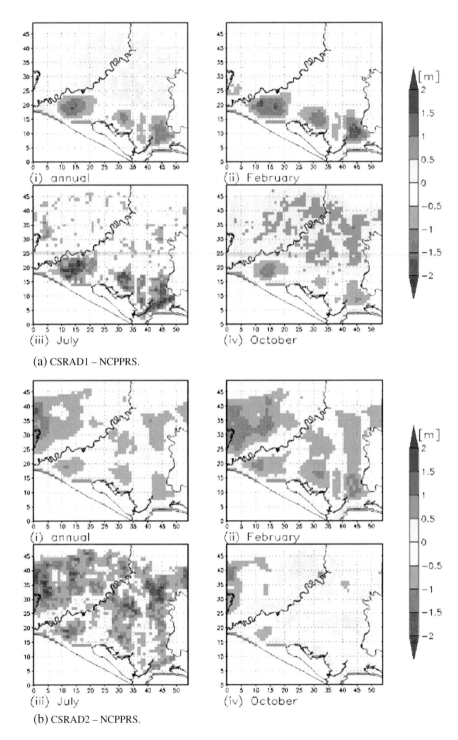

(a) CSRAD1 – NCPPRS.

(b) CSRAD2 – NCPPRS.

Fig. 6.7 Differential of groundwater level (annual mean and monthly mean of February, July and October). *Source* The authors

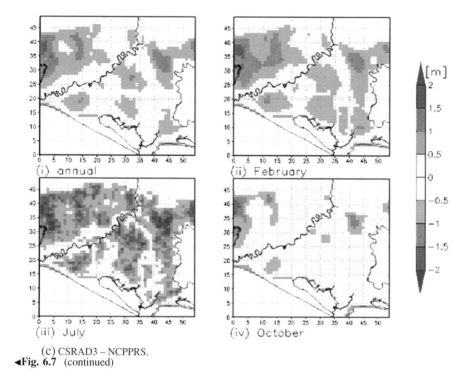

(c) CSRAD3 – NCPPRS.
◀**Fig. 6.7** (continued)

6.3.2   *Ta/Tp Ratio*

Figure 6.8 shows the ratio of the annual actual transpiration (Ta) to potential transpiration (Tp) in the control run. The Ta/Tp ratio is 1.0 without the water stress and with a decrease caused by drought or high humid soil moisture conditions. The control run varied from 0.7 to 1.0 in most of the area. The restriction of transpiration occurred mostly because of soil humidity higher than the upper threshold of the function that determines α in Eq. (6.5). Such high humid conditions appeared in the areas where groundwater level is near the ground surface.

Table 6.6 shows the average Ta/Tp ratio for all pixels with crops and throughout the simulation period (ten years). All simulations with the present management resulted in almost the same Ta/Tp ratio. The highest ratio appeared in the combination of CCSR and adaptation scenario 2. The simulations with adapted scenario 1, which increases water diversion from the river to cover a decrease in precipitation, resulted in the lowest Ta/Tp ratio.

Figure 6.9 shows the differences of the annual Ta/Tp ratio between the control run and runs with two projected climate data sets (MRI and CSR) and three

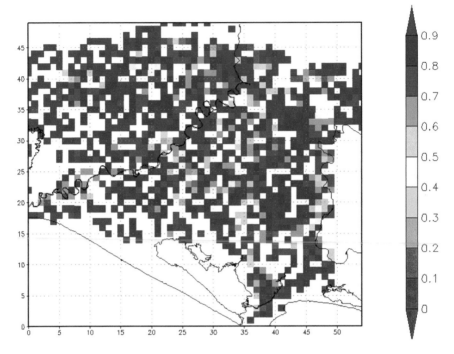

Fig. 6.8 Annual Ta/Tp in NCPPRS (ten-year mean). *Source* The authors

Table 6.6 Average T_a/T_p ratio (ten-year mean) for the all crop fields by each combination of climate data and scenario. *Source* The authors

	AD1	AD2	AD3	PRS
CSR	0.82	0.89	0.88	0.86
MRI	0.81	0.86	0.87	0.86
NCEP				0.86

adaptation scenarios (AD1–AD3). The Ta/Tp ratio in the Phase IV area was not calculated in the AD1 runs because the Phase IV area was abandoned in the AD1 scenario. The same amount of irrigation was applied in NCPPRS, MRIPRS and CSRPRS, while the latter two cases have more potential evaporation and less precipitation. No significant difference in the Ta/Tp ratio was determined among the three cases. Larger changes were seen in runs with adaptation scenarios AD2 and AD3, in which the Ta/Tp ratio increased significantly in most of the area.

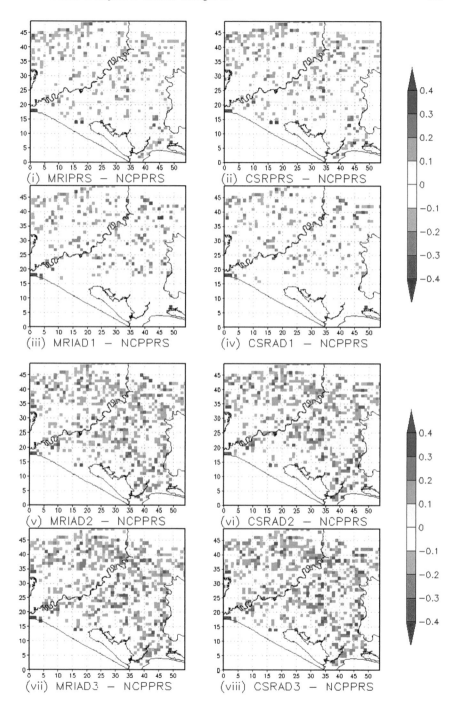

Fig. 6.9 Differential of annual mean of Ta/Tp ratio. *Source* The authors

6.4 Discussions

6.4.1 Climate Change under the Present Management

The groundwater level in the MRIPRS and CSRPRS was lower than that in the control run (NCPPRS), especially in winter, because of the decreased precipitation. Apart from that CSRPRS showed a slightly larger fall of groundwater level in the winter than MRIPRS, where there was no significant difference between the two simulation results during and just after the irrigation season. This suggested that the differences in precipitation between the projections appeared in the groundwater level only in non-irrigated periods in areas where irrigation with vast surplus was applied.

The sea level rise also only affected the groundwater level along the coastal side. Except for the lagoons and lowlands in the coastal area, the gradient of the ground surface and groundwater from the mountainside to the sea (0.15% at least) was high enough to prevent seawater from intruding by 0.8 m of seawater rise.

Figure 6.10 shows the relationship between the T_a/T_p ratio and the groundwater depth at each pixel with crops at the last time step (0.5 day) of every seven days. Although NCPPRS and CSRPRS showed no significant difference in average T_a/T_p ratio (Fig. 6.9 and Table 6.5). Figure 6.10a, b indicate that restriction of transpiration occurred with different factors. In NCPPRS, pixels with a low T_a/T_p ratio are distributed mainly in areas with a groundwater depth of less than 1 m, while in CSRPRS restriction of Ta occurred also at pixels with a groundwater depth greater

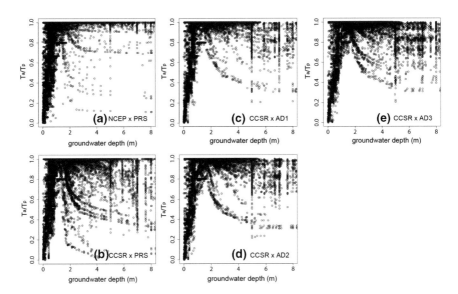

Fig. 6.10 Weekly Ta/Tp ratio and groundwater depth at each pixel with crops. *Source* The authors

than 2 m. Restriction of transpiration in CSRPRS occurred both in over-humid conditions and drought; in other words, decreased precipitation improved over-humidity in some areas while it brought drought to other areas.

These facts indicate that the decrease of precipitation on crop growth will not always be significant in areas with interaction between ground surface and groundwater.

6.4.2 Changed Climate with Adaptation

The fall in the groundwater level predicted in the coastal areas for the simulations of AD1 was due to the absence of irrigation in the Phase IV area. The significant fall in the groundwater level throughout the area in the simulations of AD2 was caused by less seepage loss. The fall of the groundwater level in AD3 was due to the irrigation water extraction from the wells. Consequently, the AD2 and AD3 simulations yielded similar groundwater levels.

The increase in the Ta/Tp ratio in the AD2 and AD3 simulations was because of the fall in the groundwater level. In runs with water management of PRS and AD1, transpiration from crops was restricted because of the high humidity prevailing in the area. A high positive correlation between Ta/Tp ratio and groundwater depth ($R^2 = 0.91$) among simulation (Tables 6.4 and 6.5) means that proper water management to reduce over-humidity is effective, and can compensate for the negative effects of climate change.

6.4.3 Overall Effects of Climate Change

It was suggested that the direct effect of climate change on hydrological conditions and crop growth in the LSIP was relatively small in comparison to the effects of changes in water and crop management. This stability was brought by the present water management with plentiful water application including losses. Even the scenarios with large decreases in water diversion from the Seyhan River (AD1 and AD2) brought positive effects to hydrological conditions and crop growth. This suggests that the selected appropriate LSIP scenarios can mitigate the impacts of climate change.

6.5 Conclusions

This study utilised two climate data sets derived from two GCMs. Simulations with the two data sets substantially showed the same direction of the hydrological changes determined in the LSIP area.

The direct effect of global warming on the hydrology of the LSIP may not be large enough to affect agricultural production in the area. The effect of the sea level rise might be limited within the range of a few kilometres from the coastline with scarce field crop areas. If changes in water and crop management are induced by global warming however, the impacts on agricultural production might be much higher.

Acknowledgements Ministry of Education, Culture, Sports, Science and Technology (MEXT), Japan carried out this study under ICCAP and RR2002, which were financially supported. We would like to express our gratitude to the Water Users' Association of the Lower Seyhan Irrigation Project, which kindly provided valuable data for simulations.

The main components and modules of IMPAM were developed in Subject 6 of the Research Revolution 2002 (RR2002) "Development of Water Resource Prediction Models," funded by the Ministry of Education, Culture, Sports, Science and Technology, Japan. This is a contribution from the ICCAP Project (Impact of Climate Changes on Agricultural Production in Arid Areas) promoted by the Research Institute for Humanity and Nature (RIHN) and the Scientific and Technical Research Council of Turkey (TÜBİTAK). This research was financially supported in part by the Japan Society for the Promotion of Science Grant-in-Aid no. 16380164. Data for spatial distribution of crop types were provided by Dr. Suha Berberoglu at Çukurova University.

References

Abe-Ouchi A, Yamanaka Y, Kimoto M (1996) *Outline of coupled atmosphere and ocean model and experiment*. Internal report, Centre for Climate System Research, University of Tokyo, Japan.

Al-Bakri J, Suleiman A, Abdulla F, Ayad J (2010) Potential impact of climate change on rainfed agriculture of a semi-arid basin in Jordan. *Physics and Chemistry of the Earth, Parts A/B/C* 36 (5–6):125–134.

Çetin M, Diker K (2003) Assessing drainage problem areas by GIS: A case study in the eastern Mediterranean region of Turkey. *Irrigation and Drainage Journal* 54(2):343–353.

Feddes RA, Kowalik PJ, Zaradny H (1978) *Simulation of field water use and crop yield*. Simulation Monographs. Pudoc. Wageningen.

Fujihara Y, Tanaka K, Watanabe T, Kojiri T (2007) Assessing Impact of Climate Change on the Water Resources of the Seyhan River Basin, Turkey, *Final Report of the ICCAP*, Research Institute for Humanity and Nature, Kyoto, Japan.

Fujihara Y, Tanaka K, Watanabe T, Naganoa T, Kojiri T (2008) Assessing the impacts of climate change on the water resources of the Seyhan River Basin in Turkey: Use of dynamically downscaled data for hydrologic simulations. *Journal of Hydrology* 353:33–48.

Hoshikawa K, Kume T, Watanabe T, Nagano T (2005) A model for assessing the performance of irrigation management systems and studying regional water balances in arid zones. *Proceedings of the 19th Congress of International Commission on Irrigation and Drainage*, Beijing, China.

IPCC working group 1 (2014) *Climate Change 2013: The Physical Science Basis*. Cambridge: Cambridge University Press; at: https://www.ipcc.ch/report/ar5/wg1/ (8 September 2017).

IPCC working group 2 (2014) *Climate Change 2014: Impacts, Adaptation, and Vulnerability*; at: https://www.ipcc.ch/report/ar5/wg2/ (8 September 2017).

Kimura F, Kitoh A (2007) Downscaling by Pseudo Warming Method, *Final Report of the ICCAP*, Research Institute for Humanity and Nature, Kyoto, Japan.

Kume T, Nagano T, Akca A, Donma S, Hoshikawa K, Berberoglu S, Serdem M, Kapur S and
    Watanabe T (2007) Impact of the Irrigation Water Use on the Groundwater Environment and
    the Soil Salinity, *The final report of ICCAP*, The research project on the impact of climate
    changes on agricultural productions system in Arid Areas (ICCAP), March 2007,
    RIHN-TUBITAK, ISBN: 4-902325-09-8.
Mary PA, William WW, Randall JH (2015) *Applied Groundwater Modeling: Simulation of Flow
    and Advective Transport*. London: Academic Press.
Nagano T, Donma S, Önder S, Özekici B (2005) Water use efficiency of the selected tertiary canals
    in the Lower Seyhan Irrigation Project Area. *Progress Report of the ICCAP*, Research Institute
    for Humanity and Nature, Kyoto, Japan.
Nagano T, Hoshikawa K, Onishi T, Kume T, Watanbe T (2009) Long-term changes in water and
    salinity management in Lower Seyhan Plain, Turkey. In Taniguchi M, Burnttt WC,
    Fukushima Y, Haigh M, Umezawa Y (eds) *From Headwaters to the Ocean: Hydrological
    Changes and Watershed Management*. London: Taylor and Francis, 313–320.
Nicholls RJ, Wong PP, Burkett VR, Codignotto JO, Hay JE, McLean RF, Ragoonaden S,
    Woodroffe CD (2007) Coastal systems and low-lying areas. In Parry M, Canziani OF
    (eds) *Climate Change 2007: Impacts, Adaptation and Vulnerability. Contribution of Working
    Group II to the Fourth Assessment Report of the Intergovernmental Panel on Climate Change*.
    Cambridge: Cambridge University Press.
Palutikof JP, van der Linden PJ, Hanson CE (eds) *Climate Change 2007 – Impacts, Adaptation,
    and Vulnerability*. Cambridge: Cambridge University Press, 315–356.
Şener S (1986) Development of drainage in Turkey over the last 25 years and its prospects for the
    future. *Proceedings of 25th International Course on Land Drainage* (ICLD), 236–246.
Selek B, Tunçok IK (2013) Effects of climate change on surface water management of Seyhan
    basin, Turkey. *Environmental and Ecological Statistics* 21:391–409. https://doi.org/10.1007/
    s10651-013-0260-5.
Umetsu C, Palanisami, Coşkun Z, Donma S, Nagao T, Fujihara Y, Tanaka K (2007) Climate
    Change and Alternative Cropping Patterns in Lower Seyhan Irrigation Project: A Regional
    Simulation Analysis, *Final Report of the ICCAP*, Research Institute for Humanity and Nature,
    Kyoto, Japan.
Van Dam JC, Huygen J, Wesseling JG, Feddes RA, Kabat P, VanWalsum PEV, Groenendijk P,
    Van Diepen CA (1997) *Theory of SWAP version* 2.0. Wageningen Agricultural University and
    DLO Winand Staring Centre. Technical Document 45.
Yukimoto S, Endoh M, Kitamura Y, Kitoh A, Motoi T, Noda A (2000) ENSO-like interdecadal
    variability in the Pacific Ocean as simulated in a coupled general circulation model. *Journal of
    Geophysical Research* 105, 13, 945-13, 963.

Chapter 7
Adaptation of Contemporary Irrigation Systems to Face the Challenges of Future Climate Changes in the Mediterranean Region: A Case Study of the Lower Seyhan Irrigation System

Rıza Kanber, Mustafa Ünlü, Burçak Kapur, Bülent Özekici
and Sevgi Donma

Abstract The Mediterranean region will be particularly affected by climate change over the 21st century. Rising temperatures and more marked drought periods will affect spatial and temporal precipitation and hence the water resources. This paper first reviews and evaluates the current and future social and environmental pressures on water resources, including climate change. The results show that pressures are not uniform across the region and sectors of water use. The changes in temperature and precipitation predicted by the general circulation models for the Mediterranean region will affect water availability and resource management, critically shaping the patterns of future crop production. The temperatures in the Mediterranean region are expected to rise by +2 to +3 °C by 2050, then by +3 to 5 °C by 2100. The water-poor countries are likely to be the most affected by 2100, and rainfall is likely to have decreased by 20–30% in the countries to the south, opposed to merely 10% in those to the north. The Mediterranean basin is thus predicted to be particularly sensitive to climate change.

This paper also evaluates the adaptation capacity of the Lower Seyhan Irrigation Project area to the future climate change as a case study. The case study reflects the

R. Kanber, Retired Professor, Department of Agricultural Structures and Irrigation Engineering, Çukurova University Adana, Turkey; e-mail: kanber@cu.edu.tr.

M. Ünlü, Professor, Department of Agricultural Structures and Irrigation Engineering, Çukurova University Adana, Turkey; e-mail: munlu@cu.edu.tr.

B. Kapur, Assistant Professor, Department of Agricultural Structures and Irrigation Engineering, Çukurova University Adana, Turkey; e-mail: bkapur@cu.edu.tr.

B. Özekici, Professor, Department of Agricultural Structures and Irrigation Engineering, Çukurova University Adana, Turkey; e-mail: ozekici@cu.edu.tr.

S. Donma, Agricultural Engineer, State Hydraulics Works, 6th Regional Directorate, Adana, Turkey; e-mail: sevgi60@yahoo.com.

© Springer Nature Switzerland AG 2019   125
T. Watanabe et al. (eds.), *Climate Change Impacts on Basin Agro-ecosystems*,
The Anthropocene: Politik—Economics—Society—Science 18,
https://doi.org/10.1007/978-3-030-01036-2_7

outcomes of the Turkish Japanese bi-lateral project entitled "Impact of climate changes on the agricultural production system in arid areas-ICCAP". The ICCAP project was launched in the Seyhan River Basin located in the east of the Mediterranean region of Turkey. The effects of climate change on temperature and precipitation have been estimated by different models of MRI-GCM and CCSR-GCM. According to the generated scenarios by these models, the surface temperature may increase by 2.0 °C to 3.5 °C respectively by 2070. The total precipitation for the whole of Turkey may decrease by 20%, while it will decrease by 25% in the LSIP area, 42–46% in Adana (located in the Lower Seyhan Plain), and by an average of 30% in the Seyhan River Basin. However, the LSIP at present seems to have a large adaptive capacity towards climatic and social changes. To sustain its productivity, it is strongly recommended to farmers and water users' associations to improve irrigation and water use efficiency by means of better maintenance of irrigation canals, better gate operations and employment of better application techniques. This would improve the equity of water allocation and distribution, avoid high water tables and conserve the soil. In the whole area, especially in the coastal zone, the appropriate management of subsurface drainage is vital for avoiding salinity and waterlogging. The use of the deep groundwater should be avoided because of the risk of salt intrusion.

Keywords Climate change · ICCAP agricultural irrigation · Water resources

7.1 Introduction

There has been a decrease in rainfall throughout the Mediterranean region over the past century. Most of the recent attention devoted by the relevant sciences to global warming or the greenhouse effect is paid to the impact of climate change and the increase in temperature. According to the IPCC (2007), temperatures are expected to rise by +2 to +3 °C in the Mediterranean region by 2050, then by +3 to +5 °C by 2100. However, some of the most severe impacts of climate change are likely to come not from the expected increase in temperature but from the changes in precipitation, evapotranspiration, runoff and soil moisture, which are crucial factors in water planning and management. An increase in the temperature is likely to reduce air relative humidity and raise the moisture-loading capacity of the atmosphere. The air will therefore have a higher saturation rate, leading to less cloud cover and hence decreased precipitation. Precipitation will be less frequent but more intense, while periods of drought will be longer and more frequent. Thus spatial and temporal precipitation patterns will be altered.

The regional effects of global climate change on the water cycle, therefore, risk decreasing water resources, emphasising their variability and reducing their exploitability (Milano 2010). The water-poor countries are likely to be the most affected by 2100, and rainfall is likely to have decreased by 20–30% in the countries to the south (particularly in the Mediterranean), opposed to merely 10% in those to the north (Giorgi/Lionello 2008). The ICCAP study conducted in the Seyhan river basin showed that water resources will decrease by 30% due to the effects of climate change in 2070 (ICCAP 2007).

Today, most of the southern rim of the Mediterranean countries imports more than 50% of their food requirements and demand for food grows faster than the rate of increase in agricultural production. Irrigation is often the driving factor for increasing agricultural production in this area. Consequently, huge efforts and budgets are allocated to the Mediterranean countries to develop their irrigation systems. Only 30% of the cultivated area in the region is irrigated, but produces 70–75% of the total agricultural production (Hamdy/Lacirigniola 1999). Water experts and politicians agree that there is an acute water shortage problem in the Mediterranean region. Four Mediterranean countries already have less than the minimum required water availability to sustain their own food production (750 m^3/inh/yr). By 2025, more countries will be in virtually the same situation concerning water shortage. These countries are essentially located on the southern rim of the Mediterranean basin. The crisis is already so acute that in Malta domestic water consumption exceeds 50% of the available water resources. In such places, the conventional water resources will be insufficient to even meet the domestic water demand at the beginning of the next century. Climate change is just one of the pressures facing water resources and their management over the next few years and decades in the Mediterranean.

This chapter describes the assessment of the effects of climate change on water resources, and their management in the selected irrigation systems of the Mediterranean basin countries. It also provides information on the adaptation of irrigation systems to the future climate changes in the Mediterranean region. Detailed information on the limiting factors of the water resources related to climate change is also discussed in this chapter.

7.2   Restrictive Factors on Water Resources in the Mediterranean

Currently there is an increasing pressure on water resources in the Mediterranean countries, derived from population dynamics, upgraded standard of living, economic and social development, and the use of water-consuming technologies. Different countries in the region have different problems with their water sectors. As stated by Hamoda (2004), water resources in the Mediterranean region are scarce and expensive, thus municipal and industrial water requirements are increasing sharply due to the population increase and rising incomes. However, the increased demands for water and the subsequent relevant approaches for solving water problems are mostly limited to improving management, as well as upgrading and modernising water delivery systems.

7.2.1 Population of the Mediterranean Region

Population growth, in many southern Mediterranean basin countries, is the major factor affecting water resources that reduces the water availability per capita (Fig. 7.1 and Table 7.1). Urbanisation increases urban demands, which are of high priority, and intensifies conflicts among users. For example, the Mediterranean population living in water-poor countries with less than 1,000 m^3/capita/yr could rise from 180–250 million by 2025. The populations living in the regions of water shortage, such as the Palestinian Territories-Gaza, Libya, Malta, Israel, Tunisia and Algeria, with less than 500 m^3/capita/yr could increase over the same period from 60–80 million inhabitants (Milano 2010). Apart from the local population, even the tourist population in the Mediterranean is significantly high compared with other parts of the world and faces an increasing trend in time. Accordingly, the EEA (2010) has reported that the tourist water consumption has lately been about 3–4

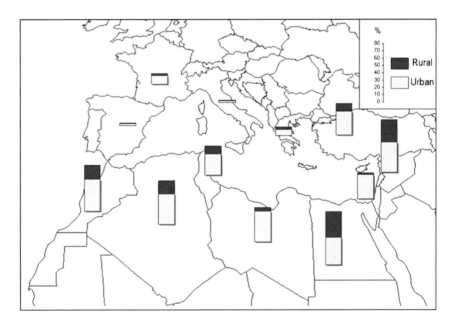

Fig. 7.1 Mediterranean countries and populations for 2025. No data is available for Lebanon, Albania, Cyprus, Malta or former Yugoslavian countries. *Source* Rearranged from Iglesias et al. (2007), Aquastat (2010)

times higher than the local demands. In some countries like Spain, Greece, France, and Turkey, the number of international tourists increases every year; and their number exceeds the total population of these countries by about one third of their population (World Tourist Organization 2010). However, touristic consumption is highly seasonal, but the tourism industry increases permanent water demand for facilities and leisure structures.

Similarly, the natural water resources in the Mediterranean region of Turkey are threatened by the tourist population, which was about 25 million in 2010. Along the Mediterranean coast of the country there are 200 golf courses, representing a demand increase of about 200 million m^3 of water. This is a small fraction of the agricultural demand at regional level. Additionally, the daily touristic water consumption is about 350–850 litres, which is about 2–4 times higher than local demand. This is documenting the serious water scarcity problem in the near future of 2025, when 200 million tourists are expected to visit Turkey (Bulut 2010). On the other hand, it is estimated that water use in the tourism sector will be 5 billion m^3 in 2023 (Evsahibioğlu et al. 2010). The other problem is the waste water production in the tourist areas of the Mediterranean coast, which was 400 million m^3/yr in 2004 and will be 1.5 billion m^3 in 2023 (Baykam 2004).

7.2.2 Environmental Factors

In general, usable water resources are always less than the potential water resources in all countries. In Turkey the total water use is less than half of the total freshwater resources (Tekinel et al. 2000; Kanber et al. 2004), and the potential use of surface water under a natural regime is only about 38% (Kanber et al. 2004). By contrast, in Egypt, Israel, Cyprus, Libya, and many other riparian countries of the Mediterranean basin, the water demand is higher than the available resources (Table 7.1).

The recurrent drought events in the region further increase the complexity of water scarcity management. Drought events in the Mediterranean have been more frequent since 1980 (Fig. 7.2) (ICCP 2007; Kitoh 2007) and have occurred two or three times in the last 20–25 years, causing social damage and severe decline in the economy of the Mediterranean region of Turkey (Şaylan/Çaldağ 2001; WWF 2007; Kanber et al. 2008). Figure 7.2 shows the time series of the annual total precipitation in Adana and the corresponding widely used drought index SPI calculated at 24-month intervals. The negative SPI values obtained after the 1990s indicate drought risk. However, drought indices do not correlate well with hydrological drought periods. This may be the outcome of the effect of the water storage capacity of Turkey, as stated by Tezcan et al. (2007) and Fujinawa et al. (2007). The Figure shows at least three periods with different precipitation trends and variability patterns. Precipitation after the 1990s has clearly followed a decreasing trend and provoked further water deficit in many areas of the country (Şaylan/Çaldağ 2001).

Table 7.1 Some water resources data in mediterranean countries. *Source* FAO (2005), Aquastat (2010)

Country	Total area (1000 ha)	Population (million)	Rainfall (mm/yr)	Internal usable water resources (km³/yr)[a]	Total usable water resources (km³/yr)[b]	Internal groundwater (km³/yr)[c]	Total water use (km³/yr)	Total water use (% renewable)	Potential total usable water resources per capita (m³/capita/yr)
Albania	2875	3.14	1485	26.90	41.70	6.20	1.710	4.10	13.268
Algeria	238 174	34.37	89	11.25	11.67	1.49	6.053	51.86	339.50
Bosnia-Herzegovina	5 121	3.77	1028	35.50	37.50	–	–	–	9939
Croatia	5 659	4.42	1113	37.70	105.50	11.00	–	–	23853
Cyprus	925	0.86	498	0.78	0.78	0.41	0.213	27.30	904.9
Egypt	100 145	81.53	51	1.80	57.30	1.30	68.200	1.19	702.8
France	54 919	62.04	867	178.50	203.70	100.00	39.950	19.61	3284
Greece	13 196	11.14	652	58.00	74.25	10.30	7.760	10.45	6667
Israel	2 077	7.05	435	0.75	1.78	0.50	1.814	101.91	252.4
Italy	30 134	59.60	832	182.0	191.30	43.00	44.270	23.14	3210
Lebanon	1 040	4.19	661	4.80	4.50	3.20	1.263	28.06	1074
Libyan Arab J.	175 954	6.29	56	0.60	0.60	0.50	4.308	718.00	95.33
Malta	32	0.41	560	0.051	0.051	0.05	0.0195	38.23	124.10
Morocco	44 655	31.61	346	29.00	29.00	10.00	12.590	43.41	917.50
Occupied Palestinian Territory-Gaza	602	4.15	402	0.812	0.837	0.74	0.418	49.94	201.8
Slovenia	2 027	2.02	1162	18.67	31.87	13.50	–	–	15816

(continued)

Table 7.1 (continued)

Country	Total area (1000 ha)	Population (million)	Rainfall (mm/yr)	Internal usable water resources (km³/yr)[a]	Total usable water resources (km³/yr)[b]	Internal groundwater (km³/yr)[c]	Total water use (km³/yr)	Total water use (% renewable)	Potential total usable water resources per capita (m³/capita/yr)
Spam	50 537	44.49	636	111.2	111.50	29.90	35.530	31.86	2506
Syrian Arab J.	18 518	21.22	252	7.13	16.80	4.84	16.690	99.34	791.4
Tunisia	16 361	10.17	207	4.19	4.59	1.49	2.837	61.80	451.9
Turkey	78 356	73.91	593	227.00	213.60	69.00	40.100	18.77	2890

[a]The values refer to both regulated and unregulated water. Real available water resources in all cases are a fraction of these values

[b]These values include transboundary water

[c]A proportion of these values in included in the total renewable resource

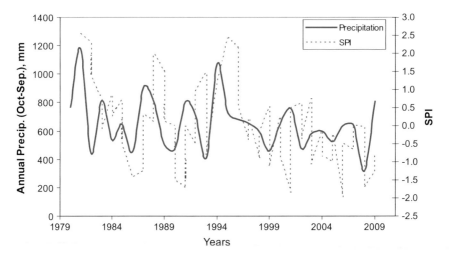

Fig. 7.2 Time series of the total annual precipitation and SPI values (24-month timescale) in Adana, Turkey. *Source* ICCAP (2007), Kitoh (2007)

7.2.3 Water Resources in Mediterranean Countries

The Mediterranean region is a transition area between a temperate Europe with relatively abundant and consistent water resources and the arid African and Arabian deserts that are very short of water except for the Nile (Giorgi/Lionello 2008). This area is a region of contrasting situations illustrated by some characteristics:

The renewable water resources (the sum of the nationwide total ground and surface waters of rainfall origin) in the Mediterranean region measure approximately 1,200 billion m^3/year, which is only about 3% of the entire renewable resources of the planet. These water resources are unevenly distributed due to the differences in climate between the northern countries and those to the south and east. 72% of the 1,200 billion m^3/year renewable water resource is in the north, 23% in the east and 5% in the south of the Mediterranean basin (Plan Blue 2009; Boucheron 2010). This major disparity in distribution, coupled with the many climatic phenomena and extreme weather events which affect the region, make the Mediterranean region one of the most vulnerable areas in the world as far as climate change is concerned. Thus renewable water resources are measured in billions of m^3/year for the northern countries, such as France, Italy, Turkey and the countries of the former Yugoslavia, whereas they are measured only in millions of m^3 in the most water-poor countries, like Malta, Libya, Cyprus, Palestinian Territories-Gaza and Jordan.

The water resources are distributed unevenly within each country. For example, in Spain, 81% of the resources are in the northern half of the country; in Morocco, the two principal drainage basins (Oum-er-Rbia and Sebou, which cover one-tenth of the territory) provide 50% of the flows; in Algeria, 75% of the renewable resources are concentrated in 6% of the territory, while in Tunisia, the north (30% of the territory) produces 80% of the resources. As stated by Tekinel et al. (2000), 31% of the renewable water resources are in the eastern and south-eastern part of Turkey, while 25.5% are in the Mediterranean region of the country. It should be noted that variable proportions of the total resources in certain countries come from outside via the trans-boundary rivers. While resources in Spain, Italy, Lebanon, Libya, Morocco and Turkey are entirely or almost entirely internal, the other States depend to a large extent on their neighbours: Egypt 98%, Syria 80%, Israel 55%, and the countries of the former Yugoslavia 45% (Boucheron 2010).

Related to populations, water resources per capita say even more about the levels of richness or poverty in the Mediterranean countries in terms of water: they range from overabundance in Albania and in the countries of the former Yugoslavia (over 10,000 m^3/year per inhabitant) to extreme water-poverty in the Palestinian Territories-Gaza and Malta, with less than 100 m^3/year per inhabitant (Hamdy/ Lacirigniola 1999; Boucheron 2010). Thus today more than 160 million of the 425 million Mediterranean people estimated by the United Nations live in countries with less than 1,000 m^3/year of water per inhabitant. Tensions between resources and needs are becoming apparent in the user nations, especially when irrigation is necessary. Of these 160 million persons, 30 million are living below the line of absolute water-poverty of 500 m^3/year per inhabitant, for example in the Palestinian Territories-Gaza, Israel, Jordan, Libya, Malta and Tunisia. The other negative aspect of water use in this area is caused by the excess use of the groundwater resources. Groundwater is severely exposed in coastal areas, where the equilibrium with seawater can easily be upset. In some countries, such as Spain, Italy, Greece, Cyprus, Israel and Libya, excess pumping of groundwater has led to the abandonment of the catchments.

Pollution, whether it is localised and caused by low standards of wastewater purification, industrial discharges and accidents or widespread due to the overuse of agricultural additives or poor waste management, is also a threat to resources, and tends to increase drinking water production costs considerably. Although less industrialised than the northern countries, the countries in the south suffer just as much from the effects of similar pollution, aggravated by inadequate purification and prevention facilities as well as by the scarcity and mediocre quality of the resources. For example, Egyptian industries dump 550 million m^3 of waste – 57% of the quantity produced – into the Nile every year.

7.2.4 Possible Climate Change

The Mediterranean climate can substantially change due to the geographical location of the Mediterranean region, which lies in a transition zone between arid and temperate and rainy climates. This makes the Mediterranean a potentially vulnerable region to climatic changes induced by increasing concentrations of greenhouse gases (Gibelin/Deque 2003). According to the Fourth Assessment Report of the

Intergovernmental Panel on Climate Change (IPCC 2007), the climate over the Mediterranean basin may become warmer and drier during the 21st century. The Mediterranean climate is affected by several tropical and subtropical systems, as illustrated and explained by numerous scientists. These factors range from the El Niño Southern Oscillation (ENSO) and tropical hurricanes to the South Asian Monsoon and Saharan dust. This leads to complex features in the Mediterranean climate variability (Alpert et al. 2006).

The climate of the Mediterranean is mild and wet during the winter and hot and dry during the summer. The westward movement of storms originating over the Atlantic and impinging upon the western European coasts (Giorgi/Lionello 2008) mostly dominates the winter climate of the Mediterranean. The winter Mediterranean climate, and most importantly precipitation, are affected by the North Atlantic Oscillation (NAO) over the region's western areas (Alpert 2004), the East Atlantic (EA) and other patterns over its northern and eastern areas. The El Niño Southern Oscillation (ENSO) has also been suggested to significantly affect winter rainfall variability over the Eastern Mediterranean (Xoplaki et al. 2004; Alpert et al. 2006; Brönnimann et al. 2007). In addition to the impact of Atlantic storms, Mediterranean storms can be produced internally in correspondence to cyclogenetic areas, such as the Lee of the Alps, the Gulf of Lyon and the Gulf of Genoa (Giorgi/Lionello 2008). In the summer, high pressure and descending motions dominate the region, leading to dry conditions, particularly over the southern Mediterranean. Summer Mediterranean climate variability has been found to be connected with both the Asian and African monsoons and with strong geopotential blocking anomalies over central Europe (Xoplaki et al. 2003; Alpert 2004; Alpert et al. 2006).

On the other hand, the Mediterranean climate is affected by local processes such as the complex physiography of the region and the presence of the large water body of the Mediterranean Sea. For example, the Alpine chain is a strong factor in modifying travelling synoptic and mesoscale systems, and the Mediterranean Sea is a major source of moisture and energy for storms (Giorgi/Lionello 2008). In addition, anthropogenic and natural aerosols of central European, African and Asian origin can reach the Mediterranean, possibly influencing its climate characteristics (Alpert 2004). Because of these factors, the Mediterranean climate is characterised by a great diversity of features, resulting in a variety of climate types and great spatial variability.

Assessments of climate change projections over the Mediterranean were initiated in 1992 (Giorgi et al. 1992) and onwards. Giorgi/Lionello (2008) have used twenty-five global climate change scenarios developed by the IPCC in order to assess the impact of climate change for 2050 and 2100 (Somot et al. 2008). According to Giorgio/Lionello (2008), at the end of the 21st century, a strong warming is expected for all seasons. This warming has been described by Deque et al. (2005) as a warming that is stronger in summer than in winter (1), stronger over land than over sea (2), warmer in the eastern part of Europe in the winter (3) and in the southern part in summer (4). As an example, temperatures are expected to rise by +2 to +3 °C in the Mediterranean region by 2050, then by +3 to 5 °C by 2100

(Figs. 7.3 and 7.4). This rise in temperature is likely to reduce relative air humidity and increase the atmosphere's capacitive moisture load (Milano 2010). The future air will therefore have a higher saturation rate, leading to less cloud cover and hence decreased precipitation. Precipitation will be less frequent but more intense, whilst periods of drought will be longer and more frequent, altering the spatial and temporal precipitation patterns. A 1–2 °C rise in air temperature accompanied by a 10% reduction in the amount of precipitation may cause a 40–70% drop in the mean annual river runoff, which will substantially affect agriculture, water supplies and hydroelectricity.

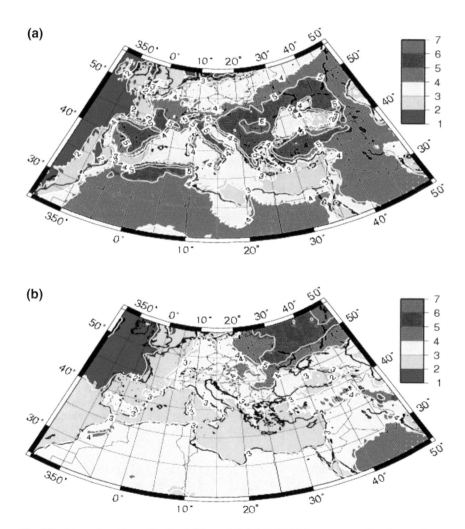

Fig. 7.3 Atmosphere-Ocean Regional Climate Model (AORCM) response to the climate change for the 30-year average 2-m temperature (in °C) between the 2070–2099 period and the 1961–1990 period **a** in winter and **b** in summer. *Source* Somot et al. (2008: 117)

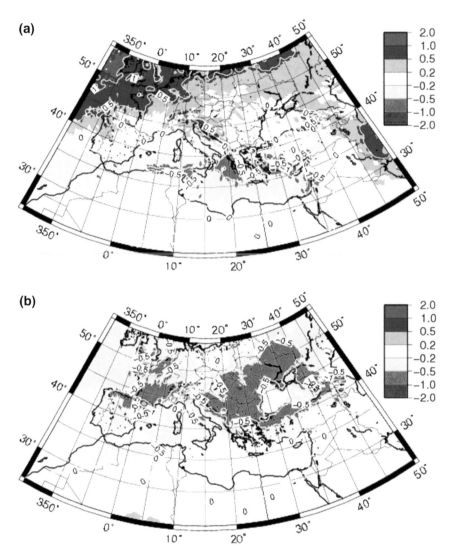

Fig. 7.4 Atmosphere-Ocean Regional Climate Model (AORCM) response to the climate change for the 30-year precipitation forecast in mm/day between the 2070–2099 period and the 1961–1990 period **a** in winter and b in summer. *Source* Somot et al. (2008: 117)

Similar work on the effects of climate change in Turkey and in the Seyhan River basin was studied by the project entitled the "Impact of Climate Change on Agricultural Production Systems in the Arid Areas (ICCAP)" for five years. This project was launched in 2002 in cooperation with the Research Institute for Humanity and Nature (RIHN) in Japan and the Scientific and Technical Research Council of Turkey (TUBİTAK) (Watanabe 2007).

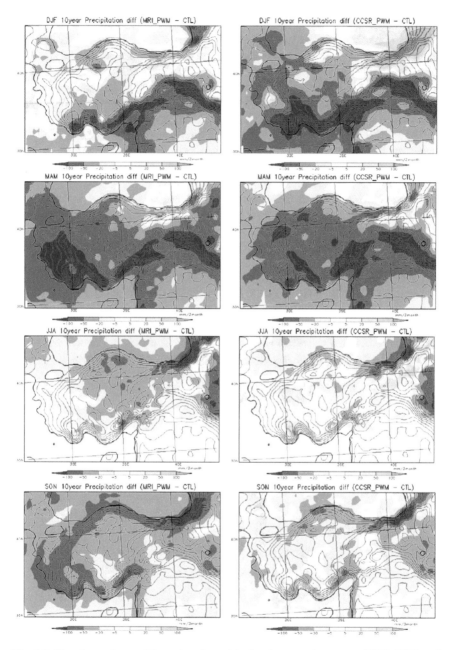

Fig. 7.5 The change in monthly-projected precipitation between the present (1994–2003) and future climate (2070–2079) by downscaling from MRI-CGCM (left) and CCSR-CGCM (right). *Source* Kimura et al. (2007: 31)

Within the research scopes of this project, researchers provided scenarios of the likely climate change in Turkey caused by greenhouse gases, in which precipitation, temperature and insulation were predicted for the period of 2070–2079. Downscaling for the ten-year climate change during the 2070s is completed for the whole of Turkey by 25 km and for the Seyhan basin by 8.3 km grid intervals. In this study, two independent General Circulation Model (GCM) projections were downscaled by only one Regional Climate Model (RCM).

Figures 7.5 and 7.6 indicate the change in monthly precipitation until 2070. The figures on the left side and right side indicate the precipitation change in the months downscaled from the Meteorological Research Institute-Coupled General Circulation Model (MRI-CGCM), and Center for Climate Science Research-Coupled General Circulation Model (CCSR-CGCM) respectively. The brown coloured areas in the figures show the decreasing projected precipitation. Moreover, a prominent decrease in the precipitation over the whole of Turkey during the winter months is illustrated. Both downscalings show that precipitation will prominently decrease in the slopes along the Mediterranean.

According to the generated scenarios, the surface temperature in Turkey may increase 2.0 °C by MRI-GCM and 3.5 °C by CCSR-GCM (Figs. 7.7 and 7.8). The total precipitation may decrease by nearly 20% except in the summer, where the difference at GCM is relatively small. The projected trend of changes in temperature and precipitation in the Seyhan River Basin is almost similar to the changes in the whole of Turkey, while precipitation is expected to decrease by about 25%.

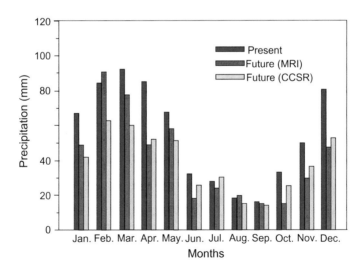

Fig. 7.6 Monthly precipitation for today (1994–2003, blue coloured) and future climate (2070–2079) in the Seyhan River Basin downscaled from MRI-CGCM (red) and CCSR-CGCM (green). Total precipitation will decrease almost 25% in Turkey, including in the Seyhan River Basin. *Source* Kimura et al. (2007: 32)

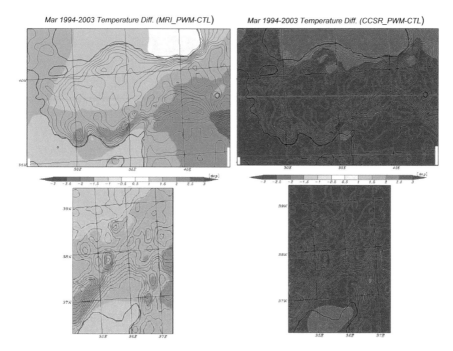

Fig. 7.7 Horizontal distribution of difference in surface temperature between the present (1994–2003) and future climate (2070–2079) in the second grid (top, Turkey) and the finest grid (below, Seyhan River Basin) of downscaling form MRI-CGCM (left) and CCSR-CGCM (right). *Source* Kimura et al. (2007: 32)

7.3   Significance of Irrigation in the Mediterranean Countries

The techniques and practices of irrigation are known to predate the Romans in the Mediterranean region, where irrigation plays a major role in the overall agricultural production. The direct impact of irrigation is represented by the increasing crop yield, productivity and farm income. The indirect effects are increasing rural employment, high economic benefit and decreasing immigration (Hamdy/ Lacirigniola 1999; Shatanawi et al. 2009). Furthermore, high-level equity in the population heterogeneity is achieved by irrigation, along with poverty and gender. Many of today's agricultural crops, such as vegetables, cereals, olives, grapes and aromatic species, were grown in some countries of the Mediterranean under irrigation (Fig. 7.9). However, the main physical constraint to irrigation development in the Mediterranean countries is usually water availability. The Mediterranean region is undergoing rapid local and global social and environmental change. In addition, competition with other sectors (domestic supply and industry) is also a

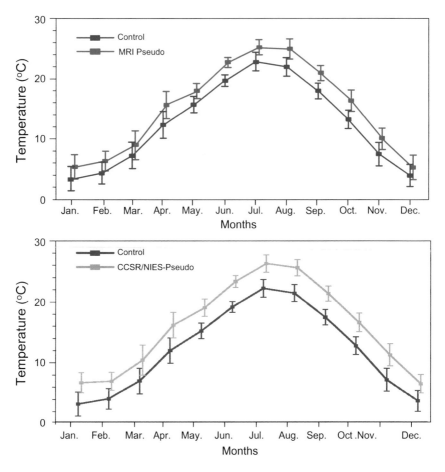

Fig. 7.8 Surface temperature for today (Control) (1994–2003) and future climate (2070–2079) in the Seyhan River Basin downscaled from MRI-CGCM (top) and CCSR-CGCM (below). Surface temperatures will increase 2.0 °C for MRI-GCM and 3.5 °C for CCSR-CGCM. *Source* Kimura et al. (2007: 33)

cause of limitation in water resources for irrigation (Hamdy/Lacirigniola 1999). All indicators point to an increase in environmental and water scarcity problems, with negative implications for current and future sustainability.

7.3.1 Management of Water Resources in the Mediterranean Basin

Water in the Mediterranean basin is not only a raw material exploitable within the limits of its availability. It is also, highly vulnerable to forms of land use and to

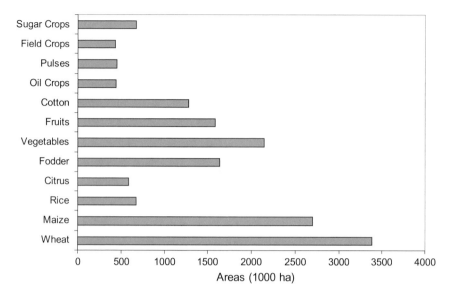

Fig. 7.9 Distribution of crops under irrigation in the Mediterranean (Aquastat 2010). Some important food and fibre crops, such as wheat, maize, rice, pulses, citrus for humans, fodder for animals and cotton, are given separately; other crops are grouped according to their properties. Sugar crops include sugar beet and sugar cane; field crops are sorghum, barley and tobacco; oil crops consist of rape, sunflower and soybean; fruits include vine, melon and bananas; vegetables and potato crops are given together under vegetables. *Source* Aquastat (2010)

numerous sources of pollution; neither is it immune to the impacts of climate change (Correia 1999).

There has recently been increasing attention to water resources management in the Mediterranean region, bearing in mind that the water resources in this area should be considered in conjunction with soil and vegetation resources. The study of Mediterranean environments and related water problems is relevant not only for this region but also for other areas in the world. Mediterranean environments are subject to very extreme conditions with respect to fluctuations in water availability and needs, causing situations of temporary or permanent scarcity and creating a need for storing water and demodulating natural regimes. Many of the approaches to deal with these problems, and some of the solutions currently adopted, may be very useful elsewhere in the future (Hamdy/Lacirigniaola 2005). The Mediterranean environment, fragile in nature, is seriously endangered by the natural trends of social and economic development of this region. Stress on the coastal areas, imbalance in the population distribution of rural and metropolitan settlements, severe water dependence, extreme sensitivity to pollution, and vulnerable equilibrium between soil and water are some of the factors to be considered in a global approach to Mediterranean environmental problems. Unquestionably, water plays a major role in this fragile environment. Not only is it an essential resource for economic growth, but also it is the most important component of the environment, with a significant impact on public health and nature conservation. Water deserves special attention in environmental studies in all regions of the world. However, in Mediterranean countries water is not simply important but essential. It is the cornerstone of all development strategies and a basic element of all planning activities (Hamdy/Lacirigniola 1994; Correia 1999; Boucheron 2010). Water resources management of the Mediterranean faces some significant constraints, such as pollution, excessive use, and silting-up of the dams (Boucheron 2010).

The failure to maintain banks in the north through terrace cultivation (which contributes to erosion), deforestation in the south and the 'artificialisation' of watercourses all contribute to irregularities in water flow and reduce renewable resources. Groundwater is severely exposed in coastal areas, where the equilibrium with seawater can be easily upset. For example, nowadays the virtually irreversible inflows of salt water caused by excessive pumping in Spain, Italy, Greece, Cyprus, Israel and Libya have led to the abandonment of catchments. The Blue Plan estimated that as early as 2010 some eleven countries would be exploiting over 50% of their resources: Morocco, Algeria, Tunisia, Libya, Egypt, Israel, Palestinian Territories-Gaza, Jordan, Malta and Syria are in the forefront, followed by Algeria, Tunisia, Cyprus and Syria. Lebanon will reach that level in 2025 (Blue Plan 2008).

The vulnerability of water resources in the Mediterranean is affected by the silting-up of the dams, which is particularly prevalent in the south rim countries. The high sediment content of floodwater shortens the effective life of reservoirs, despite the high-volume "holding ponds" or "spare capacity" provided. While the loss of the effective capacity of the Mediterranean dams is currently between 0.5 and 1% per year, it is 2% in Morocco, where the reduction in regulating capacity attributable to silting-up is equivalent today to a loss of irrigation potential of 6,000 to 8,000 ha per year. This ranges from 2 to 3% in Algeria, where the lifespan of the average-capacity reservoirs is 30–50 years, and from 1 to 2.5% in Tunisia, to such an extent that, according to experts, prevention efforts like reforestation of the basins or sediment traps can only delay the inevitable end of dam-reservoir sites. This points to a crucial 'post-dam' era, with enhanced silting-up problems for the 21st century in many Mediterranean countries (Boucheron 2010) in need of utmost attention in the management of the water resources.

7.3.2 Water Availability and Water Needs in the Mediterranean Region

The availability of water for irrigation purposes in rural areas has a significant impact on poverty and social equity in the Mediterranean countries. Water in most of the Mediterranean countries, especially in the south and the east, is inherently scarce due to the arid or semi-arid conditions. Additionally, in these countries, water shortage problems are faced due to the increasing gap between supply and demand. The gap will increase more in the future due to population growth, improved standards of living, urbanisation, industrialisation and climate change (Shatanawi et al. 2009).

The hydrological system of the Mediterranean region is affected by the climatic conditions. Changes in temperature affect evapotranspiration rates, cloud characteristics, soil moisture, storm intensity and snowfall and snowmelt regimes. However, some of the most severe impacts of climate change are likely to come not from the expected increase in temperature but from the changes in precipitation, evapotranspiration, runoff and soil moisture, which are crucial factors for water planning and management. In the meantime, changes in precipitation are predicted to affect the timing and magnitude of food and droughts, shift runoff regimes and alter groundwater recharge rates. The vegetation pattern and growth rates and the changes in soil moisture regime will also be affected according to the Blue Plan (2008).

There are some relevant variations in climatic conditions in this region, despite some prevailing common characteristics. For example, northern and north-eastern countries are clearly wetter than the south and south-eastern ones. However, most areas have a water deficit, defined as the difference between precipitation and potential evapotranspiration. This deficit is larger in the south and south east (Correia 1999). Naturally, this has a direct impact on water availability, mainly through river runoff, and contributes to a corresponding regional asymmetry. Some of the most relevant data on water availability and needs in the Mediterranean region are presented in Table 7.1 based on data adapted from FAO (Aquastat 2010). From Table 7.1, water resource managers face the dilemma of ensuring future sustainability of water resources while maintaining the strategic agricultural, social and environmental targets. Taking into account the total freshwater resources, the average annual potential water availability per capita in southern Mediterranean countries is less than 1,000 m^3 per capita and year (Table 7.1). According to scientific estimations, the proportion of people who suffer from hunger and do not have access to safe drinking water was predicted to increase by half by 2015 (Blue Plan 2008).

Irrigated agriculture is the main consumer of water in the Mediterranean (Fig. 7.10). Although northern and southern Mediterranean countries differ in relation to the rate of expanding the area of irrigated land and irrigation technologies used, the evolution of irrigation in all Mediterranean countries has been remarkable over the last half century (Iglesias et al. 2007). In general, there is little development of new irrigated areas and the investments focus on the rehabilitation of existing schemes and improvement of irrigation technologies. But, nevertheless there is a rapid increase in the water demands in all countries due to the increase in economic and social activities together with the increasing demand for tourism and pressure on the ecosystems.

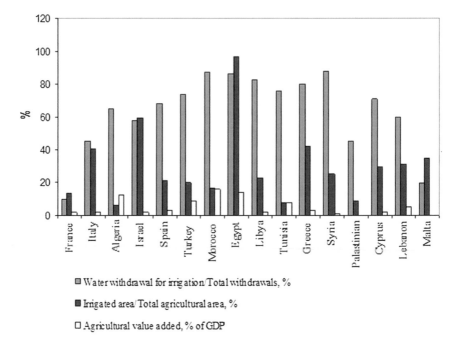

Fig. 7.10 Water use for irrigation and irrigated agricultural areas in selected Mediterranean countries. *Source* Aquastat (2010: 29)

7.4 Lower Seyhan Irrigation Project (LSIP): Case Study

7.4.1 General Characteristics

The Lower Seyhan Irrigation Project (LSIP) located on the Eastern Mediterranean coast is one of the largest irrigation projects in Turkey. It extends to the delta plain of the Seyhan River basin, with a total planned area of 175,000 ha, of which 133,000 ha has already been implemented (Fig. 7.11). The LSIP area is bound by the Mediterranean Sea on the south, by the foothills of the Taurus Mountains on the north and by the Berdan and Ceyhan rivers on the west and east respectively.

The construction of LSIP was initiated in the 1960s. The project area of 175,000 ha was divided into four areas and the construction was divided into four project phases. Phases I–III (133,000 ha) were completed by 1985. The Phase IV area, which still remains incomplete, is located in the lowest part of the project area. It has no water allocation and irrigated agriculture is practised with surplus water driven from the main canals of the completed area.

The soil in the delta is alluvial which developed from deposits of the three main rivers, namely the Ceyhan, Seyhan and the Berdan, which rise from the Taurus Mountains and flow into the Mediterranean Sea. The soils are the deepest and most fertile *Calcic Fluvisols* and *Chromic Vertisols* (Dinç et al. 1995; Çetin/Diker 2003).

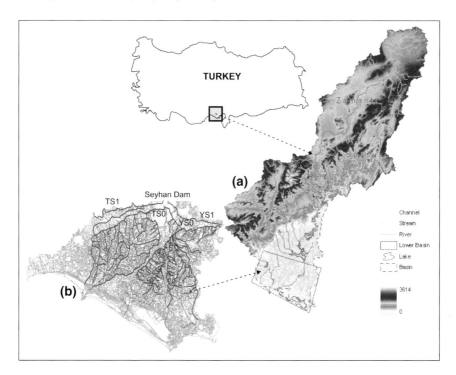

Fig. 7.11 The Seyhan River Basin (**a**) and the Lower Seyhan Irrigation Scheme (**b**). All the irrigation and drainage canals, which are of different sizes and functions, are located on the right and left banks of the Seyhan River. *Sources* Nagano et al. (2007a: 193), Tezcan et al. (2007: 64)

This plain can be regarded as one of the most important cultivated and industrial areas based on agricultural products in Turkey.

The area is divided into two parts by the Seyhan River, which flows through the plain from north to east. The part lying between the Seyhan and Berdan rivers is known as the "Seyhan Right Bank" or "The Tarsus Plain", and the part lying between the Seyhan and Ceyhan rivers is the "Seyhan Left Bank" or "The Yüregir Plain". The highest elevation in the LSIP area is about 30 m above sea level. The topography is flat and the altitude falls from the city of Adana to the Mediterranean Sea. The average slope varies from 1 to 0.1% from north towards the south (Özekici et al. 2006; Kume et al. 2007).

The climate (relatively mild and humid in the winter months) and the alluvial soil make the area highly suitable for agriculture. The average annual precipitation observed in Adana is nearly 670 mm. Most of the rain (80%) falls in the winter months. The least rainfall occurs in summer, in July and August. The mean wind velocity is 2 m/s and the annual average evaporation is 1,322 mm where 64% of this amount evaporates from May to October (Anonymous 2002).

7.4.2 *Irrigation*

The main irrigation water source of the project area is the Seyhan river, with its 6.3 billion m^3 annual flow rate from the watershed of 19,300 km^2. The quality of the water of the Seyhan river is suitable for irrigation (C_2S_1). The principal method of irrigation being implemented is gravity irrigation. The total annual water requirement for the Lower Seyhan Plain is approximately 1.8 billion cubic meters. The annual groundwater flow, which is 3 billion m^3, is not considered for irrigation purposes except for municipal use.

Water flow is controlled by two reservoirs, namely the Seyhan and Çatalan dams, in the upper stream of the Seyhan River. Irrigation water for the LSIP is being supplied by the Seyhan Dam and conveyed by various sizes and different types of irrigation canal. Irrigation efficiency in the LSIP is lower than 50%. The canal system of the LSIP consists of two conveyance canals, which are of 40.3 km (right) and 18.8 km (left) length, main, secondary and tertiary canals (Özekici et al. 2006). All types of canal provide in total 3,000 km for irrigation and 2,500 km for drainage (Fig. 7.11).

The farming system in the area can be characterised as high-input agriculture. Suitable soil, climate and topographical conditions, in addition to the rich water resources of the Seyhan River basin, allow various crops to be grown throughout the year. In response to the Mediterranean climate, farmers on the upstream part of the basin have been cultivating rain-fed winter wheat. However, in the LSIP area, agricultural crops have been irrigated mainly during the dry season from spring to autumn. Crops under irrigation are mainly wheat, maize, cotton, citrus, vegetables and watermelon. Wheat is irrigated depending on the amount of precipitation and its season. Their cultivation areas are about 20, 45, 9, 13, 4 and 6% of the total area respectively (Hoshikawa et al. 2007).

Figure 7.12 illustrates the change in the planned water release to the LSIP area in the past two decades. The amount of water has been increasing with time, mainly due to the shift in the cropping pattern and substantial increase in the irrigated area of the Phase IV area near the coast, where farm plots had not been fully consolidated (Nagano et al. 2007b). The other significant change was the transfer of the responsibility for water management from DSİ to the newly established water users associations (WUAs) in the mid 1990s. Due to the many uncontrolled problems, such as the degradation and destruction of the canals, precipitation anomalies and conflicts between WUAs, the actual amount of water release seems far more than its planned level. Although there is a lack of consistent recording of the actual diverted water, it is estimated from some data that the recent actual release is nearly 2 km^3.

Groundwater level exists only 1–2 m below the ground surface in most parts of the LSIP. Some areas of the LSIP have severe problems due to poor drainage and salt accumulation induced by the shallow water table (Çetin/Diker 2003). Therefore, implementation of the irrigation was coupled with installations of subsurface drainage and construction of drainage canal networks. By the 1980s, DSİ

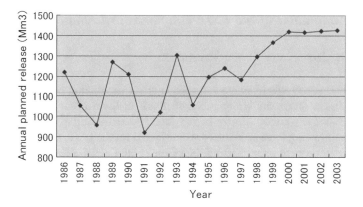

Fig. 7.12 The amount of the released irrigation water to the LSIP area during the last two decades. *Source* Nagano et al. (2007b: 212)

started monitoring the shallow groundwater level and salinity once a year over the entire irrigated area (Fig. 7.13).

The high water table that is the outcome of the inefficient irrigation practices is still problematic without the development of good drainage networks. Consequently, the area may even face severe waterlogging with the increase of irrigation.

Another potential risk is the increase in the use of deep groundwater. The deep groundwater in the area is at high risk of salt intrusion, and this may cause devastating consequences. Soil salinity decreased from 1990 to 2005, whereas the distribution pattern of salinity was still similar to that of the water table (Kume et al. 2007). Soil salinity has decreased in the LSIP area, including the coastal zone, with the application of irrigation.

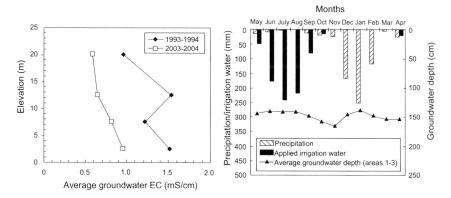

Fig. 7.13 Fluctuation of the groundwater depth (right), and groundwater EC values according to the elevation of the monitoring wells (left) in the LSIP area in 2003–2004. *Source* Kume et al. (2007: 206)

7.4.3 Adaptation of the LSIP Outcomes to Climate Change

7.4.3.1 Hydrology and Water Resources

According to the results obtained from the General Circulation Models (GCMs) of the ICCAP, the snow storage and stream runoff in the Seyhan Basin is liable to decrease under a warmer climate in the future (2070–2100) (Fig. 7.14). For the future, the following three scenarios are considered for the LSIP, (a) Land and water will be the same as at present, (b) Adaptation 1: land and water use will be under low-investment conditions; and (c) Adaptation 2: land and water use will be under high-investment conditions.

Figure 7.14 shows that, the maximum snow water equivalent (SWE) is almost 0.4 Gt in the present climate but it will decrease to 0.1 Gt under the future climate. For the Seyhan delta (irrigated area) the annual evaporation is about 800 mm, and nearly 500 mm of irrigation water must be supplied during the growing season in the hot and dry summer. As a consequence of the reduced snow cover; these areas will receive more shortwave radiation (albedo effect), and this increased energy will contribute to the increased evaporation in the spring. As reported by Fujinawa et al. (2007), the increased energy will also cause a decrease in the crop maturity period, but the amount of irrigation water requirement will increase because of the higher evaporation demand during the growing season and reduction in the soil moisture at the beginning of the growing period. According to the results given by Tezcan et al. (2007), the decrease in the mean annual snow storage is about 14.56 km^3 in the warming up period. The major decrease will occur in Aladağlar, the south-east slopes of the Erciyes and the north of the Göksu Basin. The decreased snow storage will influence the discharge of the springs in the Zamanti and Göksu Basins which feed the Seyhan River.

Similarly, future inflow will decrease markedly compared to that of the present. In addition, the decreases of inflow predicted to occur in April, May and June will

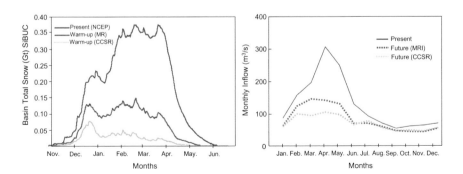

Fig. 7.14 Total snowfall in volume equivalent to water (Gt) (left) and the changes in Seyhan River flow (right) predicted from different models of MRI (red), CCSR (green) and present (blue). *Source* Fujinawa et al. (2007: 56)

be greater than those of the other months, and the peak monthly inflow will occur earlier than at present, consequently decreasing radically. Fewer flood events are estimated to occur during the warm season, when the decreased river flow may lead to water scarcity in the LSIP area (Watanabe 2007). However, Tezcan et al. (2007) have reported that the months for peak flow will be the same in both the present and warm up periods using the Mike-She simulation program.

The reliability index, which is defined as the ratio of water supply to water demand, is the indicator of the water demand if satisfied by the supply of a reservoir or the degree of water scarcity. The reliability index at present is about 0.4, which indicates low water stress; however, it ranges from 0.4 to 0.7 in the future for Adaptation 1 (high water stress) and 0.5–1.0 for Adaptation 2 (extremely high water stress) (Fujinawa et al. 2007). As a consequence, the reservoir volume in the future and Adaptation 1 will be less than at present, and in a few cases the reservoir will be void of water. The reliabilities of the dams in future and Adaptation 1 will change from 0.95 to 1.0 based on the precipitation projections of MRI and CCSR models respectively. In Adaptation 2, the reservoir is frequently empty and reliability ranges from 1.0 to 0.7 according to future data projected by the MRI and CCSR models.

On the other hand, climate change is predicted to decrease the water budget elements in the warm-up period compared to the present. The CCSR climate data reveal a greater decrease than those of the MRI data. The decrease in the actual evapotranspiration is limited by decreased precipitation (Table 7.2). Precipitation may decrease by 29.4 (MRI) and 34.7% (CCSR) in the warm-up period, which is predicted to decrease the river flow by 37.5 and 46.4% respectively. Consequently, the groundwater recharge in the whole of the Seyhan Basin will decrease by 24.7 and 27.4%. The majority of the springs in the basin will become dry due to the decline of the groundwater level below the spring level.

Groundwater resources in the LSIP area will be drastically influenced by climate change. Decreasing the recharge of the Seyhan River will cause the decrease in the subsurface recharge to the LSPP area from the higher elevations in the north of the region. The change in groundwater storage in the LSIP area due to climate change is shown in Fig. 7.15. The most significant impacts are the reduction in the recharge in the higher elevations and the increase in the abstraction due to the limitations of surface water resources. Decline in the head will also cause saline water intrusion to the LSIP area. In the case of a 50% increase in the groundwater abstraction in the

Table 7.2 Decrease of the water budget elements in the warm-up period with respect to the present conditions (mm). *Source* Tezcan et al. (2007: 69)	Water budget elements	MRI	CCSR
	Precipitation	29.4	34.7
	Actual evapotranspiration	16.9	16.9
	River flow	37.5	46.4
	Discharge to the Medit. Sea	50.0	54.2
	Recharge	24.7	27.4
	Spring discharge	50.0	50.0

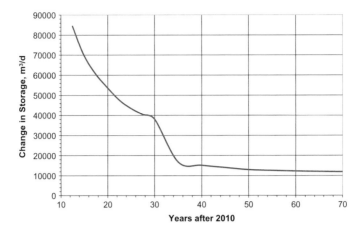

Fig. 7.15 Impact of climate change on groundwater storage in the Lower Seyhan Delta Plain. *Source* Tezcan et al. (2007: 65)

warm-up period of 2080, the seawater intrusion will reach 10 km inland (Tezcan et al. 2007). At the same time, in the coastal zone of the LSIP area, the groundwater salinity will reach 25% of the seawater composition.

7.4.3.2 Crop Productivity

The Seyhan River Basin and especially the LSIP area is an important food production area for Turkey and the European Union. The Seyhan River Basin is suitable for different types of rain-fed crops in the north and for irrigated crops in the southern part. The major crops grown in the LSIP area are maize, wheat, cotton, citrus and vegetables. Most of the winter crops, such as wheat, are grown in the rain-fed area north of the LSIP.

Wheat is the most dominant crop in the LSIP area and is grown as a winter crop under rain-fed conditions. Sometimes it is irrigated depending on the precipitation. In recent years, the total wheat production in the Adana province is about 1.3 million tons on average, produced on 0.31 million ha cultivated area, 0.42 million tons of which is obtained from the LSIP area. The second widespread crop of the LSIP area is maize, which is grown in the summer under irrigation. Its total production is 18% of the total wheat production. Accordingly, the two most widespread crops, namely wheat and maize, of the LSIP, are considered for climate change predictions.

Wheat sown in due time and late in the winter is grown under irrigation (drip) and rain-fed conditions, whereas irrigated maize is grown as the first crop and second crop after wheat. During the experiments, some crop growth parameters, such as ET, yield, plant height, LAI, HI and WUE, are measured and used for the

Table 7.3 Water balance components for wheat and maize. *Sources* Ünlü et al. (2007: 100)

Items	Wheat			Maize		
	1994–2003 observed	2070–2100 estimated	Diff. (%)	1994–2003 observed	2070–2100 created	Diff. (%)
Precip. (mm)	531.1 ± 184.7	510.6 ± 209.5	−4.0	47.5 ± 23.8	24.0 ± 18.0	−47.6
Evapo. (mm)	373.9 ± 30.6	333.0 ± 24.2	−10.9	399.9 ± 29.8	−1.7	393.1 ± 33.9
Irrig. (mm)	74.8 ± 72.8	67.4 ± 1.32	−9.9	371.1 ± 49.4	+3.3	383.3 ± 44.2
Biomass (t/ha)	16.5 ± 1.39	14.33 ± 1.37	−13.1	27.0 ± 1.53	−25.1	20.6 ± 1.34
Yield (t/ha)	5.11 ± 0.73	4.83 ± 0.7	−5.5	15.19 ± 1.29	−31.0	10.48 ± 1.0
Growing period (day)	183.4 ± 6.3	167.4 ± 5.4	−8.7	115.2 ± 3.6	−8.1	105.9 ± 0.7

calibration of the SWAP Model (Soil-Water-Atmosphere-Plant). Wheat and maize growth, yield and other growth parameters are estimated by the SWAP Model for future climatic and soil conditions of the 2070s to 2100s. Water balance components for the wheat-growing period are depicted in Table 7.3.

Table 7.3 shows the effect of the higher predicted temperatures and less rainfall on wheat production in the future climate. Higher temperatures are predicted to cause shorter growing periods and lower water balance components in the wheat-growing season than those of the future. The higher concentration of CO_2 is predicted to enhance photosynthesis and increase crop yields. However, the higher temperatures of the summer months will increase the water requirements for irrigation. The yield of the conventional rain-fed wheat depends on the rainfall in the winter from November to May (Kanber et al. 2007).

Similarly, the second crop maize yield will decrease because of the reduced grain weight caused by the short growing period. Future grain yield decreases due to the increased temperatures are to be expected more drastically under rain-fed conditions. Grain yield reductions due to the increased temperatures will occur even under irrigated conditions. Moreover, grain weight increases will not be sufficient to compensate for the reduced grain numbers. This indicates that high temperatures will also have an adverse effect on grain growth.

7.4.3.3 Irrigation and Drainage

Land use in the 2070s: Climate change will influence evapotranspiration due to the decrease in precipitation, especially in the wintertime. However, the spatial distribution of ET is different in the whole of the basin. ET values increase in the north of the basin, but decrease in the south, especially in the areas of crop production and natural vegetation during the warm-up period. The reduced ET in these areas is due to the lack of available soil water in the root zone to evaporate. In these zones, irrigation water requirement will be greater in the future because of the decrease in precipitation.

Table 7.4 Yearly irrigation water requirement (IWR) for the major crops in LSIP during the warm up period of 2070. *Source* Nagano et al. (2007a: 194)

Crops	Present[a] 1990s	MRI-GCM		CCSR-GCM	
		IWR (mm)	Diff. (mm)[b]	IWR (mm)	Diff (mm)
Fruit	762.1	848.6	+86.4	778.8	+16.6
Citrus	661.4	749.0	+87.6	724.4	+63.0
Maize	569.0	611.0	+42.0	594.2	+25.2
Soybean	539.0	559.9	+20.9	546.2	+7.2
Cotton	524.2	583.0	+58.8	569.3	+45.1
Maize II	391.4	385.9	−5.5	380.3	−11.1
Vegetable	229.2	302.0	+72.8	289.2	+60.0
Melon	195.9	195.2	−0.7	239.6	+43.7

[a]From the DSİ
[b]Differences between present and future values were estimated from the pseudo-warming experiment

Table 7.4 shows the crop water demand in the 2070s according to climate change. According to the table, the water demand for fruit crops would significantly increase due to water shortages in the early spring. Vegetables would have a greater water demand for the same reason.

Similarly, irrigation efficiencies, cultivated area, and the amount of water release from the Seyhan dam to the LSIP area will be affected by the future climate change (Table 7.5). Nagano et al. (2007a) reported that efficiencies of conveyance and application will be the same as in the present in the S1 scenario (passive and low investment scenario predicting a decline in the maintenance level of the canals compared to the present in non-irrigated Phase IV area conditions). However, the relevant efficiencies are estimated to increase in the S2 and S3 scenarios (S2 and S3: pro-active and high investment scenarios), where the maintenance of the canals will be improved compared to the present conditions seeking to irrigate the Phase IV area i.e., the lowlands of the LSIP with the highest groundwater uptake. On the whole, an average of 150 mm of new (future) groundwater use is assumed to occur in the LSIP area. In these scenarios the same soil physical properties and the geology and terrain characteristics of the LSIP area were used for both models of MRI and CCSR. The actual water release for LSIP increases 7% in S1, whereas it decreases 22% in S2 and S3 for MRI-GCM. These results are also used for the CCSR-GCM scenarios of the future. The decrease in water availability estimated for IWr is nearly 50% of the present in the S2 and S3 scenarios, whereas it is predicted to remain constant in the total service area of the LSIP in the warm-up period of 2070. However, the water availability for a unit area will decrease drastically in the future for both models. All figures on the water management estimated by CCSR-GCM are lower than those by MRI-GCM.

Table 7.6 shows the results of the simulation for the future land use. Citrus would remain constant around 20% and, in the case of scarce water supply, watermelon would emerge. However watermelon is usually cultivated only once

Table 7.5 Water availability in the LSIP under climate change and water development scenarios. *Source* Nagano et al. (2007a: 195)

	Present 2002	MRI-GCM (2070s)			CCSR-GCM (2070s)			Unit
		S1	S2	S3	S1	S2	S3	
Ec	0.8	0.6	0.8	0.8	0.6	0.8	0.8	
Ea	0.6	0.6	0.7	0.7	0.6	0.7	0.7	
Ei	0.48	0.36	0.56	0.56	0.36	0.56	0.56	
IWr	1424	1523	1112	1112	1294	854	854	10^6 m^3
Waa	683.5	548.1	622.7	622.7	465.8	478.5	478.5	10^6 m^3
Area	1168.83	1168.83	1168.83	1168.83	1168.83	1168.83	1168.83	1000 da
WA/da	585	469	533	683	398	409	559	m^3/da
AreaT			1450.98	1450.98		1450.98	1450.98	1000 da
WA/da*			429	579		330	480	m^3/da

Note Ec, is conveyance efficiency; Ea, application efficiency; Ei, total efficiency (Ec × Ea); IWr, actual water release for LSIP; Waa, actual water available for LSIP (Ei × IWr); Area, total service area of LSIP; AreaT, total service area with Phase IV area complete; S, water development scenarios, WA/da*, actual available water for AreaT

Table 7.6 Simulated cropping pattern for future climate in 2070 and social scenarios. *Source* Nagano et al. (2007a: 190)

	Base case	MRI-GCM			CCSR-GCM		
		S1	S2	S3	S1	S2	S3
Available Water (mm)	585	469	429	579	398	330	480
Citrus	22.0	22.1	22.1	21.9	21.9	18.3	21.8
Cotton	59.3	24.0	15.1	48.3	4.3		26.0
Vegetable	7.0	4.4	3.6	6.4	3.0	3.2	4.7
Watermelon, and Maize		41.3	51.7	12.9	64.0	78.5	38.8
Fruit	11.6	8.3	7.5	10.4	6.8		8.6
Gross revenue (TL/da)	717.9	706.9	702.6	715.6	696.4	670.0	707.9
Shadow price of water		0.101	0.117	0.056	0.164	0.137	0.116
Idle water (mm)	23.5						

every five years to avoid replant failure, thus requiring use of a particular crop rotation approach when setting up simulation studies based on the weighed average of watermelon cultivated for one year and maize cultivated for four years (Nagano et al. 2007a).

Regarding the predicted future land use change due to the decrease in the available water together with the projection of the present revenue-water demand relation, cotton seems to become the major crop in the less water deficit conditions. However, in severe water deficit conditions the combination of watermelon with maize would become the major crop. Citrus would also have a stable land-wise distribution unless a severe water deficit (<350 mm) occurs.

Fluctuation of groundwater depth and salinity: For the simulations on groundwater depth and salinity, the groundwater depth at the northern boundary was assumed to be fixed at 5 m below the ground surface by Nagano et al (2007a). On the other hand, at the southern boundary on the coast, the present climate was set to 0 m and to 0.8 m for the projected climate for the simulation. The level of groundwater table in simulations with projected climate data was lower than in the present conditions (Fig. 7.16). The water table within 3 km from the coast was mostly affected by the higher seawater level (0.8 m) and was more sensitive to the degree of management (Nagano et al. 2007a).

According to the simulations, the level of groundwater depths may increase in the future climate compared to the present climatic conditions. So, the shallow water table in the LSIP is projected to be lower in the 2070s. This will be caused by the decreased precipitation and the amount of the irrigation water. In general, the risk of higher water table occurrences seems less likely to occur due to the projected

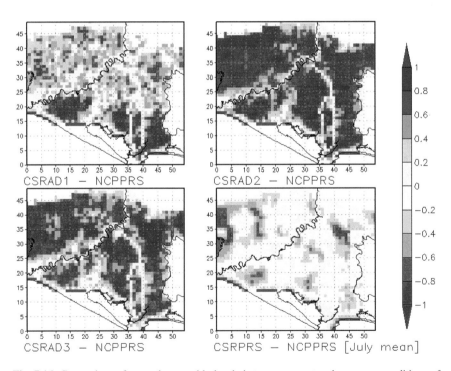

Fig. 7.16 Comparison of groundwater table levels between present and warm-up conditions of the future with different adaptation scenarios (the case of CCSR runs in July average; Scenario 1: top left, Scenario 2: top right, Scenario 3: bottom left and present land use: bottom right). *Source* Nagano et al. (2007a: 194)

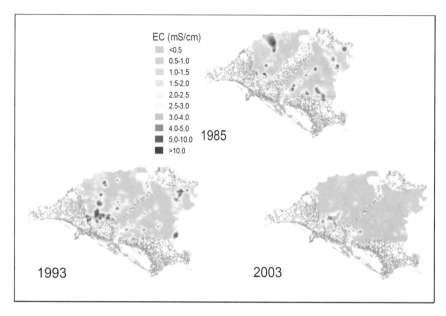

Fig. 7.17 Electrical conductivity of the shallow water table during the different years in July in the LSIP area. *Source* Nagano et al. (2007b: 213)

decrease in precipitation and the water supply. Waterlogging is only partially projected to occur along the coast, where irrigation management has in the present and will in the future have considerable influences on the shallow water table. The influence of irrigation management on the shallow water table will most likely be higher than the change predicted to occur due to the climate.

Sodium contents dominantly contribute to increasing salinity in the LSIP area, where measurements were conducted during the peak irrigation season in July. Although there may have been dilution effects by the increased irrigation, it is quite definite that the shallow water table has been consistently decreasing over the years (Fig. 7.17).

Salinity was severe when rain-fed agriculture was widespread before the implementation of the irrigation system. In those days, the summer dryness was the major driving force for the rising salt in the soil. However, after the implementation of the farmer-based excess irrigation practices, the soil water flux in the summer was reversed, consequently leaching the soluble salts in the soils. Additionally, the low sodium Seyhan Dam waters have been an irrevocable advantage for leaching salinity in the area (Nagano et al. 2007b).

Irrigation performance: The actual irrigation practices of the farmers/water user associations in the area of LSIP are somewhat different from the regulations established by the DSİ in the past. The contemporary regulation-based problems are the outcomes of the lack of or insufficient communication between the farmers and water distribution technicians concerning the allocation of water within the tertiary canals and the insufficient control of irrigation water by technicians. Additionally, the inconsistencies between the initial design of the management programmes and gradually diversifying land use also brought conflict over water allocation. Moreover, the amount of irrigation demand is projected to increase from 100 to 170 mm and the duration of irrigation to extend from early spring to late autumn due to the considerable decrease in precipitation in the winter period via climate change.

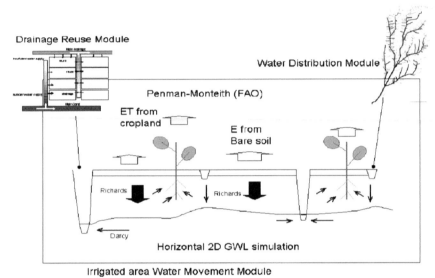

Fig. 7.18 Concept of calculation of the IMPAM (upper), and its important hydrological elements, input (I) and output (O), in irrigated agriculture (lower). *Sources* Hoshikawa et al. (2007: 220), Nagano et al. (2007c: 654)

The high water table is projected to be lower in the 2070s following the transition from gravity irrigation to drip irrigation in citrus gardens, which will enhance the fall in the water level and help users adapt to the future water deficit. Ultimately, the Nagano et al. (2007c) study conducted for irrigation performance in the area highlights the need for an integrated water distribution programme that could include the canal networks, mixed land use, crop growth and water balance.

The Irrigation Management Performance Assessment Model (IMPAM) (Fig. 7.18) was applied to the LSIP for simulating the effects of climate change on the flexibility of the irrigation system, systematic assessment of itsperformance, and diagnosis for sustaining productivity. The model was primarily calibrated to the water budget structure of a specific irrigation district and later the adaptive capacity of the district towards the projected climate change for the 2070s was tested. On the other hand, this calibrated model was used for learning and developing decisions for the future vulnerabilities of the present irrigation management programme and assessing the effect of possible adaptations concerning climate change (Hoshikawa et al. 2007).

From the outcomes of the IMPAM, irrigation efficiencies (around 40%) and irrigation water losses are estimated to decrease in the future. The annual total of irrigation water intakes and drainage exceed 2,500 mm and 1,500 mm respectively in the upper part of the LSIP. It is also determined that the majority of this large amount of irrigation intake was the water that was not lost from the canal as leakage or dropped to the drainage system as tail water.

7.5   Conclusions

The ultimate statement of this chapter derived from the above mentioned research and modelling studies is, briefly, that agriculture in the LSIP area may be affected by the future predicted climate change. In the 2070s, the precipitation in Adana is projected to decrease by 42–46%. The decreased precipitation would mainly occur in the winter period. There would be need for irrigation in the early spring for the tree crops and vegetables to cope with the drier winter. Additionally, the irrigation demand would increase and the irrigation period would become longer under the assumed climate change. For this reason, the LSIP may have to change crop and irrigation management to adapt to situations with less water availability. The LSIP has two large-scale reservoirs in the upstream as water sources. The decreasing river discharge based on climate change would necessitate the determination of a resource-wise adaptive capacity. However, the 30,000 ha of additional irrigation development in the Phase IV area will cause an extra water deficit in the future.

Moreover, the irrigation and drainage infrastructure is already aged and deteriorated so that the carrying capacity would farther deteriorate without proper maintenance. Low conveyance and application efficiencies are both contributing to high water table problems. Therefore, the management of the canal system would be the key factor determining the system-wise adaptive capacity of the LSIP. For conserving the adaptive capacity against possible climate change, it is strongly recommended that irrigation efficiencies, both application and conveyance, must be improved in the LSIP. In this context, improved irrigation efficiencies would point to the enhanced maintenance of the canals, gate operations, and application techniques. This will lead to the equity of allocation and the prevention of the high water table, along with appropriate soil conservation in the long term. This will also be useful in the coastal area seeking un-irrigated management and, in turn, the consequent absence of drainage networks. However, the coastal area may face higher water tables due to a sea level rise as foreseen by the changing climate projections. This may suggest the need for good drainage networks or, more appropriately, a land-water management programme based on the natural properties of the wetland area that would consider halophyte management. Use of deep groundwater for conventional agriculture would increase the risk of salt water intrusion from the sea and cause devastating outcomes.

Additionally, for enhancing an adaptive capacity for the LSIP area to climate change, effective modifications in the strategies of water use and irrigation system management should be accomplished. The gravity water delivery system should be transformed to a closed delivery system, and serious precautions must be taken immediately to prevent erosion and sedimentation in the water basin and water storage structures. This effort would subsequently provide high level effectiveness for the water supply and water use.

Highly efficient drip and sprinkler irrigation networks should be widely implemented in the LSIP area to irrigate cash crops and increase the efficiency of water use. However, other adaptations can be developed by the increased use of groundwater as a source of irrigation to cope with the hazards of the forthcoming climate change. Nevertheless, in many parts of the plain, the deep aquifers are saline and their degree of recharge is still questionable. Moreover, excess exploitation of the groundwater in the coastal area would surely increase seawater intrusion, causing devastating outcomes. This suggests the need for good drainage networks and the use of river waters with increased irrigation efficiencies.

Consequently, as reported by Nagano et al. (2007a), the LSIP at present seems to have large adaptive capacity towards climatic and social changes. To sustain its productivity, it is strongly recommended that farmers and water user associations improve irrigation and water use efficiencies by means of better maintenance of canals, better gate operation, and employment of better application techniques. This would improve equity of water allocation, avoid high water tables, and conserve the soil in the long term. In the whole area, especially the coastal zone, good management of subsurface drainage is vital to avoid salinity and waterlogging. The use of deep groundwater should be avoided because of the risk of salt build-up.

Acknowledgments This research was conducted as one part of the ICCAP (Impact of Climate Change on Agricultural Production System in Arid Areas) Project. It is a collaboration research between Research Institute of Humanity and Nature of Japan and TUBİTAK (The Scientific and Technical Research Council of Turkey). Finally, the authors extend their thanks to the guest editors for their contributions during the review process of our chapter.

References

Alpert P (2004) Rainfall and Temperature Trends to Extremes over the Mediterranean and Tropical effects. ICCAP-Cappadocia workshop 21–23 November 2004.

Alpert P, Baldi M, Ilani R, Krichak SO, Price C, Rodo X, Saaroni H, Ziv B, Kishcha P, Barkan J, Mariotti A, Xoplaki E (2006) Relations Between Climate Variability in The Mediterranean Region and the Tropics: ENSO, South Asian and African Monsoons, Hurricanes and Saharan Dust. In Lionello P, Malanotte-Rizzoli P, Boscolo R (eds) *Mediterranean Climate Variability.* p 149–177, Amsterdam: Elsevier.

Anonymous (2002) Lower Seyhan Irrigation Project. Regional Directorate of State Hydraulic Works, p 8 Adana (Unpublished, data).

Aquastat (2010) Global İnformation System on Water and Agriculture. Food and Agriculture Organization of The United Nations.

Baykam AR (2004) *The Environmental Atlas of Turkey.* Ministry of Environment and Forestry, Dir of Environmental Impact Assessment, Office of Environmental Archiving, 143–157 Ankara (in Turkish).

Blue Plan (2008) Water: Virtual Water: Which perspective for the Mediterranean Water Management and Distribution? *Environment and Development in the Mediterranean.* N° 8 April 2008.

Boucheron JM (2010) *Water Resources in the Mediterranean* (report).

Brönnimann S, Xoplaki E, Casty C, Pauling A, Luterbacher J (2007) ENSO influence on Europe during the last centuries. *Climate Dynamics* 28:181–197.

Bulut A (2010) The tourism and water resources of Turkey. The Antalya Medical Board; at: http://www.antalyatabip.org.tr (14 December 2010).

Çetin M, Diker K (2003) Assessing drainage problem areas by GIS: A case study in the Eastern Mediterranean Region of Turkey. *Irrigation Drainage* 52(4):343–353.

Correia FN (1999) Water Resources in the Mediterranean Region. International Water Resources Association. *Water International* 24(1):22–30.

Deque M, Jones RG, Wild M, Giorgi F, Christensen JH, Hassell DC, Vidale PL, Rockel B, Jacob D, Kjellström E, de Castro M, Kucharski F, van den Hurk B (2005) Global high resolution versus Limited Area Model scenarios over Europe: results from the PRUDENCE project. *Climate Dynamics* 25:653–670.

Dinç U, Sari M, Şenol S, Kapur S, Sayin M, Derici MR, Çavuşgil V, Gok M, Aydın M, Ekinci H, Ağca N, Schlichting E (1995) The Çukurova Region Soils. Publication of Agricultural Faculty, University of Çukurova, No. 26 (in Turkish with extended English summary), Adana

EEA (2010) The European *Environment-State and Outlook 2010: Consumption and Environment.* European Environment Agency.

Evsahibioğlu AN, Aküzüm T, Çakmak B (2010) Water management, water use strategies and international water rights. *7. Technical Congress of the chamber of Agricultural Engineers Book of Proceedings* 1, 119–134, Ankara (in Turkish).

FAO (2005) *FAO's global informations system of water and agriculture.* Food and Agriculture Organizations (2005) Databases; at: http://www.fao.org/.

Fujinawa K, Tanaka K, Fujihara Y, Kojiri T (2007) *The impacts of climate change on the hydrology and water resources of Seyhan River Basin, Turkey.* The Research Project on the Impact of Climate Changes on Agricultural Production System in Arid Areas. ICCAP Publication 10, Kyoto, Japan, pp 53–58.

Gibelin AL, Deque M (2003) Anthropogenic climate change over the Mediterranean region simulated by a global variable resolution model. *Climate Dynamics* 20:327–339.

Giorgi F, Marinucci MR, Visconti G (1992) A 2 × CO_2 climate change scenario over Europe generated using a limited area model nested in a general circulation model. II: climate change scenario. *Journal of Geophysical Research* 97:10011–10028.

Giorgi F, Lionello P (2008) Climate change projections for the Mediterranean region. *Global Planet Change* 63:90–104.

Hamdy A, Lacirigniola C (1994) Water Resource Management in the Mediterranean Basin. Proc. In *International Conference on Land and Water Resources Management in the Mediterranean Region. Vol. I*, CIHAEM/IAM-B, Bari, Italy, pp 1–28.

Hamdy A, Lacirigniola C (1999) *Mediterranean Water Resources: Major Challenges Towards the 21st Century*. Bari, CIHAEM/IAM-B.

Hamdy A, Lacirigniola C (2005) *Coping with Water Scarcity in the Mediterranean: What, Why, and How*. Bari, CIHAEM/IAM-B.

Hamoda MF (2004) Water strategies and potential of water reuse in the south Mediterranean countries. *Desalination* 165:31–41.

Hoshikawa T, Nagano T, Kume T, Watanabe T (2007) Evaluation climate changes on the Lower Seyhan Irrigation Project, Turkey. The Research Project on the Impact of Climate Changes on Agricultural Production System In *Arid Areas*. ICCAP Publication 10, Kyoto, Japan, pp 217–226.

ICCAP (2007) Final report of Turkish Group on the ICCAP: Impact of climastic change on agricultural production system in arid areas (Turkish). ICCAP Pub. No. 12, Kyoto, p 181.

Iglesias A, Garrote L, Flores F, Mone M (2007) Challenges to Manage the Risk of Water Scarcity and Climate Change in the Mediterranean. *Water Resources Management* 21:775–788.

IPCC (2007) Climate Change 2007: *Impacts, Adaptation and Vulnerability. Contribution of Working Group II to the Fourth Assessment Report of the Intergovernmental Panel on Climate Change, Published for the Intergovernmental Panel on Climate Change*. Cambridge: Cambridge University Press.

Kanber R, Çullu MA, Kendirli B, Antepli S, Yılmaz N (2004) Irrigation, Drainage and Salinity. 6. Technical Congress of the Chamber of Agricultural Engineers, *Proceedings* 213–252 (in Turkish).

Kanber R, Watanabe T, Ünlü M, Tekin S (2007) Outcomes of the ICCAP-Turkish Team. ICCAP Workshop, Jan. 30, 2007. Kyoto, Japan.

Kanber R, Kapur B, Ünlü M, Koç DL, Tekin S (2008) Agricultural drought and contemporary irrigation technologies. *Türktarım J* 178:14–18 (in Turkish).

Kimura F, Kitoh A, Sumi A, Asanuma J, Yatagai A (2007) Downscaling of the global warming projections to Turkey. *The final report of ICCAP*. The research project on the impact of climate changes on agricultural production system in arid areas. ICCAP Publication 10, Kyoto, Japan, p 21–32.

Kitoh A (2007) Future climate projections around Turkey by global climate model. *The final report of ICCAP*. The research project on the impact of climate changes on agricultural production system in arid areas. ICCAP Publication 10, Kyoto, Japan, pp 39–42.

Kume T, Nagano T, Akça E, Donma S, Hoshikawa K, Berberoğlu S, Serdem M, Kapur S, Watanabe T (2007) *Impact of irrigation water use on the groundwater environment and soil salinity*. The research project on the impact of climate changes on agricultural production system in arid areas. ICCAP Publication 10, Kyoto, Japan, pp 205–210,

Milano M (2010) The foreseeable impacts of climate change on the water resources of four major Mediterranean catchment basins. HSM, *Plan Blue*, Reginal Activity Centre, p 6.

Nagano T, Donma S, Hoshikawa K, Kume T, Umetsu C, Akça E, Önder S, Berberoğlu S, Özekici B, Watanabe T, Kapur S, Kanber R (2007a) *The integrated assessment of the impact of climate change on Lower Seyhan Irrigation project*. The research project on the impact of climate changes on agricultural production system in arid areas. ICCAP Publication 10, Kyoto, Japan, pp 193–204.

Nagano T, Donma S, Kume T, Berberoğlu S, Hoshikawa K, Akça E, Kapur S, Watanabe T (2007b) Long-term changes of level and salinity of shallow water table in the Lower Seyhan Plain, Turkey. The research project on the impact of climate changes on agricultural production system in arid areas. ICCAP Publication 10, Kyoto, Japan, pp 211–216.

Nagano T, Hoshikawa K, Donma S, Kume T, Önder S, Özekici B, Kanber R, Watanabe T (2007c) Assessing Impact of Climate Change on the Large Irrigation District in Turkey with Irrigation Management Performance Assessment Model. *Proc. in the International Congress on River Basin Management,* Chap. II, 22–24 March, 2007, pp 651–664, Antalya.

Özekici B, Donma S, Önder S, Nagano T (2006) Evaluation of Lower Seyhan Irrigation Project. Communications to the International Conference on Renewable Energies and Water Technologies. 7-7 Oct. RES 1-7, Almeria, Spain.

Plan Blue (2009) *State of the Environment and Development in the Mediterranean.* United Nations Environment Programme/Mediterranean Action Plan (UNEP/MAP) – *Plan Bleu,* P.O. Box 18019, p 200, Athens, Greece.

Şaylan L, Çaldağ B (2001) *Drought and overcome recommendations.* Chamber of Meteorological Engineers, 14–15 (in Turkish).

Shatanawi M, Naber S, Shammout M (2009) Mainstreaming agricultural and water polices for social equity and economical efficiency. In technological perspectives for rational use of water resources in the Mediterranean Region. Edired By M. El Moujabber, L. Mandi, G.T. Liuzzi, I. Martin, A. Rabi and R. Rodriguez. *Options Mediterranean.* Series A: Mediterranean Seminars, Number 88, pp 209–220.

Somot S, Sevault F, Deque M, Crepon M (2008) 21st Century Climate Change Scenario for the Mediterranean using a coupled Atmosphere-Ocean Regional Climate Model. *Global Planet Change* 63:112–126.

Tekinel O, Kanber R, Çetin M (2000) Development and use of water resources. *6. Technical Con. of the Chamber of Agr. Eng.* Ankara (in Turkish).

Tezcan L, Ekmekçi M, Atilla Ö, Gürkan D, Yalçınkaya O, Namkhai O, Soylu EM, Donma S, Yılmazer D, Akyatan A, Pelen N, Topaloğlu F, Irvem A (2007) Assesment of climate change impacts on water resources of Seyhan River Basin. The Research Project on the Impact of Climate Changes on Agricultural Production System in Arid Areas. ICCAP Publication 10, pp 59–72, Kyoto, Japan.

Ünlü M, Koç M, Barutçular C, Koç L, Kapur B, Tekin S, Kanber R (2007) *Effects of Climate change on evapotranspiration, and crop growth under Çukurova Conditions,* Turkey. ICCAP Workshop, Jan. 30–31, Kyoto, Japan.

Watanabe T (2007) Summary of ICCAP: *Framework, outcomes and implication of the project. The final report of ICCAP. The Research Project on the Impact of Climate Changes on Agricultural Production System in Arid Areas.* ICCAP Publication 10, Kyoto, Japan, pp 1–14.

World Tourist Organization (2010). Tourism Outlook 2010; at: http://www.onecaribbean.org/content/files/UNWTOTOURISMOUTLOOK2010.pdf (22 November 2010).

WWF (2007) Drought: The subtle catastrophe on earth. Evaluation Report. Turkish WWF. Istanbul (in Turkish).

Xoplaki E, Gonzalez-Rouco JF, Luterbacher J, Wanner H (2003) Mediterranean summer air temperature variability and its connection to the large-scale atmospheric circulation and SSTs. *Climate Dynamics* 20:723–739.

Xoplaki E, González-Rouco JF, Luterbacher J, Wanner H (2004) Wet season Mediterranean precipitation variability: influence of large-scale dynamics and trends. *Climate Dynamics* 23:63–78.

Part III
Climate Change Impacts
on Land Use and Vegetation

Chapter 8
Impacts of Agriculture on Coastal Dunes and a Proposal for Adaptation to Climate Change: The Case of the Akyatan Area in the Seyhan Delta

Kemal Tulühan Yılmaz, Didem Harmancı, Yüksel Ünlükaplan, Hakan Alphan and Levent Tezcan

Abstract The overall aim of this paper is to propose a sustainable management alternative for the dunes in the research site in order to adopt the concept of the expected impacts of climate change. For this aim we analysed the agricultural land use along with income generation and sharing versus ecosystem losses using a case study of the Akyatan (Kapıköy) coastal area, located in the Lower Seyhan Plain, Turkey. The coastal agriculture practised on sand dunes is more profitable than that in the lower plain, but it causes irreversible damage to the dune environment. The existing land rental procedure was considered improper since it causes the extensive conversion of the sand dunes to agricultural fields, which in turn causes the loss of the natural heritage. This conversion results in the loss of the barrier function of the dune ridges which could provide protection for sea level rise. A land use proposal, including dune reclamation by means of afforestation and cultivation of indigenous dune plants, suggests alternative income generators for the local farmers. The goals of the dune reclamation action plan are to: (i) foster re-vegetation of indigenous woody plant cover, (ii) establish stone pine or carob orchards, and (iii) introduce certain perennial plant species (e.g. aromatic plants and fragrant geophytes) which will generate income for the locals. This approach will help to protect the native bio-diversity and create alternative income generation instead of the existing land management, which is degrading the coastal dune ecosystem.

K. T. Yılmaz, Professor, Çukurova University, Department of Landscape Architecture, Faculty of Agriculture, Adana 01330, Turkey; e-mail: tuluhan@cu.edu.tr

D. Harmancı, forestry engineer, Adana Regional Directorate of Forestry, Adana, Turkey; e-mail: didemharmanci@ogm.gov.tr

Y. Ünlükaplan, lecturer and researcher, Çukurova University, Department of Landscape Architecture, Faculty of Agriculture, Adana, Turkey; e-mail: yizcan@cu.edu.tr

H. Alphan, Professor, Çukurova University, Department of Landscape Architecture, Faculty of Agriculture, Adana, Turkey; e-mail: alphan@cu.edu.tr

L. Tezcan, Professor, Hacettepe University (UKAM), International Research Center for Karst Water Resources, Ankara, Turkey; e-mail: tezcan@hacettepe.edu.tr

© Springer Nature Switzerland AG 2019
T. Watanabe et al. (eds.), *Climate Change Impacts on Basin Agro-ecosystems*,
The Anthropocene: Politik—Economics—Society—Science 18,
https://doi.org/10.1007/978-3-030-01036-2_8

Keywords Afforestation · Coastal zone management · Sand dunes
Sea level rise · Seyhan delta · Sustainable agriculture · Welfare economics

8.1 Introduction

8.1.1 Sea Level Rise and Threats to Coastal Areas

Rising sea level results in a spatial shift of coastal geomorphology, manifested through the redistribution of coastal landforms comprising subtidal bed forms, intertidal flats, saltmarshes, shingle banks, sand dunes, cliffs and coastal lowlands (Pethick/Crooks 2000). This evolution in geomorphology will determine not only the quality and quantity of associated habitats and the nature of their ecosystem linkages, but also the level of vulnerability of wildlife, people and infrastructure in coastal areas. Consequently, effective management of the coast has been difficult. A society that is unaccustomed to change and adaptation inhabits the coast. To manage for change, the functional geomorphic values of coastal landform need to be recognised and incorporated into planning strategies (Crooks 2004).

The International Panel on Climate Change (IPCC) estimates that the global average sea level will rise between 0.6 and 2 feet (0.18–0.59 m) in the next century (IPCC 2007). EPA (2010) reports that sea level is rising along most of the US coast and around the world. Later IPCC (2013) updated the earlier estimation, and, according to the RCP 8.5 scenario, the rise in sea level was projected to be around 0.45–0.82 m. In the last century, the sea level rose 5–6 in. more than the global average along the Mid-Atlantic and Gulf Coasts, because coastal lands there are subsiding. Assessments from accurate surveys demonstrate that over the last century unconsolidated sections of the ocean coastline have been moving inland between 50 and +300 m, depending on compartment position. In some places it has resulted in the complete loss of coastline public reserve (Helman 2007). Estimates for fifty-year periods between 1850 and 2100 on the east coast of Australia show a slowly accelerating rate of sea level rise that will continue into future centuries until warming is slowed. The increasing rate of rise will result in a corresponding increasing rate of inland coastal migration. Between 1820 and 1945 (125 years) sea level rose some 90 mm, and between 1945 and 2008 it rose another 90 mm. It is claimed that sea level is likely to continue rising at an increasing rate, and it is projected to rise another 90 mm by 2040 and another 90 mm by 2055 (Helman/Tomlinson 2009). Prediction of future sea level rise and the resulting shoreline retreat are, however, among the most important tasks facing coastal and global change scientists, particularly given the population concentration in coastal zones (Cooper/Pilkey 2004). Rising sea levels inundate wetlands and other low-lying lands, erode beaches, intensify flooding, and increase the salinity of rivers, bays, and groundwater tables. Some of these effects may be further compounded by other effects of a changing climate. Additionally, measures that people take to protect private property from rising sea level may have adverse effects on the environment and on public uses of beaches and waterways. Some property owners and state and local governments are already starting to take measures to prepare for the consequences of rising sea level (EPA 2010).

8.1.2 Expected Effects on Beaches and Coastal Dunes

The effect of future sea level rise on coastal dunes may be expected to vary spatially and temporally, and will depend on such factors as the rate of rise, local sediment supply and wind energy conditions. Many dune systems occur within coastal process settings where longshore sediment transport is important, and where frontal dune erosion in one area releases sediment for shoreline progradation and new dune development down-drift. Particularly around estuaries, patterns of shoreline and frontal dune erosion and progradation are often complex due to the dynamic behaviour of banks and channels. The presence of coastal protection structures such as sea walls and revetments will also complicate the response of the dunes to sea level rise (Saye/Pye 2007). Van der Meulen (1990) claimed that geomorphological changes to the dune system under conditions of rising sea level will, in turn, affect the groundwater hydrology, which in turn will influence vegetation patterns and aeolian processes in the inland dunes (Pye 2001).

It is widely known that there is already widespread erosion on the world's beaches. Bird (1985) claims that nearly 70% of the earth's sandy beaches retreated and less than 10% advanced between 1880 and 1980, the remainder having remained stable or shown no net change.

It is stated that coastal dunes behind eroding beaches will show increased cliffing of the backshore (the shoreward, usually dry zone of the beach) by storm waves as sea level rises. As the coastal dune fringe is cut back, blowouts will be initiated, and some of these will grow into large transgressive dunes as sand is excavated and blown landward. If, at the same time, the climate becomes drier and windier, coastal dunes that are at present stable and vegetated may become unstable as the vegetation cover is weakened and sand mobilised; dunes that are already active will become more mobile. A wetter and calmer climate could aid vegetation growth and dune stabilisation (Bird 2010).

8.1.3 Expected Effects on Salt Marshes, Estuaries and Lagoons

A rising sea level will submerge existing intertidal areas, and as the nearshore water deepens, stronger wave action will initiate or accelerate erosion on the seaward margins of salt marshes. Changes can be seen in salt marshes on coasts that are already submerging. On the Atlantic seaboard of the United States salt-marsh islands in Chesapeake Bay have diminished in area or disappeared as a result of sea level rise, and submergence has resulted in salt-marsh plants invading backing meadow land (Bird 2010). It is expected that estuaries and lagoons will generally become wider and deeper as sea level rises; their shores will be submerged and eroded. Tides will penetrate farther upstream and tide ranges may increase. Entrances to coastal lagoons are likely to be enlarged and deepened, increasing the inflow of seawater during rising tides and drought periods. This will raise the salinity of lagoons that were previously relatively fresh. Erosion of enclosing barriers may lead to breaching of new lagoon entrances and could eventually reopen the lagoons as marine embayment. On the other hand, new lagoons may be formed by seawater incursion or rising water tables in low-lying areas behind dunes on coastal plains. It is reported that this has occurred around parts of the Caspian coast as the sea level has risen since 1977 (Bird 2010).

8.1.4 Agricultural Land Use and Dune Management Along the Turkish Mediterranean Coast

Turkey has the greatest extent of dunes among European countries, with 845 km long dune coasts covering almost 10% of the total coastline (8,333 km). Rapid destruction of the dunes was reported, including 110 individual areas, most of which were owned by the Treasury and some of which were designated as National Parks or Nature Reserves.

At present, coastal areas, particularly lowlands, are threatened by intensive land use, including agricultural activities, as well as sea level rise and coastal erosion in the Eastern Mediterranean part of Turkey. An increasing proportion of arable land is also lost to urbanisation and other infrastructures, along with the increase of agricultural expansion on to marginal lands such as coastal dunes. There has always been a mismatch between land quality and land use, resulting in land resource exploitation in the coastal countries of the Mediterranean, especially the eastern flank of the area (Kapur et al. 1999).

The major destruction in the Eastern Mediterranean coastal dunes is caused by levelling to create agricultural fields. From 1990 to 1993, excluding the Göksu Delta, 7,036 ha of dunes have been transformed to agricultural fields (Bal/Uslu 1996). In certain cases, including the Kapıköy dune area (see Fig. 8.1), afforestation has been proposed for stabilising mobile dunes and preventing coastal erosion along the Mediterranean coast. Stone pine (*Pinus pinea*), which is native to Anatolia, has been used for these plantations and covers an area of 884 ha on the sand dunes along the Turkish Mediterranean coast.

Since the coastal areas are considered public property by the Coast Law in Turkey, as in many parts of the world, private use of these lands is illegal. However, in some cases, state-owned land allocation for private use is the underlying concern

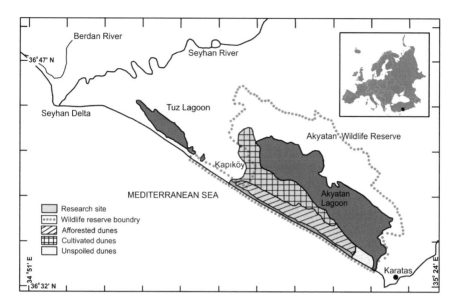

Fig. 8.1 Location and boundaries of the research site. *Source* The authors

in land use conflict between local community benefits and nature conservation demand. The sandy soils – Arenosols – of the coast are ideal sites for certain crops, such as watermelon and peanuts, and are often preferred for intensive farming because of their excellent drainage characteristics. Turkey ranks second to China in watermelon yield, with a cropland of about 150,000 ha and harvest of 4 million tons/year. Watermelon is one of the important cash crops as well as cotton, paddy rice, melon, strawberry, citrus and cereals in the Çukurova plain. Çukurova has an outstanding importance for watermelon cropping not only due to its yield capacity but also to its very early season. Cropland under low polythene tunnels doubled from 1985 to 1992 and increased to 17,100 ha in Adana. It reached 20,287 ha in 2001 and the watermelon yield increased by 2.8 times and exceeded 655,000 tons in the Adana province (Hatırlı et al. 1993).

8.1.5 Economic Assessment of Coastal Values

Most particular to environmental issues, property rights remain poorly defined because even if they were clearly defined, they would be difficult to enforce (Holcombe 2006). Property rights convey the right to benefit or harm one or another. They also specify how people may be better or worse off, and therefore who must pay whom to modify the actions taken by people (Demsetz 1967). Property rights therefore have an important role to play in resolving environmental issues. Coase (1960) states that in the absence of transaction costs, parties bargain and by the establishment of property rights the externalities become internalised. In this approach, it does not matter to whom the right is given.

Local people's willingness-to-pay (WTP) and to protect the coastal environment was measured by Ünal (2003) for the site used in this study. The mean value for the WTP was reported to be 100 €/ha per year per household in the study site. This low amount of WTP is derived from the lack of awareness of the local people both of the natural value and the option value of the coastal ecosystem. Having considered the low WTP, any ecotourism scheme, as an alternative income-generating activity that would be based on local visitors, is not applicable to the site. In addition, the attitude of the "free rider", who wants to benefit from public goods and services without paying for them, plus local people's preference revelation mechanisms and reflection ability should also be considered. However, relevant studies reveal that WTP differs depending on visitor profile. Egashira (2006) stated that WTP values obtained from Asian, North American, and West European tourists differed for historic heritage and the natural environment. It was reported that Asian visitors agreed to pay the highest value for the natural environment while European tourists were willing to pay the highest amount for historical heritage.

In the natural state, vegetated sand dune ridges act as a barrier to block rising seawater and safeguard the fertile lower plain as well as natural habitats located in the landward extent of the ridges. The coastal agriculture practised on sand dunes causes irreversible damage to the dune environment. The existing land rental procedure is considered improper since it causes the extensive conversion of sand dunes to agricultural fields, which in turn causes the loss of the natural heritage. This conversion results in loss of the barrier function of dune ridges which could provide coastal defences against sea level rise. The focal point of this paper was based on re-structuring land use and reclamation of degraded sand dunes, considering expected sea level rise according to the climate change scenarios.

8.2 Materials and Methods

8.2.1 Materials

8.2.1.1 Research Site

The study site Kapıköy – a sample site for Mediterranean deltaic environments – is located at the south-western coast of the Akyatan lagoon in the Seyhan River delta (Fig. 8.1). The delta consists of extensive sand dunes, two lagoons, namely Tuz and Akyatan, and salt marshes. This coastal area, which was nominated as one of the Important Bird Areas in Turkey, is a key migratory stopover site for the Palearctic-African route. On the left bank of the Seyhan River, including the Tuz lagoon, an area of about 5,768 ha was declared the "Seyhan River and Tuz Lake Wildlife Reserve". Further east, the Akyatan lagoon and the surrounding wetland-sand dune complex is also managed as a wildlife reserve and was declared a RAMSAR site due to its considerable potential for waders and waterfowl. In the delta, including the right and left banks of the river, a total dune area of about 3,500 ha was afforested and 3,200 ha cultivated (Yılmaz 2002).

Kapur et al. (1999) reported that 46% of the coastal sand dunes in the delta had been cleared for cultivation between 1923 and 1999. It was also concluded that the agricultural activities have caused severe degradation on the sand dunes during the last five decades around the Tuz Lagoon due to the six-fold increase in the cultivated area during this period (Yılmaz et al. 2003; Berberoglu et al. 2003). Due to the intensive migration from other parts of the country and the rapid conversion of arable lands to residential and industrial areas in the upper parts of the lower Seyhan Plain, the demand for cultivated land has increased further over the last decades. This phenomenon has resulted in an agricultural encroachment upon the marginal lands of coastal dunes and temporary wetlands around the Akyatan and Tuz lagoons. The most rapid land conversion was observed in the 1990s. There has been a close relationship between the expansion of cultivation and decision-making in land use for the coastal area. The rate of land conversion from dunes to agricultural fields accelerated tenfold from 1993 to 2002, through the designation of the Karataş district for tourism development.

The banks of the delta have been the subject of earlier attempts at afforestation and two large areas of *Stone Pine/Acacia/Eucalyptus* plantations can be found on the landward side of the Akyatan Lagoon and left bank of the Berdan River estuary (Yılmaz et al. 2004).

The impact mechanism of afforestation on the dunes differs from agriculture, as the shelter provided by the canopies of planted trees encourages the regeneration of woody perennials while preventing invasive weeds in the under storey. Despite its negative effect on the native dune flora, afforestation generates social benefits for the local community via the utilisation of income-generating non-timber forest products. Çakan et al. (2005) reported that the greatest number of taxa in the area are threatened by farming and changing agricultural practices, followed by the effects of small populations and restricted habitats, recreational activities, and inappropriate forestry practices. Studies revealed strong evidence that cultivation is a major cause of the destruction of dunes (which are regularly levelled for this purpose) where watermelons and peanuts are cultivated. The floristic analysis, carried out in the sampling plots selected from undisturbed dune sites and abandoned fields, showed that significant changes were detected in the floristic composition of the vascular plant cover. These changes occurred with an increase in the weeds and decrease (species richness and coverage) in the native psammophytes (Aytok 2001; Yılmaz 2002).

As a potential threat, impacts of expected sea level rise along the Turkish coast were studied by Kuleli (2010). It was stated that the Mediterranean region, including the research site, is the most vulnerable to land loss among the coastal areas in Turkey. Potential coastal land loss from inundation in the research site and adjacent areas has been evaluated by the authors, and threatened coastal areas are illustrated in Fig. 8.2.

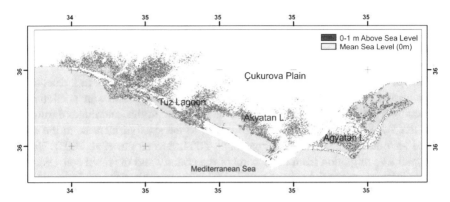

Fig. 8.2 Potential coastal land loss from expected inundation in accordance with the RCP 8.5 scenario. *Source* The authors

8.2.1.2 Water Resources

Tezcan et al. (2007) reported that the main water resources for the coastal zones of the Seyhan plain are composed of deep groundwater aquifers. There are at least two distinct aquifer systems in the region. The upper one is between the depths of 150 and 250 m, and the lower one is deeper than 300 m (minimum thickness of 150 m). Both systems are confined and artesian, and fed by higher-elevation sources in the upper Seyhan basin, and according to the stable isotope analysis conducted, the turnover time of these water resources is over twenty years. The upper aquifer is in hydraulic contact with the sea, and no evidence is found for the connection of the deeper system with the sea. These groundwater systems are under pressure, which creates a higher piozometric head than the sea level and in turn negative natural conditions for seawater intrusion. The extraction of the groundwater through the deep bore holes in the coastal zone causes seawater intrusion in late summer, but during winter and springtime it is refreshed by the recharge of the water sources of the higher elevations. The salinisation in the lower aquifer system has never been observed. A hydraulic connection between the two aquifer systems is possible, since the separating layer is not completely impervious. In some parts, the separating layer is very thin and the conductivity is high enough for the exchange of the groundwater between the two layers. The available well logs, some discrete groundwater level measurements, and the major ion hydrochemistry of a few wells are the only sources for the definition of the hydraulic system. Systematic monitoring of the quality and level of the groundwater, and extraction through the wells are not available.

Methods	Observed behaviour	Hypothetical
Direct	Market price	Contingent valuation
	Simulated markets	
Indirect	Travel cost	Contingent ranking
	Hedonic property values	
	Hedonic wage values	
	Avoidance expenditures	

Table 8.1 Economic methods for measuring environmental and resource values. *Source* Tietenberg (2001: 36)

8.2.2   Methods

The land rental procedure and the economic benefit of watermelon cropping was investigated in this study. Thus, the total revenues, total cost of inputs, and the rental cost on treasury lands were analysed at twelve test areas (Harmancı 2005). "Contingent Valuation", which is a "direct observation method", was used for deriving values regarding watermelon cropping by asking local farmers.

As a traditional approach, economists decompose the total economic values conferred by resources into three main components: (1) "Use value", the value derived from the actual use of a good or service; (2) the "Option value", which reflects the value people place on a future ability to use the environment; and (3) the 'Non-use' value, also referred to as "passive use" values, which are values that are not associated with the actual use, or even the option to use a good or service (Wierenga 2003). The sum of these three values yields the "Total Willingness to Pay" (Tietenberg 2001), which might be regarded as the ultimate economic approach in the sustainable use of the coastal resources (Table 8.1).

Some indigenous shrub and bulbous plant species which have tolerance to sand burial, were considered as "optional values" of the sand dune environment that can be utilised for alternative income generation by the locals. For calculating the total revenue of cultivation of indigenous plant species, European market prices were taken into consideration.

The land ownership pattern (State, forestry and private) was analysed based on the cadastral information provided by the authority. Both information sources were used to analyse the agricultural land use in the research area, and a proposal for sustainable management was suggested for the coastal ecosystem.

8.3   Results

Kapıköy is a village with 300 inhabitants of 60 families. The majority of the families (80%) with medium-size households (average six individuals) live on agriculture. The average size of cropland cultivated by the villagers varies from 1.5 to 2 ha. Hence, the largest portion of the agricultural enterprise (40%) is small scale farms. Only four families cultivate large fields of more than 25 ha.

Table 8.2 Profitability assessment of watermelon cropping in different conditions. *Source* The authors

(Euro/hectare)	Location		
	Adana	İmamoglu (Plain)	Kapıköy (Dune)
Total revenues (crop price)	5,080	6,930	6,930
Total cost of inputs	3,010	4,560	2,890
Rental cost	580	580	150
Profit	1,490	2,430	4,040

Profit = Total Revenues (Crop Price) − Total Costs (Total Costs of Inputs + Rental Cost)

Despite the strict prohibitions stated in the Coastal Law concerning physical changes to the coastal structure, dune ridges have been excavated for sand extraction as construction materials. Additional sand material has been used for elevating dune fields above the salinisation, which is an expected outcome of the water table rising in the dunes (and naturally stabilised dunes that were levelled for cultivation). This destructive misuse of coastal dunes has been legalised by the application of the rental and allocation procedure by the local authority of the National Treasury. The land ownership pattern in the study site showed: 29% of the total area (1,554 ha) is treasury lands; 38% (2,088 ha) is owned by the forestry authority; and the remainder (33%, 1,780 ha) is private land. Almost half of the property (768 ha) of the coastal area owned by the Treasury in Kapıköy is registered as sand fields and dunes by the local authority. The state owned land of about 300 ha has been allocated for cultivation to the local farmers. Sand dune areas rented by the locals cover 146 ha, i.e., a large part (49%) of the cultivated land owned by the State is part of the sand dune habitats. The amount of the rental fee for the study site was evaluated as 145 €/ha in 2005. Local farmers cultivate only 100 ha of land for watermelon cropping in Kapıköy, while 200 ha have been used by the investors from other regions. As a direct observation method, we also investigated total costs, crop prices, and generated profits for a unit cropland (Euro/hectare) in different locations, including the study site, for 2005. Those variables are given in Table 8.2 for Adana, İmamoğlu (plain), and Kapıköy (dune) to compare the profitability of watermelon cropping in different conditions. Irrigation costs in the plain and on the dunes differ (5 €/ha and 17.3 €/ha respectively).

It is obvious that watermelon cropping on the dunes is more profitable than that in the plain conditions. The basic reasons for this are the low costs of agricultural inputs (namely for tillage), high prices due to the advantage of early harvests (namely due to the high water use efficiency in the sandy soils and optimal rhizosphere aeration together with the climatic benefits – stable and moderate to high temperatures throughout the growth period), and low rental costs provided by the local authority.

8.4   Discussion

The expected sea level rise is threatening the lower Seyhan plain and adjacent coastal areas. McGlashan (2003) gives examples of coastal lowlands that have been abandoned to marine invasion in Britain: Freiston in Lincolnshire and Porlock Bay in Somerset. More broadly, some coastal structures have been moved inland in response to the threat of coastline recession. It was claimed that as the sea rises, great modifications will have to be made to cities, ports, and low-lying areas along the world's coastlines. Preparation of suitable management plans to ensure conservation and sustainable use of dune systems also requires assessments to be made over the likely impacts of future changes in climate, sea level, sediment supply and human activities, both within the dune systems themselves and in adjoining onshore and offshore areas. The geomorphological response of sand dune systems to future changes in sea level and climate, and the consequences for their status as ecological habitats, are of key interest for future management strategies.

It is clear that measures are urgently needed to (a) mitigate the negative impacts of agriculture; (b) prevent illegal land conversion; and (c) prevent uncontrolled use of groundwater; thus achieving sustainable use of coastal resources in the study site. For this aim, a comprehensive ecological evaluation is essential, but, on the other hand, socio-economic considerations aiming to re-organise land use pattern for wise use of these resources should not be neglected.

8.4.1   Economic Basis of Dune Management

Two systems are recommended to harmonise the decision-making procedures for coastal dunes. These were (1) the dune system in which the natural values and processes prevail, and (2) a dune management system based on relevant research and politics (Van der Meulen/Salman 1996). For the dune system as natural heritage it is accepted within the basic rules of the global economy that a utilisation approach will not limit potential use of resources for future generations while satisfying the needs of the existing community. This means that our society is responsible for sustaining this system, which can be considered a capital as well as natural heritage for future generations (Pearce/Moran 1994). Natural heritage has a diverse and good capital which has beneficial effects on the revitalisation of the rural economy and/or the improvement of living conditions. However, a comparison of the present economic level with the optimal conditions for the future would require an analysis of the income-generating potential of the dune environment.

The value of a sand dune environment can be classified into use and non-use values. The former can be categorised into direct, indirect and optional use values. The direct use value is obtained as the product that can be consumed and includes crop production in the agricultural case, on a natural sand dune environment. Indirect use value cannot be applied to consumption, and includes many types of recreation, such as sight-seeing, birdwatching, and angling. Option value is the value that can be obtained by saving the natural environment, which is not used today, but has the potential for use in the future. The non-use value includes (a) the bequest value which can be obtained by saving the benefits of the natural resources for future generations and not contemporary society, i.e., standing for the desire for preservation of the heritage or the natural environment; and (b) the existence value, which shows the specific sites where the natural heritage exists. A good example of this would be the protection of plant and bird diversity, along with the original geomorphologic structure as well as the scenic value of the dune environment.

8.4.2 Natural Product Management

The predominant factor affecting sand dune mobilisation is wind erosion, while the other factors that influence the mobility and stability of coastal sand dunes are related to human activity. Wind above a certain wind velocity can erode sand to such an extent that it prevents seeds from germinating in the sand and stabilising it. The drift potential (DP) of the wind (Fryberg 1979), which refers to the sand transport equation, may indicate the wind magnitude. Similarly in Tsoar's (2002) threshold value, vegetation starts covering the sand dunes as the wind power decreases below 500 DP. Dunes start shifting with stronger winds at higher threshold values.

Therefore, the basic issue for sand dune management is to safeguard the natural state or the successional pathways. To achieve this strategy, the conservation of natural vegetation should be considered a priority.

In the research site patches of most of the remnant dunes have been destroyed by intensive cultivation and have only remained in a limited number of localities in an agriculture-dominated coastal landscape. Indigenous shrub species, such as *Rubus sanctus*, *Cionura erecta*, *Vitis sylvestris* and *Vitex agnus-castus*, are present on dune ridges. Among the above species, *Vitex agnus-castus* (Chaste tree) is a crucial element since it tolerates sand burial, and helps stabilise dune ridges. Since its products are well-known components in the natural healthcare market, the Chaste tree has an important economic value. Active components are extracted from the dried, ripened fruits and the root bark of this species. Several new luteolin-like flavonoids, iridoid glycosides, aucubin, eurostoside, agnuside, some triterpenoids, and an alkaloid, vitricine, have been isolated from the root bark. Both free and conjugated forms of progesterone and hydroxyprogesterone have been isolated from the leaves and flowers. Existence of testosterone and epitestosterone were reported in the flower parts. It was also stated that essential oils and Androstenedione were extracted from the leaves. Sand lily (*Pancratium maritimum*) is a perennial and bulbous plant (geophyte) species and occurs in the Mediterranean coastal dunes. This native species contains several alkaloids, particularly galanthamine in its bulbs and other vegetative parts. It has attractive and sweetly scented flowers and has been used widely as an ornamental plant. Recent

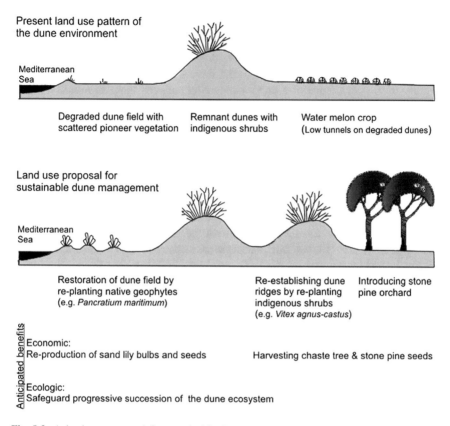

Fig. 8.3 A land use proposal for sustainable dune management in the study site. *Source* The authors

studies revealed successful reproduction of *P. maritimum* under field conditions (Tatlı 2004). Chaste tree and sand lily, which are indigenous plant species, can be considered "optional values" of the sand dune environment that can be utilised for alternative income generation by the locals. A simplified scheme of land use proposal for sustainable dune management is given in Fig. 8.3.

8.4.3 Water Management

The other primary concern should be the management of the water resources, which are crucial for sustaining agriculture as well as the sandy habitats occupied by crops. The extensive utilisation of the groundwater on the coastal zone may cause several consecutive problems in the region, both for the groundwater resources and the land use. The irrigation of the watermelon fields is by deep wells mostly located around the lagoons of the Tuzla site between the Seyhan River bed and the Akyatan lagoon. The increasing number of deep wells and the extracted amount of water will cause the decline of the groundwater pressure/level in both aquifers, and seawater intrusion will be inevitable. Once the hydraulic balance of the upper aquifer is distorted, saline water may intrude to the deeper system, where there is the possibility of the occurrence of a direct connection between these aquifers. The only way to avoid this hazard is to keep the groundwater level always much higher than the sea level, and keep the piezometric level of the lower system greater than the upper one. Climate change scenarios predict sea level rises of 40–90 cm in the area, which is expected to create a susceptible environment for the future groundwater resources. The most important impacts are the possible decrease in the recharge at higher elevations and increase in the abstraction due to the limitations in surface water resources.

Simulation studies indicate that the intrusion distance is controlled by the groundwater level in the plain. In the case of a 50% increase in groundwater abstraction in warmer conditions, the distance of the seawater intrusion will reach 10 km inland by the end of 2080 (Tezcan et al. 2007). Below the sand dunes is impervious clay, which prevents the deep percolation of the irrigation excess water to the groundwater resources. On the other hand, the permeability and the porosity of the sand dunes are enormously high. The rising water level over the impervious clay zone will most likely transform the land to a swamp if the excess water is not drained. The extent of the swampy areas to be found on the coast seems to be a clue for this transformation or the resilience of the pre-existing wetland conditions.

8.5   Conclusions

In conclusion, we emphasise that the application of the land rental procedure which is encouraging illegal occupation of the State-owned dune areas and results in destabilisation and salinisation of the dunes, should be reviewed. Increased agricultural activities have caused substantial degradation of the research site. This degradation is an externality as it rises from the "property right" problem. Considering the land ownership pattern, the dominance of the treasury lands causes over-production and prevents efficient determination of rental fees and product prices. In order to correct this market failure, public and private solutions should be employed by paying special attention to the distribution of land ownership. The existing land use pattern does not generate sufficient income to the locals compared with the amount of total land allocated for cropping. The portion of treasury land subjected to allocation to local farmers allows them to cultivate only a third of the total area. The profitability of watermelon cropping on the dunes is relatively high in comparison with the plain conditions, but marketing conditions for this crop are not stable. Therefore, the annual fluctuation of crop price is an important factor in reducing the overall income of the farmers. Imports from Iran, where the watermelon crop is harvested earlier than that in the Lower Seyhan plain, caused a significant decrease in the market value. A similar situation for groundnut crop, cultivated in the plain, has been experienced recently due to imports from China.

Having considered severe conversion from dunes to agricultural fields, which results in a loss of the natural heritage, it is obvious that the local community has been exploiting this economic capital, which will be essential for future generations. This issue should be taken into consideration not only for existing or bequest values of the coastal environment, but also from the point of "option value" which can generate income for the long term environmental management.

Hence, income sharing should be taken into account to reach realistic consideration of the demands of the local community in the future as well as the present. However, it is not realistic, in the short term, to re-organise the land use pattern. Due to the lack of alternative income generation strategy for the local community, it is difficult to reduce the intensity of agricultural activities by abandoning the watermelon fields gained from the dunes. But at least two-thirds of the dune area allocated for cropping can be abandoned, since this land is not currently used by the locals. The authority can achieve this mitigation measure since 29% of the total area (1.554 ha) is treasury lands. Degraded treasury lands should be restored by replanting indigenous plant species as is shown in Fig. 8.2, and local farmers should be encouraged to be involved in maintaining and harvesting noon-wood products.

The actual market price of ten seeds of sand lily (*Pancratium maritimum*) is about 1.5 € in the European market. Considering the average seed yield of sand lily (100 seeds), the total revenue for 1 ha has been calculated as 150,000 €. However the net income of the local farmers will be much lower than the total revenues when the total cost of inputs is calculated. It is obvious that in the case of having ideal market conditions, the total revenue of sand lily is much higher than that of the watermelon crop. Stone pine plantations are another optional investment instead of watermelon cropping, but establishing an orchard is a long-term investment since it can require fifteen years to reach maturity.

Sand lily reproduction and stone pine orchard establishment and indigenous shrub re-planting are recommended for degraded sites along the foredunes and the inner dune fields respectively. Therefore the areas of specific habitats and biodiversity that need to be protected, such as unspoiled dunes along the seashore and mud flats, salt flats and salt marshes surrounding the lagoon, will not be subjected to the proposed applications. Protected nesting and spawning areas for *Caretta caretta* and *Chelonia mydas*, endangered marine turtle species, overlap the unspoiled foredunes. Negative impacts of the proposed application are not expected on *Caretta caretta* or any other endangered species.

Recently a massive population of reed cat (*Felis chaus*) has been detected in and around the afforestation area by the Akyatan lagoon. It was reported that this is one of the healthiest populations in the country. Despite negative impacts of afforestation on native flora, it is obvious that woody habitats support resilience of this species. Therefore, establishing stone pine orchards will not cause any negative impacts on reed cat, which is an endangered species. Further, it has been suggested that stone pine orchards should be established on heavily degraded dunes, with no harmful impact on native dune flora either. Afforested dunes support some colonial bird populations as well as reed cat. Tall trees located between the seashore and the lagoon serve as a standing post for cormorants (*Phalacrocorax carbo*). Briefly, no negative impacts from the recommended land use proposal for sustainable dune management are expected in the study site.

Co-ordination among the governmental institutions that currently work independently is essential for implementing existing legislation as well as creating alternative income generation (e.g. non-wood products, eco-tourism) instead of agriculture. An eco-tourism scheme should be evaluated, a profile of the potential visitors (foreign visitors) identified and the WTP evaluation reconsidered for the study site.

A long-term interdisciplinary wetland management plan was put into force in 2012 to mitigate the negative impacts of human interference on the research site where complex structures are represented by both landscape features and land ownership patterns. The management plan was prepared and harmonised with the ecosystem dynamics while safeguarding the benefits to the local community at a reasonable level. The focal point of the management plan was based on sustaining wetland values; however, the reclamation of degraded sand dunes was also an aim. For achieving holistic environmental management, re-structuring the land use pattern and land allocation for cropping is an essential issue, as well as protection of bio-diversity and habitats, covering wetlands, salt marshes and dunes.

Acknowledgements This study was partly supported by the Academic Research Project Unit of Çukurova University and TÜBİTAK. The authors wish to thank ICCAP project (Impact of Climate Changes on Agricultural Production in Arid Areas) promoted by the Research Institute for Humanity and Nature (RHIN, Japan). We are also grateful to the representatives of Adana Regional Directorate of Forestry for their contribution and Professor Dr. Selim Kapur and Professor Alan Feest for reviewing the manuscript.

References

Aytok Ö (2001) *Investigating the impacts of agricultural activities on dune ecosystems of Tuzla coastal dunes.* Master's thesis. Çukurova University, Institute of natural and applied sciences, Department of Landscape Architecture.

Bal Y, Uslu T (1996) Beach erosion in the Eastern Mediterranean. In Kapur S, Akca E, Eswaran H, Kelling G, Vita-Finzi C, Mermut AR, Öcal AD (eds) *First International Conference on Land Degradation*, Adana (Adana: Çukurova University).

Berberoğlu S, Alphan H, Yılmaz KT (2003) Remote sensing approach for detecting agricultural encroachment on the Eastern Mediterranean coastal dunes of Turkey. *Turkish Journal of Agriculture and Forestry* 27(3):135–144.

Bird ECF (1985) *Coastline Changes: A Global Review*. Chichester, UK: Wiley.

Bird ECF (2010) Sea level future, implications, effects of climate change, effects on cliffed coasts, effects on beaches, effects on coastal dunes.

Coase R (1960) The problem of social cost. *Journal of Law and Economics* 7:19–20.

Cooper JAG, Pilkey OH (2004) Sea-level rise and shoreline retreat: time to abandon the Bruun Rule. *Global Planet Change* 43:157–171.

Crooks S (2004) The effect of sea-level rise on coastal geomorphology. *Ibis* 146(Suppl.1):18–20.

Çakan H, Yılmaz KT, Düzenli A (2005) First comprehensive assessment of the conservation status of the flora of the Çukurova Deltas, Southern Turkey. *Oryx* 39(1):17–21.

Demsetz H (1967) Toward a theory of property rights. *American Economic Review* 57(2):347–359.

Egashira R (2006) Analysis of heritage value on tourists for sustainable tourism and conservation: using tourists' differences and contingent valuation method. In *14th International Conference on Cultural Economics*, 6–8 July, Vienna.

EPA (U.S. Environmental Protection Agency) (2010) Coastal zones and sea level rise, climate change health and environmental.

Harmancı D (2005) *Analysis of coastal agricultural activities in terms of resource usage planning: KAPIKÖY*. Master's thesis. Çukurova University, Institute of Natural and Applied Sciences, Department of Landscape Architecture.

Hatırlı S, Soysal AM, Yurdakul O (1993) Marketing conditions and problems of watermelon cultivation in agricultural enterprises located in Adana province. *Journal of Agricultural Faculty*, Çukurova University Publication, Pub No: 91.

Helman P (2007) *Two hundred years of coastline change and future change, Fraser Island to Coffs Harbour, East Coast Australia*, Ph.D. Thesis, Southern Cross University.

Helman PR, Tomlinson RB (2009) Coastal vulnerability principles for climate change. In Galvin S, Edmondson T (eds) *Proceedings of the Queensland Coastal Conference 2009*. Gold Coast: SEQ Catchments Ltd.

Holcombe R (2006) Public sector economics. *The Role of Government in the American Economy*. Upper Saddle River, New Jersey: Pearson Education.

IPCC (2007) *Climate Change 2007: Impacts, Adaptation and Vulnerability*. In Parry ML, Canziani OF, Palutikof JP, van der Linden PJ, Hanson CE (eds) *Contribution of Working Group II to the Fourth Assessment Report of the Intergovernmental Panel on Climate Change*. Cambridge: Cambridge University Press.

IPCC (2013) Climate Change 2013: *The Physical Science Basis*. Stocker T, Qin D (Co-Chairs) Contribution to the Fifth Assessment Report of the Intergovernmental Panel on Climate Change. Working Group I Technical Support Unit, Switzerland.

Kapur S, Eswaran H, Akça E, Dinç O, Kaya Z, Ulusoy R, Bal Y, Yılmaz KT, Çelik İ, Özcan H (1999) Agroecological management of degrading coastal dunes: a major land resource area in Southern Anatolia. In Özhan E (ed) *The Fourth International Conference on the Mediterranean Coastal Environment*, Antalya.

Kuleli T (2010) City-based risk assessment of sea level rise using topographic and census data for the Turkish zone. *Estuaries and Coasts* 33:640–651.

McGlashan DJ (2003) Managed relocation: an assessment of its feasibility as a coastal management option. *Geographical Journal* 169(1):6–20. https://doi.org/10.1111/1475-4959. 04993.

Pearce D, Moran D (1994) *The Economic Value of Biodiversity*, IUCN. London: Earthscan.

Pethick JS, Crooks S (2000) Development of a coastal vulnerability index: a geomorphological perspective. *Environmental Conservation* 27:359–367.

Pye K (2001) Long-term geomorphological changes and how they may affect the dune coasts of Europe. In Houston JA, Edmondson SE Rooney PJ (eds) *Coastal Dune Management-Shared Experience of European Conservation Practice*. Liverpool: Liverpool University Press, pp 17–23.

Saye SE, Pye K (2007) Implications of sea level rise for coastal dune habitat conservation in Wales, UK. *Journal of Coastal Conservation* 11:31–52. https://doi.org/10.1007/s11852-007-0004-5.

Tatlı M (2004) The capability of domestication and effects of different nitrogen applications on yield and yield components of sea daffodil (*Pancratium maritimum* L.). Master's thesis. Çukurova University, Institute of natural and applied sciences, Department of Field Crops.

Tezcan L, Ekmekçi M, Atilla Ö, Gürkan D, Yalçınkaya O, Namkhai O, Soylu ME, Donma S, Yılmazer D, Akyatan A, Pelen N, Topaloğlu F, İrvem A (2007) Assessment of climate change impacts on water resources of the Seyhan River Basin. In Kanber R, Ünlü M, Tekin S,

Kapur B, Koç DL, Güney İ (eds) *Impact of Climate Changes on Agricultural Production System in Arid Areas (ICCAP)*, ICCAP Project Turkish Group Final Reports Adana, pp 1–20.

Tietenberg T (2001) *Environmental Economics and Policy*. New York: The Addison-Wesley Series in Economics.

Tsoar H (2002) Climatic factors affecting the mobility and stability of sand dunes. In Lee JA, Zobec TM (eds) *Proceedings of ICAR5/GCTE-SEN Joint Conference*, International Center for Arid and Semiarid Land Studies, Texas Tech University, Lubbock, Texas, USA Publication 02-2.

Ünal N (2003) *A case study on the prediction of social benefits and costs expected from a conservation proposal*. Akyatan. Master's thesis. Çukurova University, Institute of Natural and Applied Sciences, Department of Landscape Architecture.

Van der Meulen F (1990) European dunes: consequences of climate change and sea level rise. In Bakker W, Jungerius PD, Klijn JA (eds) Dunes of the European coasts: geomorphology-hydrology-soils. *Catena Supplement* 18:209–223.

Van der Meulen F, Salman AHPM (1996) Management of Mediterranean coastal dunes. *Ocean Coast Manage* 30(2–3):177–195.

Wierenga M (2003) *A Brief Introduction to Environmental Economics*, ELAW. Washington, DC: Environmental Law Alliance Worldwide, Eugene.

Yılmaz KT (2002) Evaluation of the phyto-sociological data as a tool for indicating coastal dune degradation. *Israel Journal of Plant Sciences* 50(3):229–238.

Yılmaz KT, Çakan H, Szekely T (2003) Ecological importance and management needs of coastal areas in the Eastern Mediterranean region/Turkey. In Özhan E (ed) *MEDCOAST 03, Proceedings of the Sixth International Conference on the Mediterranean Coastal Environment, Ravenna, Italy*, pp 877–888.

Yılmaz KT, Çakan H, Feest A, Düzenli A, İzcankurtaran Y (2004) Assessment of biodiversity for coastal eco-tourism. In Micallef A, Vassallo A (eds) *First International Conference on the Management of Coastal Recreational Resources, Malta*, pp 227–232.

Chapter 9
Estimating Spatio-temporal Responses of Net Primary Productivity to Climate Change Scenarios in the Seyhan Watershed by Integrating Biogeochemical Modelling and Remote Sensing

Süha Berberoğlu, Fatih Evrendilek, Cenk Dönmez and Ahmet Çilek

Abstract Climate change will have a significant impact on ecosystem functions, particularly in the Mediterranean. The aim of this study is to estimate responses of terrestrial net primary productivity (NPP) to four scenarios of regional climate change in the Seyhan watershed of the Eastern Mediterranean, integrating bio-geochemical modelling and remote sensing. The CASA model was utilised to predict annual fluxes of regional NPP for baseline (present) (2000–2010) and future (2070–2080) climate conditions. A comprehensive data set including percentage of tree cover, land cover map, soil texture, normalised difference vegetation index, and climate variables was used to constitute the model. The multi-temporal metrics were produced using sixteen-day MODIS composites at a 250-m spatial resolution. The future climate projections were based on the following four Representative Concentration Pathways (RCPs) scenarios defined in the 5th Assessment Report of The Intergovernmental Panel on Climate Change: RCP 26, RCP 4.5, RCP 6.0 and RCP 8.5. The future NPP modelling was performed under CO_2 concentrations ranging from 421 to 936 ppm and temperature increases from 1.1 to 2.6 °C. Model results indicated that the mean regional NPP was approximately 1185 g C m^{-2} yr^{-1}. Monthly NPP ranged from 10 to 260 g C m^{-2} for the baseline period. The total annual NPP was, on average, estimated at 3.19 Mt C yr^{-1} for the baseline period and 3.08 Mt C yr^{-1} for the future period. NPP in the Seyhan watershed appears to be sensitive to changes in temperature and precipitation. The CASA provide

S. Berberoğlu, Professor, Çukurova University, Department of Landscape Architecture, Adana 01330, Turkey; e-mail: suha@cu.edu.tr.

F. Evrendilek, Professor, Abant Izzet Baysal University, Department of Environmental Engineering, Bolu, Turkey; e-mail: fevrendilek@ibu.edu.tr.

C. Dönmez, Associate Professor, Çukurova University, Department of Landscape Architecture, Adana, Turkey; e-mail: cdonmez@cu.edu.tr.

A. Çilek, Ph.D., Çukurova University, Department of Landscape Architecture, Adana, Turkey; e-mail: cilek@cu.edu.tr.

© Springer Nature Switzerland AG 2019
T. Watanabe et al. (eds.), *Climate Change Impacts on Basin Agro-ecosystems*,
The Anthropocene: Politik—Economics—Society—Science 18,
https://doi.org/10.1007/978-3-030-01036-2_9

promising results for a better understanding and quantification of ecological and economic implications of regional impacts of climate change on biological productivity across the complex and heterogeneous watersheds of Turkey.

Keywords Climate change · IPCC · Modelling · MODIS · NPP
RCPs · Turkey

9.1 Introduction

Terrestrial net primary productivity (NPP) is a key component of the global carbon (C) cycle. Accurate estimations of NPP and its sensitivity to regional changes in temperature and precipitation are of great interest in understanding the amount and rate of changes in C emissions/removals over time and space due to climate change. Increased capability of estimating spatial and temporal responses of NPP to various future climate scenarios paves the way for policy-makers to cope with the changing climate (Evrendilek 2014). In other words, such dynamic predictions are needed not only for measuring regional C budgets but also for mitigating greenhouse gas (GHG) emissions, particularly, in heterogenous ecosystems (Wang et al. 2011). This is especially true for Turkey, where climate change can affect its rich vegetation biodiversity and productivity in various ways (Erşahin et al. 2016).

Accurate NPP estimates in complex environments call for analysing scenarios of projected climate change, spatial and temporal changes in C sinks (Houghton 2005), and conversion dynamics of C sinks to C sources (Smith et al. 2001; Morales et al. 2007; Tang et al. 2010; Zhao/Running 2010). The Intergovernmental Panel on Climate Change (IPCC) has developed different climate scenarios within its 5th Assessment Report. These scenarios are called "Representative Concentration Pathways (RCPs)" because they were developed to be representative of possible future scenarios of emissions and concentrations. The RCPs focus on the trajectory of GHG concentrations over time to reach a particular radiative forcing at 2100 (Australian Government 2014).

Remote sensing-based indices, field measurements, and biogeochemical models are typically used to estimate spatial and temporal dynamics of terrestrial NPP at regional level since in-situ NPP measurements are costly, time-consuming, and provide point samples instead of areal measurements of standing stock, biomass growth, and litterfall. Due to these limitations, process-based terrestrial biogeo-chemistry models, including vegetation models, have been widely used in estimating spatial and temporal responses of forest biomass and NPP at coarse spatial scales to the projected climate changes (Tang et al. 2010).

The aims of this study were to (1) estimate above-ground NPP for the Seyhan watershed, integrating the biogeochemical model of Carnegie Ames Stanford Approach (CASA) and land cover products of Moderate Resolution Imaging Spectroradiometer (MODIS); and (2) model changes in above-ground NPP based on four RCP climate change scenarios of the IPCC under the following CO_2

concentrations: 421 ppm (RCP 2.6), 538 ppm (RCP 4.5), 670 ppm (RCP 6.0), and 936 ppm (RCP 8.5) by the year 2100.

9.2 Materials and Method

9.2.1 Study Area

The Seyhan Watershed is located along the Taurus mountain chain in the Eastern Mediterranean region of Turkey (Fig. 9.1). The region covers an area of approximately 26,000 km^2 and comprises pure and mixed conifer forests that are classified as a Mediterranean evergreen cover type and estimated from tree cores to be up to about 100 years old. The dominant tree species are Crimean pine (*Pinus nigra*), Lebanese cedar (*Cedrus libani*), Taurus fir (*Abies cilicica*), Turkish pine (*Pinus brutia*), and juniper (*Juniperus excelsa*) (Dönmez et al. 2011). The region has a mountainous terrain with a diverse microclimate and vegetation. The typical climate of the watershed is a semi-arid Mediterranean climate whose mean annual rainfall and air temperature are about 800 mm and 19 °C respectively.

9.2.2 Data Processing

The meteorological data used in modelling were obtained on a daily basis from 48 meteorological stations of the State Meteorological Works and included rainfall, solar radiation, and air temperature. Daily climate data were aggregated on a monthly basis and input into the CASA model as the main driving variables. Future climate data were obtained from the WorldClim research group and are based on four RCP climate projections from global climate models (GCMs). These projections are the most recent GCM climate projections used in the Fifth Assessment of the IPCC report. The future climate data were developed with respect to the RCP scenarios within the Coupled Model Intercomparison Project Phase 5 (CMIP5). The GCM output was downscaled and calibrated (bias corrected) as the baseline current climate (Hijmans et al. 2005; WorldClim 2013).

Remote sensing data used in this study were derived from MODIS. The sensor has the Terra and Aqua satellites on board with thirty-six spectral bands. The following spectral bands are primarily designed for analyses of vegetation and land surface: blue (459–479 nm), green (545–565 nm), red (620–670 nm), near infrared (841–875 nm and 1,230–1,250 nm), and shortwave infrared (1,628–1,652 nm and 2,105–2,155 nm) (Gobron et al. 2000; Wang et al. 2013). MODIS sixteen-day composite data at a 250-m spatial resolution were obtained from the National Aeronautics and Space Administration (NASA). These images were geometrically corrected according to WGS 84 coordinate system.

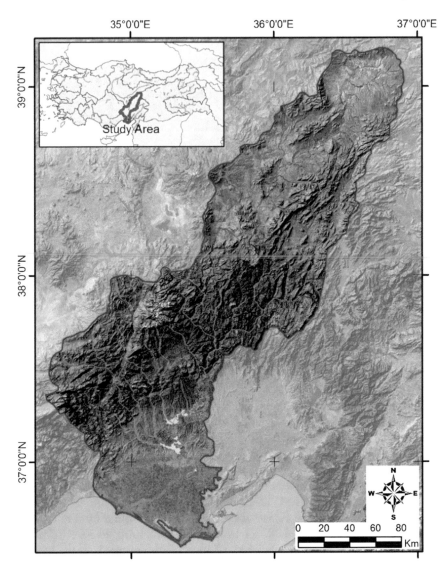

Fig. 9.1 Location of the Seyhan watershed. *Source* The authors

9.3 Methodology

9.3.1 Model Description

The CASA model was used to estimate the present and future spatio-temporal dynamics of monthly NPP in the study region as a function of absorbed photosynthetically active (400–700 nm) solar radiation (APAR) and mean light use efficiency (ε) (Potter et al. 2003). The fundamental relationship in the CASA model is thus:

$$NPP = APAR \times \varepsilon$$

$$NPP = f(NDVI) \times PAR \times \varepsilon \times g(T) \times h(W)$$

where the APAR (in megajoules per square metre per month) is a function of the normalised difference vegetation index (NDVI), and downwelling photosynthetically active solar radiation (PAR in megajoules per square meter per month), while ε (in grams of C per megajoule) is a function of the maximum achievable light utilisation efficiency. The light use efficiency is adjusted by such reduction functions as temperature effect [g(T)] and water stress [h(W)]. Although the previous versions of the CASA model (Potter et al. 1993, 2004) used NDVI to estimate FPAR (APAR/PAR ≈ NDVI), the current model version relies on canopy radiative transfer algorithms (Knyazikhin et al. 1998) in order to generate improved FPAR products as inputs to C flux calculations. In addition, the percentage of tree cover, land cover map of the region, and soil texture were used to run this model.

Climate Data: Monthly precipitation, air temperature and solar radiation were used as the climate data sets. These variables were based on ten years (2000–2010) of records from the meteorological stations of the Seyhan Watershed. Climate variables were spatially-interpolated together with the Digital Elevation Model (DEM) using the co-kriging method, and mapped on a monthly basis and incorporated into the modelling process. The future climate maps obtained from CMIP5 in grid format were incorporated into the CASA model to constitute future simulations of NPP.

Land Cover Map: The MODIS land cover product was used to classify the vegetation types with the CoORdination of Information on the Environment (CORINE) land cover scheme. Initially, the output was comprised of thirteen land cover classes at a 250-m spatial resolution (Fig. 9.2). Accuracy analysis was carried out comparing the classification map and the ground truth data collected from field campaigns.

Soil Texture Map: The soil texture data are based on the FAO soil texture classification system with five classes. For the dominant soil type in a soil unit, the designations 'coarse', 'medium', 'fine', and 'very fine' or combinations of them were assigned based on the relative amount of clay, silt, and sand present in the top 30 cm of soil (Potter et al. 2003). These classes were derived from the pedotransfer rules and expert opinions, and the regional soil maps at the 25,000 scale were utilised for this study.

NDVI: MODIS NDVI images were extracted for the study area. The MODIS/ Terra NDVI images (L3 Global) produced at sixteen-day intervals and at a 250 m spatial resolution were used as one of the main inputs to the CASA model. The images contained transformations of the red (620–670 nm), near infrared (841–876 nm) and blue (459–479 nm) bands to enhance vegetation signals and allow for

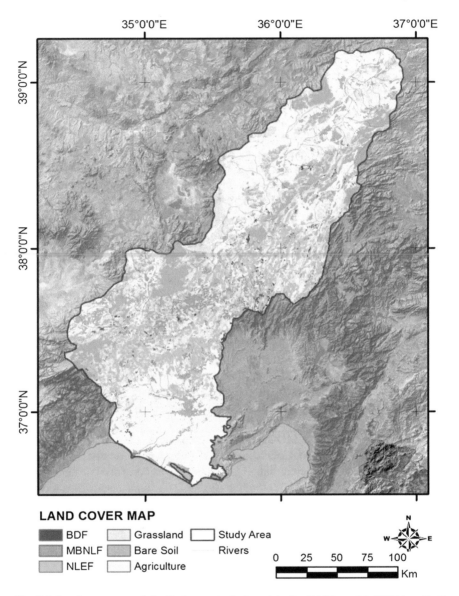

Fig. 9.2 Land cover map of the Seyhan watershed used in the CASA model (*BDF* broadleaf deciduous forest, *MBNLF* mixed broadleaf deciduous and needleleaf forest, and *NLEF* needleleaf forest). *Source* The Authors

precise inter-comparisons of spatial and temporal variations in photosynthetic activity (NASA 2013).

Percent Tree Cover Map: A Regression Tree (RT) algorithm was used to predict the percent tree cover of the Seyhan watershed. The RT algorithm produces a rule-based model for predicting a single continuous response variable from one or more explanatory variables (Loh 2002). Regression trees are built through a process known as binary recursive partitioning, an iterative process of splitting data into subsets called nodes. At each node, the algorithm investigates all possible splits of all explanatory variables (Tottrup et al. 2007). Partitioning the data is based on reducing the deviance from the mean of the target variables. A search is conducted

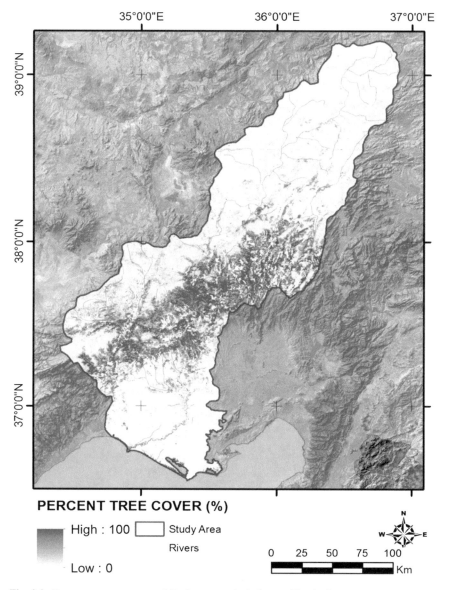

Fig. 9.3 Percent tree cover map of Seyhan watershed. *Source* The Authors

over all predictors and possible split points so that the reduction in deviance is maximised (Breiman et al. 1984). The percent tree cover map of the Seyhan watershed derived from RT is shown in Fig. 9.3.

The accuracy of the percent tree cover map was defined using correlation coefficient (r) values. The r value of the percent tree cover for the study was 0.80, which showed that the RT model was performed within a reasonable accuracy to express the distribution of different tree species within the complex terrain of the study region.

9.4 Results and Discussion

Eleven-year data sets of MODIS and climate variables were used to estimate variability patterns of monthly NPP in relation to climate change in the Eastern Mediterranean region of Turkey. The NASA-CASA model based on FPAR and Light Use Efficiency (LUE) was adopted for simulations over the Seyhan watershed. Modelling results showed that forest NPP for the baseline (2000–2010) condition varied spatially and temporally across the study region. The mean NPP differed significantly across all the months and ranged from 10 to 260 g C m^{-2}. Monthly NPP variations estimated by the CASA model are shown in Fig. 9.4. The total annual NPP were mapped at a 250-m grid cell size (Fig. 9.5).

NPP response to future climate change was explored using projections of the CASA model run over the Seyhan watershed for the period of 2070–2080. Model simulations were intended to reflect monthly and annual variability of NPP under the four climate scenarios (RCP 2.6, RCP 4.5, RCP 6.0, and RCP 8.5) defined by IPCC by which the mean long-term rainfall and air temperature values varied

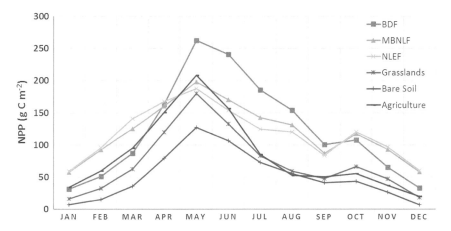

Fig. 9.4 Monthly NPP variations (g C m^{-2}) in the Seyhan watershed simulated by the CASA model. *Source* The Authors

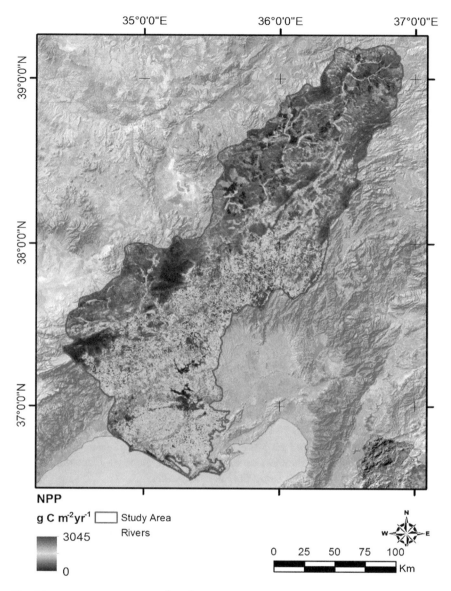

Fig. 9.5 Total NPP map (g C m^{-2} yr^{-1}) of Seyhan watershed derived from the CASA model. *Source* The Authors

substantially. The long-term mean annual temperature indicated an increase of 2 °C for the period of 2070–2080 relative to the baseline (present) conditions between 2000 and 2010. Monthly variations in NPP for different land cover classes were estimated under the present and future climate conditions, thus quantifying changes

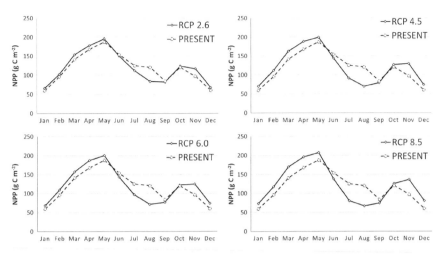

Fig. 9.6 Monthly variations of needleleaf forest NPP (g C m^{-2}) under four RCP scenarios. *Source* The Authors

between the baseline and future periods for each land cover class, including forests, grasslands, and agricultural areas.

Our CASA model simulations indicated a significant increase in NPP for the spring season for the 2070–2080 period. A dramatic decrease in NPP was simulated with RCP 6.0 and RCP 8.5 during the summer. Monthly variations of needleleaf forest NPP under the RCP scenarios are shown in Fig. 9.6. An increase in NPP occurred in the spring and winter seasons due to projected temperature increases for each of the RCP scenarios. In the summer, a decrease in needleleaf forest NPP occurred due to an increased air temperature in the months of June to September. An increase in air temperature is also expected from October to December for the future period when simulated using the WorldClim data.

The total NPP was estimated at 1,179 g C m^{-2} yr^{-1} under RCP 4.5 and 1,345 g C m^{-2} yr^{-1} under RCP 8.5 for Seyhan watershed. A significant increase in mean NPP is projected under the RCP 8.5 scenario. Monthly variations of needleleaf forest NPP under the RCP scenarios are shown in Fig. 9.6. A slight decrease in broadleaf deciduous forest NPP occurred under RCP 4.5 by 50 g C m^{-2} yr^{-1} (Fig. 9.7).

Mixed broadleaf deciduous and needleleaf forest stands cover wide areas in the study area where their biological productivity plays an important role in the regional C budget. The total annual mixed broadleaf deciduous and needleleaf forest NPP was estimated at 1,445 g C m^{-2} yr^{-1} under RCP 2.6. An increase in NPP by 13 g C m^{-2} yr^{-1} was predicted for the RCP 4.5 scenario. Monthly mixed broadleaf deciduous and needleleaf forest NPP was also estimated under all the scenarios (Fig. 9.8).

The monthly grassland NPP was estimated to vary between 10 and 200 g C m^{-2} (Fig. 9.9). Agricultural production provides a significant contribution to Turkey's

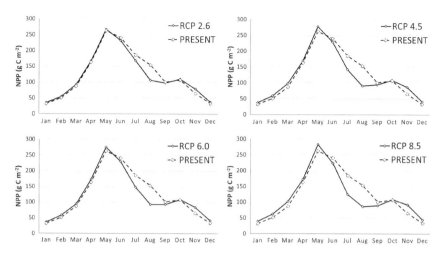

Fig. 9.7 Monthly variations of broadleaf deciduous forest NPP (g C m^{-2}) under four RCP scenarios. *Source* The Authors

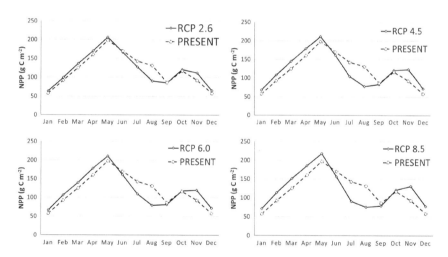

Fig. 9.8 Monthly variations of mixed broadleaf deciduous and needleleaf forest NPP (g C m^{-2}) under four RCP scenarios. *Source* The Authors

economy, and NPP indicates the performance of agricultural productivity. According to our model simulations, a minor increase in agricultural NPP was estimated, with the greatest increase under RCP 8.5. Monthly variations in agriculture NPP are shown in Fig. 9.10.

Our assessment of impacts of the climate change scenarios on NPP and its spatio-temporal variations was based on predictions produced by the CASA model. Model simulations were interpreted for each land cover class over the Seyhan

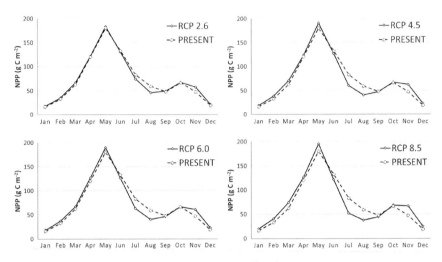

Fig. 9.9 Monthly variations of grassland NPP (g C m^{-2} yr^{-1}) under four RCP scenarios. *Source* The Authors

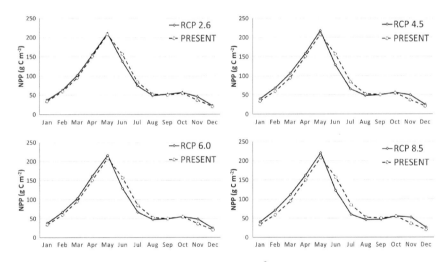

Fig. 9.10 Monthly variations of agriculture NPP (g C m^{-2}) under four RCP scenarios. *Source* The Authors

watershed. Predicted values of annual and total NPP for the land cover classes are shown in Table 9.1.

The mean annual NPP simulations of broadleaf deciduous forest under the RCP scenarios on average indicated a decrease by 48 g C m^{-2} yr^{-1}. By contrast, the total NPP of the needleleaf forest and mixed broadleaf deciduous and needleleaf forest stands showed an increase of up to 50 g C m^{-2} yr^{-1}. Mixed broadleaf deciduous and needleleaf forest stands exhibited slight differences in NPP under the future

Table 9.1 Estimations of NPP responses to present and future climate scenarios for the Seyhan watershed (g C m^{-2} yr^{-1}) (BDF broadleaf deciduous forest, *MBNLF* mixed broadleaf deciduous and needleleaf forest, and *NLEF* needleleaf forest), (*RCP* representative concentration pathways). *Source* The Authors

Land cover classes	2000–2010	2070–2080			
	Present	RCP 2.6	RCP 4.5	RCP 6.0	RCP 8.5
BDF	1482.15	1445.63	1432.98	1430.83	1427.36
MBNLF	1432.81	1445.04	1454.13	1442.64	1469.54
NLEF	1412.36	1430.19	1445.69	1432.56	1464.60
Grasslands	867.74	868.97	870.60	866.31	873.69
Bare soil	618.67	610.88	604.55	605.69	602.84
Agriculture	1005.38	1007.13	1014.12	1007.44	1019.07

projections. The most pronounced impact of the projected climate change scenarios was observed on NPP in needleleaf forest based on the CASA simulations. The greatest increase between the present and future climate conditions was under the RCP 8.5 scenario. Grasslands and agricultural areas showed a critical response to changes in temperature and precipitation according to the CASA simulations.

The total NPP simulations for the present and future climate conditions indicated an increase of 0.238 Mt for the needleleaf forest and of 0.044 Mt for grassland, additionally agricultural NPP increased by 0.089 Mt. NPP increase in agriculture is expected by the period of 2070–2080 due to the effects of CO_2 fertilisation. However, the present study did not take into account changes in irrigation, soil nutrients, land use/cover and agricultural management practices. In the future studies, interaction effects of the driving variables (e.g. CO_2 fertilisation, changes in minimum and maximum daytime versus night-time temperatures, and management practices) should be quantified not only on all the ecosystem compartments (e.g. soil organic C pools, and soil water content) but also on both ecosystem structure (e.g. biodiversity, and herbivory) and function (e.g. nutrient and water cycles).

The differences between the present and future NPP under the RCP scenarios were mapped at a 250-m spatial resolution and are shown in Fig. 9.11.

The changes in the spatial distribution of the NPP for the present and future climate scenarios were predicted using the CASA model. NPP presented significant differences in its spatial distribution across the study region. An increase ranging from 403 to 489 g C m^{-2} yr^{-1} was projected under the four regional climate scenarios. The increase was predicted mostly in the northern part where the broadleaf deciduous forest is located. All the scenarios indicated an increase in the northern section and a great decrease in the southern section. The spatial distribution maps of differences in NPP between the present and future climate scenarios revealed that the higher elevation belts possessed the largest increase primarily due to increased temperature and decreased precipitation. The decrease in annual NPP was estimated to vary from 510 to 557 g C m^{-2} yr^{-1}.

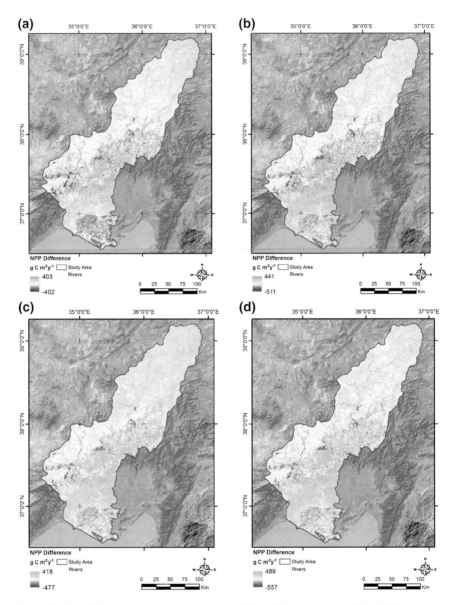

Fig. 9.11 The differences between the baseline (2000–2010) and future (2070–2080) periods: **a** RCP 2.6, **b** RCP 4.5, **c** RCP 6.0, and **d** RCP 8.5, of the NPP maps. *Source* The Authors

9.5   Conclusions

The climate projections should be considered to be projections developed on the basis of specific assumptions over the next sixty years. Different factors, including future developments and emission loads are involved in the evaluation of climate change projections. The climate scenarios indicated an increase in the annual temperature, particularly in the winter season, of up to 2 °C.

Addressing the projected climate changes, the present (2000–2010) and future (2070–2080) regional climate scenarios of RCP 2.6, RCP 4.5 RCP 6.0 and RCP 8.5 and their impact on spatio-temporal dynamics of total, annual and monthly NPP in the Seyhan Watershed were modelled integrating the CASA model and time-series MODIS data.

The study provided an insight into forest NPP responses to future climate change in various scenarios in a semi-arid Mediterranean region. The modelling results of this study indicated that the projected forest NPP in the Seyhan Watershed will differ under the greenhouse gas emission scenarios. Moreover, the results also showed that forest NPP would positively respond to the RCP scenarios within high greenhouse gas emissions (CO_2 concentrations of 421 and 936 ppm in the atmosphere by 2080).

Our results also indicated an increase in total NPP with slight differences in its spatial distribution, and emphasised the importance of high resolution input data for regional impact analysis with respect to climate change. Our model simulations revealed increased NPP, particularly at higher altitudinal zones. NNP of main land cover types in this zone, such as BDF and NLEF, showed an increase, whereas agriculture decreased significantly. Variation in forest NPP appears to be primarily a function of temperature. As far as the increased temperature is concerned, the central part of the study region can have drier conditions that in turn decrease NPP. A distinct decrease in forest NPP was simulated which results from lower precipitation and higher temperature projected for the lowland regions, especially during the summer. However, it should be noted that increased temperature, CO_2 fertilisation, and changes in precipitation regime would interactively determine ecosystem responses to climate change. Their interaction effects in the modelling studies should be considered not only simultaneously but also on all the biogeochemical processes among all the ecosystem components in the orientation of preventive and mitigative measures towards ecosystem sustainability.

The NPP decrease in colder temperate in the upper parts of the region may be associated with the rising temperature. An increase in temperature may reduce the water effectiveness through enhancing evapotranspiration, which can affect the forest growth negative. The NPP decrement in the lower parts of the region for warm temperate conditions can be explained by increasing water stress and deficit which resulted increased evapotranspiration.

The results gained from this study imply that the potential carbon sequestration and climate change response of the Eastern Mediterranean forests of Turkey are strongly related, with changes in heterotrophic respiration and disturbances from natural forces and human activities along with NPP changes.

The uncertainty of NPP modelling over the space was dependent on spatial variation of climate variables and spatial resolution of the variables. A few steps are required in order to reduce these uncertainties of NPP over space. First, climate projections need to be downscaled at finer spatial scales as the input data for modelling. Given the importance of parameters in spatially distributed estimations with biogeochemical models, the uncertainty caused by parameterisation should be minimised through generating spatial layers of the parameters at fine spatial scales. Using remotely sensed data, particularly hyperspectral images, to derive parameter layers could be possible. However, such an approach may be impractical because of the often limited availability of hyperspectral data. Hence, spatial information of the required parameters may be produced by field campaigns, inventory data of the governmental organisations and other related remote sensing platforms, such as airborne systems.

Higher spatial data contributed to increasing the model capability of capturing local changes in NPP and its distribution. In particular, MODIS data hold great potential for predicting NPP with the CASA model owing to its appropriate spatial and spectral resolutions. CASA was the appropriate model for dealing with vegetation heterogeneity in the study region where sparse canopy cover and high species diversity are exhibited.

The response of the Eastern Mediterranean terrestrial ecosystem to climate change will contribute to the general understanding of the vulnerability of ecosystems to climate change in a local scale. The results obtained from this study can provide a basis for environmental decision making and increase adaptability of physical planning strategies to the regions.

Acknowledgements We would like to acknowledge the research project grants (project no: 110Y338) from the Scientific and Technological Research Council (TÜBITAK) of Turkey.

References

Australian Government (2014) Department of the Environment, 2014. *Representative Concentration Pathways (RCPs) Fact Sheet.*

Breiman L, Friedman JH, Olshen RA, Stone CJ (1984) *Classification and Regression Trees*, Chapman and Hall, New York.

Dönmez C, Berberoğlu S, Curran PJ (2011) Modelling the current and future spatial distribution of net primary production in a Mediterranean watershed. *International Journal of Applied Earth Observation and Geoinformation* 13(3):336–345.

Erşahin S, Bilgili BC, Dikmen Ü, Ercanlı I (2016) Net primary productivity of Anatolian forests in relation to climate, 2000–2010. *Forest Science* 62(6):698–709.

Evrendilek F (2014) Modeling net ecosystem CO_2 exchange using temporal neural networks after wavelet denoising. *Geographical Analysis* 46:37–52.

Gobron N, Pinty B, Verstraete MM, Widlowski JL (2000) Advanced vegetation indices optimised for up-coming sensors: Design, performance, and applications. *IEEE Transactions on Geoscience and Remote Sensing* 38(6):2489–2505.

Hijmans RJ, Cameron SE, Parra JL, Jones PG, Jarvis A (2005) Very high resolution interpolated climate surfaces for global land areas. *International Journal of Climatology* 25:1965–1978.

Houghton RA (2005) Aboveground forest biomass and the global carbon balance. *Global Change Biology* 11:945–958.

Knyazikhin Y, Martonchik JV, Myneni RB, Diner DJ, Running SW (1998) Synergistic algorithm for estimating vegetation canopy leaf area index and fraction of absorbed photosynthetically active radiation from MODIS and MISR data. *Journal of Geophysical Research* 103:32257–32276.

Loh WY (2002) Regression trees with unbiased variable selection and interaction detection. *Statistica Sinica* 12:361–386.

Morales P, Hickler T, Rowell DP, Smith B, Sykes MT (2007) Changes in European ecosystem productivity and carbon balance driven by regional climate model output. *Global Change Biology* 13:108–122.

NASA (2013) Land processes distributed archive center web site, 2013; at: https://lpdaac.usgs.gov.

Potter CS, Randerson JT, Field CB, Matson PA, Vitousek PM, Mooney HA, Klooster SA (1993) Terrestrial ecosystem production: A process model based on global satellite and surface data. *Global Biogeochemical Cycles* 7:81–841.

Potter C, Klooster S, Steinbach M, Tan P, Kumar V, Shekhar S, Nemani R, Myneni R (2003) Global teleconnections of climate to terrestrial carbon flux. *Journal of Geophysical Research-Atmospheres*. https://doi.org/10.1029/2002JD002979.

Potter CS, Klooster S, Steinbach M, Tan P, Sheikarand S, Carvalho C (2004) Understanding global teleconnections of climate to regional model 13 estimates of Amazon ecosystem carbon fluxes. *Global Change Biology* 10:693–14 703.

Smith B, Prentice IC, Sykes MT (2001) Representation of vegetation dynamics in the modeling of terrestrial ecosystems: Comparing two contrasting approaches within European climate space. *Global Ecology Biogeography* 10:621–637.

Tang G, Beckage B, Smith B, Miller PA (2010) Estimating potential forest NPP, biomass and their climatic sensitivity in New England using a dynamic ecosystem model. *Ecosphere* 1(6):1–20.

Tottrup C, Rasmussen MS, Eklundh L, Jönsson P (2007) Mapping fractional forest cover across the highlands of mainland Southeast Asia using MODIS data and regression tree modelling. *International Journal of Remote Sensing* 28(1):23–46.

Wang F, Xu YJ, Dean TJ (2011) Projecting climate change effects on forest net primary productivity in subtropical Louisiana, USA. *AMBIO* 40:506–520. https://doi.org/10.1007/s13280-011-0135-7.

Wang L, Gong W, Ma Y, Zhang M (2013) Modeling regional vegetation NPP variations and their relationships with climatic parameters in Wuhan, China. *Earth Interactions* 17(4):20.

WorldClim Climate Layers Web Page (2013); at: http://www.worldclim.org.

Zhao M, Running SW (2010) Drought-induced reduction in global terrestrial net primary production from 2000 through 2009. *Science* 329:940–943.

Chapter 10
Prediction of Vertical and Horizontal Distribution of Vegetation Due to Climate Change in the Eastern Mediterranean Region of Turkey

Junji Sano, Shigenobu Tamai, Makoto Ando
and Kemal Tulühan Yılmaz

Abstract Climate change affects land use patterns all over the world. This study describes the present condition of vegetation in the Seyhan and Ceyhan basins, in the Eastern Mediterranean Region of Turkey, to estimate the impacts of climate change on the species composition and vegetation productivity. There is a great variety of forests ranging from evergreen coniferous forests to deciduous broad-leaved trees along an altitudinal gradient. There are single-species forests of both evergreens and deciduous forests, and also mixed forest formations. The most frequently occurring evergreen forest consists of *Pinus brutia*. Other needle-leaf forest trees are *Pinus halepensis* in the coastal regions, and *Abies cilicia* and *Cedrus libani* in the higher parts of mountains. The most commonly occurring deciduous forest trees are comprised of various *Quercus* species, such as *Q. infectoria* and *Q. cerris*, which are widespread in this region. The other common trees in the mid-altitude and lower mixed and broad-leaved forests are *Carpinus, Fraxinus, Styrax* and some maquis species, such as *Arbutus andrachne* and *Quercus coccifera*. However, anthropogenic (man-made) destruction has greatly reduced their importance in this region. We estimated that the vegetation would be strongly impacted by global warming and a drier climate in the future, based on the data produced by the climate group of ICCAP (Kimura et al. 2007). The contemporary and 2070s biomass and productivity of this area were estimated from satellite images and field research. The present vegetation was remarkably changed from the potential by anthropogenic pressure, especially at the lower level elevations. Furthermore, recent climate changes were discovered to have had a powerful effect

J. Sano, Professor, Tottori University, Forest Ecology and Ecosystem Management Laboratory, Faculty of Agriculture, Tottori 680-8553, Japan; e-mail: jsano@muses.tottori-u.ac.jp.

S. Tamai, Professor Emeritus, Tottori University, Arid Land Research Center, Tottori, Japan; e-mail: tamai@alrc.tottori-u.ac.jp.

M. Ando, Associate Professor, Kyoto University, Field Science Education and Research Center, Kyoto, Japan; e-mail: ando@kais.kyoto-u.ac.jp.

K. T. Yılmaz, Professor, Çukurova University, Department of Landscape Architecture, Faculty of Agriculture, Adana, Turkey; e-mail: tuluhan@cu.edu.tr.

© Springer Nature Switzerland AG 2019
T. Watanabe et al. (eds.), *Climate Change Impacts on Basin Agro-ecosystems*,
The Anthropocene: Politik—Economics—Society—Science 18,
https://doi.org/10.1007/978-3-030-01036-2_10

on the evergreen coniferous forests. All vegetation types are predicted to shift from their present distribution areas to northern or higher altitudes, the consequence being that the distribution of the steppe areas will increase and in turn the evergreen coniferous forests will decrease. The biomass distribution in this area will decrease, compared with that in the present, but net primary production will increase by the 2070s. In conclusion, the natural forests should be conserved and reforestation enhanced to mitigate the climate change of the future.

Keywords Anthropogenic impact · Biomass · Deciduous broad-leaved forest Evergreen forest · Global warming · Present and potential vegetation Productivity · Steppe · Succession

10.1   Introduction

Warming of the climate system is unequivocal, and many of the observed changes are unprecedented whether decades or millennia are taken into account (IPCC 2013). Climate change affects the land use pattern across the world (Allen et al. 2010). It is important to clarify the impact of climate change on vegetation distribution and carbon budget (Bachelet et al. 2001). Mediterranean ecosystems may be particularly vulnerable to degradation in warmer and drier climates (Henne et al. 2015). Impacts of climate change on Mediterranean vegetation have been examined with regard to phenology (Kramer et al. 2000), productivity (Dönmez et al. 2011) and vegetation (Osborne et al. 2000). However, the effect of climate change on vegetation change and productivity associated with altitude is not clear in this region.

There are various vegetation types along climatic and topographic gradients in Turkey (Altan 2000). In the Eastern Mediterranean region of the country they contain grasslands above the timberline, evergreen and deciduous forests, scrublands, river beds, lagoons, coastal saltmarshes, coastal woodlands, and sand dunes (Fig. 10.1) (Yılmaz 1998). The vertical distributions of the dominant tree species in the Mediterranean usually follow: (1) Elevation: 0–800 m a.s.l., Evergreen broad-leaved *Quercus* species such as *Q. coccifera and Q. ilex,* and *Pinus brutia* and *P. halepensis,* (2) Elevation: 800–1,500 m a.s.l. (above sea level), Deciduous broad-leaved tree species such as *Quercus cerris, Q. infectoria, Fraxinus ornus* and *Styrax officinalis,* (3) Elevation: 1,500 m a.s.l. or more: Coniferous species such as *Abies cilicica, Pinus nigra* ssp. *pallasiana, Juniperus* spp. and *Cedrus libani* (Polunin/Huxley 1990).

Yılmaz (1993) showed the potential natural plant cover and actual vegetation type with the dominant species and geological aspects of the Amanos mountains, located in the Eastern Mediterranean region (Fig. 10.2). The sand dune vegetation located along river deltas and maquis (scrub) vegetation cover large areas from the coast to ca. 400 m a.s.l. *Pinus brutia* forests are found on serpentine rocks from 400 to 800 m. *Quercus cerris* occurs over 800 m on limestone. *P. brutia* forests

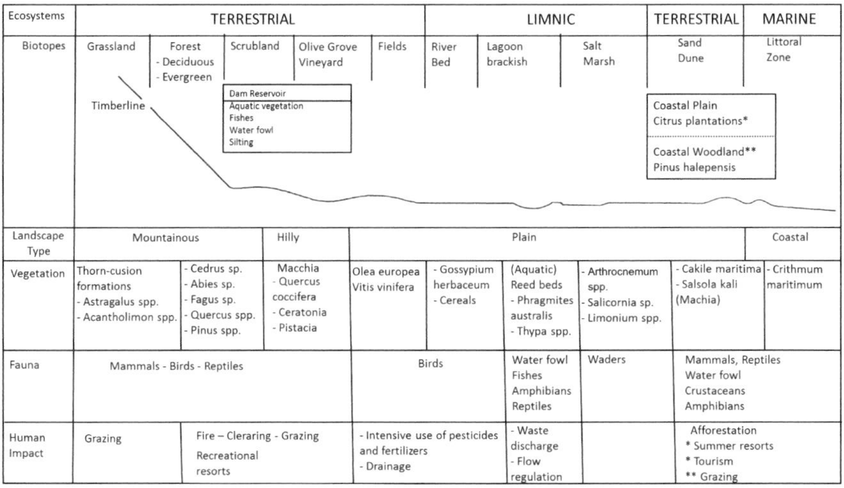

Ecosystems	TERRESTRIAL					LIMNIC			TERRESTRIAL	MARINE
Biotopes	Grassland	Forest - Deciduous - Evergreen	Scrubland	Olive Grove Vineyard	Fields	River Bed	Lagoon brackish	Salt Marsh	Sand Dune	Littoral Zone
	Timberline		Dam Reservoir Aquatic vegetation Fishes Water fowl Silting						Coastal Plain Citrus plantations* Coastal Woodland** Pinus halepensis	
Landscape Type	Mountainous		Hilly			Plain				Coastal
Vegetation	Thorn-cusion formations - Astragalus spp. - Acantholimon spp.	- Cedrus sp. - Abies sp. - Fagus sp. - Quercus spp. - Pinus spp.	Macchia - Quercus coccifera - Ceratonia - Pistacia	Olea europea Vitis vinifera	- Gossypium herbaceum - Cereals	(Aquatic) Reed beds - Phragmites australis - Thypa spp.	- Arthrocnemum spp. - Salicornia sp. - Limonium spp.		- Cakile maritima - Salsola kali (Machia)	- Crithmum maritimum
Fauna	Mammals - Birds - Reptiles				Birds	Water fowl Fishes Amphibians Reptiles	Waders		Mammals, Reptiles Water fowl Crustaceans Amphibians	
Human Impact	Grazing		Fire – Clearing - Grazing Recreational resorts		- Intensive use of pesticides and fertilizers - Drainage	- Waste discharge - Flow regulation			Afforestation * Summer resorts * Tourism ** Grazing	

Fig. 10.1   Model of the ecosystem complex of the Eastern Mediterranean region. *Source* Yılmaz (1998: 88)

gradually decrease and *P. nigra* forests with *Q. cerris* appear at elevations from 800 to 1,500 m. *Fagus orientalis* forests are found as relict communities from 1,500 to 2,000 m (the forest limit) on the sandstone at the northwest exposed slopes of the Amanos Mountains.

Climate change is expected to affect the vertical and horizontal distribution of the vegetation of this region via the changing distribution of each species. This region has been affected by past human activities since at least the early Neolithic period (Yılmaz 1998). Not only the mountainous part of the region but also the Seyhan and Ceyhan plains were covered with dense oak forests in the eighteenth century. People occupied dense settled villages, seasonally, utilising the uplands in their vicinity based on agriculture and pastoralism. The nomadic activities, which continued between the mountains and the plain, had a detrimental effect on the natural vegetation of the region. The most common livestock in this region are goats, which graze grasses and young trees of the forest floor. People in this region obtain wood for several uses, including the illegal gathering of fuel wood in the winter. Thus, it is difficult to find original vegetation especially in the plains and the low elevation area of the mountains (Yılmaz 1998).

The purpose of this study is to describe the present condition of vegetation of the Seyhan and Ceyhan basins and to estimate the impacts of climate change on the species composition and vegetation productivity, which would be severely affected by climate change, especially on the higher regions of the Eastern Mediterranean region of Turkey.

POTENTIAL NATURAL VEGETATION   ACTUAL VEGETATION TYPES
FORMATIONS   AND GEOLOGICAL FORMATIONS

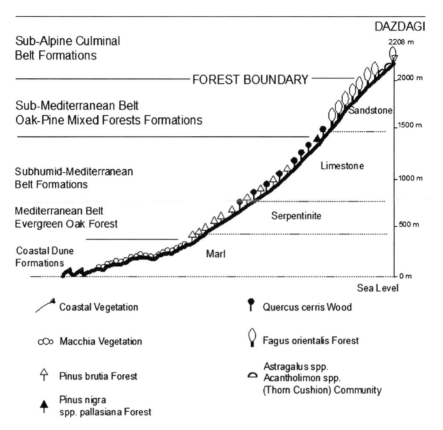

Fig. 10.2 Potential plant cover and actual species distribution at the north-west exposed slopes of the Amanos Mountains on the geological base. *Source* Yılmaz (1993: 73; 2001: 45)

10.2 Study Area and Methods

Turkey's forestland is found on the mountains bordering the Black Sea, Marmara, the Aegean and the Mediterranean, and is located in an altitude of 0 to 2,000 m. The central and eastern parts of the country are much less densely forested. Small concentrations of needle-leaf forests (*Pinus nigra* and *P. sylvestris*) are found in some protected localities of central Anatolia. In both regions, however, the most common forest trees are species of *Quercus*, which includes many evergreen and deciduous species in the Mediterranean region (Roda et al. 1999). There are eighteen species are included in the genus *Quercus* in Turkey (Baytop 1994). They are *Q. aucheri, Q. brantii, Q. cerris, Q. coccifera, Q. ilex, Q. infectoria, Q. ithaburensis, Q. libani, Q. petraea, Q. robur, Q. vulcanica, Q. pontica,*

Q. macranthera, Q. hartwissiana, Q. frainetto, Q. pubescens, Q. trojana and *Q. virgiliana.* In total, 23 taxa, including sub-species and varieties, are represented in the genus *Quercus.* In the Eastern Mediterranean there are five *Quercus* species which, because of their relatively high drought tolerance (Abrams 1990), will probably increase in domination under climate change.

After the preliminary research of species identification and vegetation distribution along an altitudinal gradient, from the Mediterranean coast (ca. 0 m a.s.l.) to the mountain region (ca. 1,500 m a.s.l.), we set up seven plots, namely the Yumurtalık, Çatalan, 2 Karatepe, and 3 Aladağ in 2003, and an additional seven plots – 2 Çatalan, 4 Aladağ, and 1 Adana – in 2004. We divided them into three sites of the area (Nos. 1, 2, and 3 in Fig. 10.3). Yumurtalık is the lowest site facing the Mediterranean Sea (No. 1), and the Adana, Çatalan and Karatepe are the middle elevation sites (No. 2). The Aladağ site was the highest of the research sites located over ca. 800 m a.s.l. (No. 3). The stand characteristics of the research plots in 2003 and 2004 are shown as Tables 10.1 and 10.2 respectively.

We set up replicated plots in each site. Then we identified tree species and measured DBH (diameter at 1.3 m high) and tree height in order to estimate dominance and biomass. Also we took hemispherical photographs to estimate canopy covers, using a Nikon Cool-Pix 950 camera with Fisheye converter,

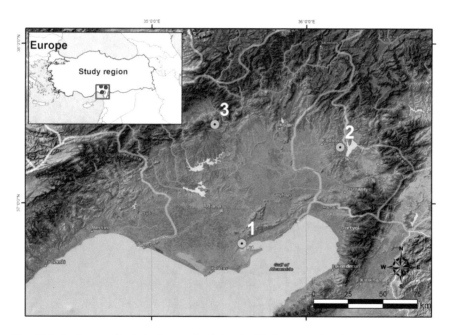

Fig. 10.3 Study area. No. 1, the lowest site facing the Mediterranean Sea, including Yumurtalık; No. 2, the middle elevation sites including Adana, Çatalan and Karatepe; No. 3, the highest site in this area, including Aladağ. *Source* The authors

Table 10.1 Stand characteristics of research plots in 2003. *Source* The authors

Plot	1 Yumurtalık	2 Çatalan	3 Karatepe 1	4 Karatepe 2	5 Aladağ 2	6 Aladağ 3	7 Aladağ 1
Dominant species	*Pinus halepensis*	*Pinus brutia*	*Pinus brutia*	*Arbutus andrachne*	*Pinus brutia*	*Abies cilicica*	*Cedrus, Abies*
Size m × m	50 × 40	20 × 20	30 × 20	15 × 6	50 × 30	40 × 40	30 × 20
Inclination	2	10	21	21	10	26	12
Direction	N50W	N40W	N45E	N30W	N65W	N60W	S35E
N	36°44′49.2	37°12′04.4	37°17′45.4	37°15′48.4	37°33′32.9	37°28′06.4	37°36′20.8
E	35°37′40.4	35°15′22.4	36°15′02.7	36°13′35.5	35°23′31.7	35°19′10.1	35°29′17.3
Altitude	3	151	253	559	793	1223	1532

Table 10.2 Stand characteristics of research plots in 2004. *Source* The authors

Plot	8 Çatalan 1	9 Çatalan 2	10 Aladağ 1	11 Aladağ 2	12 Aladağ 3	13 Aladağ 4	14 Adana
Dominant species	*Pinus brutia*	*Pinus brutia*	*Pinus nigra*	*Pinus nigra*	*Cedrus libani*	*Abies cilicica*	*Quercus coccifera*
Size m × m	20 × 20	20 × 20	20 × 40	20 × 20	20 × 40	20 × 40	10 × 10
Inclination	18	22	18	20	10	15	32
Direction	S70W	S55W	N60E	N70W	N80E	N80W	N80E
N	37°16′47.9	37°16′03.4	37°37′28.6	37°37′31.7	37°36′28.8	37°36′25.9	37°03′51.1
E	35°11′16.6	35°11′37.6	35°28′13.7	35°28′43.2	35°28′53.7	35°28′51.0	35°21′18.2
Altitude	263	329	1951	1840	1403	1379	102

analysed by LIA32 for Windows (Yamamoto 2000). Also we took wood cores to determine tree ages and growth patterns.

We used two approaches for the purpose, (1) making a present vegetation map by using satellite photographs (LANDSAT ETM+, resolution 30 m, 8 bands, 13 June 2000) with 100 ground control points, and (2) estimation of potential vege- tation by using the Thornthwaite p/e Index (PEI) and the Warmth Index (WI) calculated from the climate data at present and the future (2070) provided from the Climate Sub-group of ICCAP (Kimura et al. 2007). The definition of each index is as follows;

Thornthwaite p/e Index (Thornthwaite 1931, 1948):

$$(p/e)i = 0.164[pi/(ti + 12.2)]^{1.11} \tag{10.1}$$

$$\mathrm{PEI} = 10\sum(p/e)i \tag{10.2}$$

T10: PEI < 16: Perarid
T20: 16–32: Arid
T30: 32–64; Semi-arid
T40: 64–; Humid

Warmth Index (WI) (Kira 1976):

$$\mathrm{WI} = \sum(ti - 5)ti \ > \ 5\,^\circ\mathrm{C} \tag{10.3}$$

W1: 15–45: sub-arctic zone, evergreen coniferous forest
W2: 45–85: cool-temperate zone, broadleaved deciduous forest
W3: 85–180: warm-temperate zone, evergreen forest

where, 'i' means month from January to December, 'p' is annual precipitation and 't' is monthly averaged air temperature.

We combined these two indices for classification of potential vegetation.

T10W3 (13): Desert
T20W3 (33): Steppe
T30W2 (32): Woodland a
T30W3 (33): Woodland b
T40W1 (41): Evergreen coniferous forest
T40W2 (42): Broadleaved deciduous forest
T40W3 (43): Maquis

Then present and future patterns of potential vegetation were estimated.

10.3   Environmental Factors

We measured several environmental factors in the plots set at each altitude in 2003 (Table 10.3). The temperature was measured at a height of 1.5 m from the ground, while soil temperature and humidity were measured at 5 cm below the ground as environmental factors in each plot. Soil temperatures were discovered to gradually decrease from the lower to the higher parts of this region, together with air temperatures. Soil temperatures were lower than air temperatures in each plot. Soil humidity (water contents in the soil) was relatively high at the sites of the lower parts, and relatively low at the sites of the higher parts of the region. The border of this difference of soil humidity was found at 600–700 m a.s.l. The canopy cover was extremely low at Yumurtalık because of the scarce distribution of canopy trees (*Pinus halepensis*), and relatively low at the sites of lower to middle elevations, and relatively high at the sites of higher elevation areas.

The influence of light intensity with the climate change on vegetation was not clarified until now. However, vegetation changes with climate change (Matthews et al. 2004), which may affect the light environment, and this change in turn might be sufficient to change the vegetation cover.

10.4   Species Composition

Tree species composition with relative basal area (BA%) in each plot is shown in Table 10.4 based on our field research. There were sixteen species in the plots. The dominant tree species were *Pinus halepensis*, *Pinus brutia*, *Arbutus andrachne* and *Abies cilicica* in the coastal region, lowland, maquis, and highland respectively. *Pinus brutia* and *Quercus coccifera* were found at relatively large areas from the low to the highlands. *Arbutus andrachne*, *Quercus infectoria* and *Styrax officinalis* were limited to mid-altitude regions. Maquis shrubs (Maki) included *Arbutus andrachne* and *Quercus coccifera*. *Carpinus betulus*, *Carpinus orientalis* and *Quercus cerris* were found only at the lower part of the high altitude region, while *Cedrus libani* and *Pinus nigra* were found only at the higher parts of the high latitude region.

We thus defined the vertical distribution associated with the altitude of flora by several conifers (*Pinus, Abies, Cedrus* and *Juniperus*) and *Quercus* species. The vertical distribution of the main tree species is as follows (see also Table 10.4).

2,000–3,000 m in altitude: Alpine pastures
Astragalus spp. (Milk vetch). e.g. *A. angustifolius*
Acantholimon spp. (Prickly thrift). e.g. *A. glumaceum*
6,00–2,000 m in altitude: Aladağ
Abies cilicica (Toros göknarı)
Cedrus libani (Lübnan Sediri)
Juniperus oxycedrus (Katran Ardıçı. Prickly juniper)

Table 10.3 Environmental factors in each plot in 2003. *Source* The authors

Plot	Yumurtalık	Catalan	Karatepe 1	Karatepe 2	Aladag 2	Aladag 3	Aladag 1
Date	20030823	20030829	20030827	20030828	20030825	20030825	20030824
Altitude	3	151	253	559	793	1223	1532
Air temperature (time) (°C)	NA	36.7 (14:00)	33.4 (16:05)	31.2 (14:40)	31.1 (12:27)	24.3 (18:10)	23.2 (19:05)
Soil temperature (°C)		(−5 cm)					
Mean	NA	30.20	27.67	26.13	28.10	23.09	21.11
SD	–	1.99	0.98	1.29	2.49	0.98	1.24
CV	–	6.61	3.55	4.95	8.84	4.23	5.85
Max	–	34.00	30.00	29.40	35.00	24.60	23.60
Min	–	26.80	26.10	23.60	24.60	20.80	18.70
Soil humidity (%)		(−5 cm)					
Mean	NA	41.45	43.31	42.47	27.06	28.74	29.46
SD	–	2.37	4.66	3.68	5.75	8.07	8.26
CV	–	5.71	10.76	8.66	21.26	28.09	28.04
Max	–	46.30	51.20	48.70	39.70	39.50	42.90
Min	–	38.40	29.70	36.00	14.30	12.60	12.60
Canopy cover (%)							
Mean	49.024	72.587	71.745	78.139	64.862	84.502	80.718
SD	12.111	2.642	3.638	4.957	3.593	3.082	3.870
CV	24.704	3.640	5.070	6.344	5.540	3.648	4.794
Max	62.190	76.021	76.735	86.375	71.074	88.626	87.878
Min	24.698	69.642	67.141	73.160	57.988	81.607	77.497

Table 10.4 Species composition and dominance (basal area) of trees in the order of altitude. *Source* The authors

Plot	1	14	2	3	8	9	4	5	6	13	12	7	11	10
Altitude m	3	102	151	253	263	329	559	793	1223	1379	1403	1532	1840	1951
Pinus halepensis	100.0													
Quercus coccifera		78.6		2.1	0.0		20.3							
Cistus creticus		3.1												
Pistacia terebinthus		1.6												
Phillyrea latifolia		2.4			0.3									
Pinus brutia		14.3	98.9	95.2	99.6	100.0		100.0						
Fontanesia phillyrioides			0.5											
Olea europea			0.2											
Arbutus unedo					0.1									
Arbutus andrachne				1.8			57.7							
Myrtus communis					0.1									
Quercus infectoria				0.3			13.6							
Styrax officinalis				0.5			2.8							
Fraxinus ornus							5.6							
Carpinus betulus									3.3					
Carpinus orientalis									2.0					
Quercus cerris									1.9					
Abies cilicica									92.5	35.5	2.4	43.9		
Juniperus oxycedrus									0.3			14.3		
Juniperus excelsa										11.8				
Cedrus libani										24.9	87.2	35.1		
Pinus nigra										27.9	10.4	6.8	100.0	100.0
BA (m2/ha)	8.8	1.2	16.0	42.4	31.1	8.1	20.6	36.5	36.4	40.1	40.8	43.0	73.3	46.6

Quercus cerris (Türk meşesi. Turkish oak, deciduous)
Pinus nigra (Karaçam. Austrian black pine. European black pine)
Pinus brutia (Kızılçam)
Less than 600 m in altitude: Karatepe and Çatalan
Pinus brutia (Kızılçam)
Quercus infectoria
Quercus coccifera (maquis)
Arbutus andrachne (maquis)
0 m in altitude: Yumurtalık
Pinus halepensis (Aleppo pine) and maritime maquis, halophyte communities.

10.5 Vegetation Change Caused by Climate Change

We determined the vegetation distribution and some environmental factors affecting the vegetation change using satellite images and topographic maps. The present vegetation map obtained from the unsupervised classification method is represented in Fig 10.4. We found eleven classes for land use patterns. The most abundant class was grassland (23.9%), and wild grass or crops (22.6%), soil 1 (16.3%), *Pinus brutia* (10.5%), water (9.8%) and soil 2 (8.2%), followed by mixed forests (3.9%).

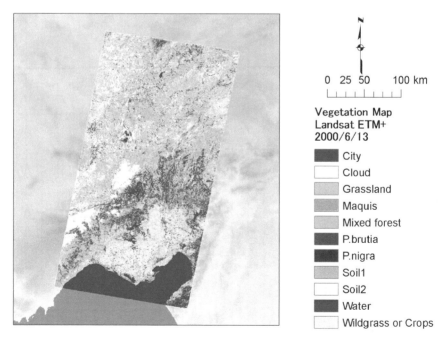

Fig. 10.4 Vegetation map of the Seyhan and Ceyhan river region on 13 June 2000 (Landsat ETM +). *Source* The authors

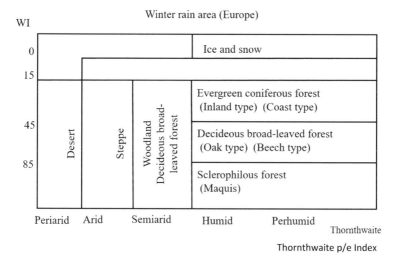

Fig. 10.5 Classification of potential vegetation from the Thornthwaite p/e Index and WI. *Source* The authors

Maquis and *Pinus nigra* communities had low distributions (0.8 and 0.3% respectively). *Pinus nigra* and mixed forests were found at higher elevations, and maquis shrubs were at lower parts, whereas the other classes (grassland and *Pinus brutia*) were relatively evenly distributed. These facts suggested the impact of human intervention especially on lowland areas in this region similar to the central and western Europe (Vera 2000).

The potential patterns of vegetation distribution using climate data at present and in the future were estimated in this study. The present and future patterns of the Thornswaite p/e Index and WI (warmth index) were estimated and the potential vegetation maps were produced from the combined classification of these two indices (Fig. 10.5). Relative areas of steppe, woodland a, woodland b, evergreen coniferous forests, broadleaved deciduous forests and maquis were 1, 10, 26, 1, 45, 18% respectively in the present patterns (Fig. 10.6). Those of desert, steppe, woodland a, woodland b, broadleaved deciduous forests and maquis in the future were different (Fig. 10.7). The difference should indicate the strong impact of global warming and drier climate in the future prospects for the potential vegetation after climate changes in this region (Table 10.5). Climate changes as well as human-induced disturbances are increasingly threatening the natural ecosystems (Evrendilek/Wali 2004). By comparing the present vegetation map (Fig. 10.6) and estimated vegetation map (Fig. 10.7), we can predict the vegetation change caused by climate change for the 2070s. We found the lower parts of this region to be affected by anthropogenic disturbances, such as tree-cutting and grazing (Atik et al. 2007). Ultimately, long-term monitoring and sustainable management of the natural resources are required for the welfare of the future generations (Kılıç et al. 2006).

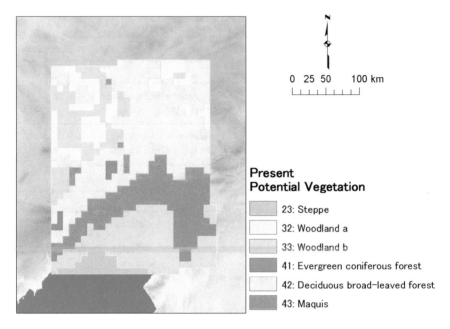

Fig. 10.6 Estimated potential vegetation map at present. *Source* The authors

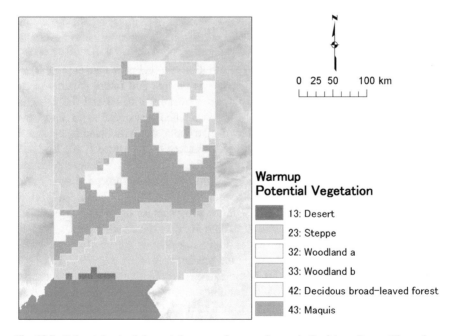

Fig. 10.7 Estimated potential vegetation map after warming up in the future. *Source* The authors

Table 10.5 Estimated succession of vegetation in the research area using satellite photographs and field researches. *Source* The authors

Elevation	Present	After 30 yrs	After 50 yrs	Climax
0–50 m	*Pinus halepensis, Pinus brutia* with maquis	*P. brutia: recession* maquis *Phryigana*	maquis *Phryigana P.brutia*	*Quercus infectoria*
50–800 m	*P. brutia Q. infectoria Q. coccifera*	*Q. coccifera* maquis *P. brutia*	*Q. coccifera* maquis *P.brutia*	*Q. infectoria*
800–1,000 m	*P.brutia* mixed with *Sorbus, Carpinus, Cornus* and *Acer*	Same as the present	Same as after 30 yrs	Same as after 50 yrs
1,000 m	*P. nigra, Abies cilicica, Cedrus libani*	*A. cilicica*	*A. cilicica* higher northface: *P. nigra* rocky south or west face: *Juniperus*	*A. cilicica*

10.6   Changes in Biomass

Climate change affects the growth of trees in this region (Sabate et al. 2002). We estimated the vegetation in the 2070s from the WI and Thornswaite p/e Index by using the climate data provided by the Climate Sub-group of ICCAP (Kimura et al. 2007) in this area. The vegetation of the 2070s was at least somewhat shifted to a climax under the estimated climate conditions. To recover the original vegetation from the degraded, the required period seems to be 60–80 years. In addition to the time of recovery, some 20–30 years will most likely be required to establish the new vegetation that would develop in the changed physical elements of the climate. Moreover, the Mediterranean climate will most likely require about 100 years to encounter a change to another vegetation type in certain areas. According to the vegetation dynamics analysed in this study, the present vegetation will remain intact in one third of the area, whereas the rest of the other vegetation types will change in the Steppe and Woodland areas. These are smaller than those of the potential one, i.e., the climax stage. On the other hand, areas of Evergreen coniferous forests will disappear.

We created two scenarios to estimate the biomass in the 2070s, both of which were assumed to have the same biomass per unit area in each vegetation type as those in the present (Case I) and those of 1.5 fold higher than the present biomass per unit area (Case II).

10.6.1   Case I

The biomass in the present and in the 2070s was estimated and is shown in Fig. 10.8. The biomass of the evergreen coniferous forest of the present, most of

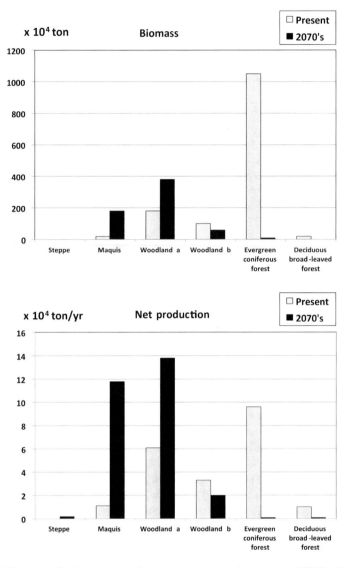

Fig. 10.8 Biomass and net production of each vegetation type in present and 2070s. *Source* The authors

which was dominated by *Pinus nigra*, was remarkably high, though its area was very small, and it depended on a higher average biomass per unit area. The biomass value of the "woodland a" of the present was due to a relatively high average biomass of this vegetation type in this area. Conversely, the biomass value of "woodland b" was lower, though its area was the largest among the five vegetation types of the present. The biomass of the evergreen coniferous forest in the 2070s was remarkably low compared with that of the present, which was due to the decrease in the area in the 2070s. The higher biomass of the woodland and maquis reflected the increase in their distribution areas. The biomass of the total of the five vegetation types in the 2070s was 45% of that of the present. The decrease of the total biomass in the 2070s was due to the increase of the Steppe area, which was lower in the biomass per unit area. This was accompanied by a decrease in the area of the Evergreen deciduous forests with higher biomass. The net primary production patterns among the vegetation types were almost similar to the biomass patterns (Fig. 10.8). The net primary production patterns among the vegetation types were almost similar to the biomass patterns. However, the difference in the net primary production of this area among vegetation types was found to be smaller than the biomass. For example, the biomass of the woodland a in the 2070s was about twofold higher of that of the maquis, but the difference of the net production between them was only 1.1 fold higher.

The total net primary production of this area in the 2070s was estimated to be 1.3 fold higher than that of the present. The biomass of this area in the 2070s decreased, while the net primary production increased, compared with those of the present. Consequently, the study revealed that the biomass of the future may gradually increase in the area, assuming that drought conditions are not too severe. The net primary production of the maquis in the 2070s increased and was 11 fold higher of that of the present. Increase of the production of this vegetation type mostly depended upon an increase in the area (five times) and its productivity.

10.6.2 Case II

The biomass and net primary production of the 2070s were estimated when the net production of each vegetation type except evergreen coniferous forests increased by about 50% of that of the present,. Productivities of species in evergreen coniferous forests, where we actually measured in this area, were higher and almost the same as the productivity of the similar areas studied by Cannel (1982) elsewhere. In this case, we did not increase the productivity of the evergreen coniferous forests to estimate its biomass and productivity (Fig. 10.9). The proportion of the biomass and net production among vegetation types in Case II were scarcely different from those in Case I because of the smaller area occupied by the evergreen coniferous forests. total biomass of and net primary production of Case II in this area were 67% and 200% of those of the present, respectively.

From remote sensing data, the carbon dynamics of this region were estimated to be highly variable, whereas the net primary production (NPP) rose relative to the present climate conditions (Dönmez et al. 2011; Berberoğlu et al. 2015). We had the same trend in the lower elevation parts of this study area, such as in the maquis and woodland a areas. However, the evergreen coniferous forests would markedly decline (Figs. 10.8 and 10.9) in the higher elevation areas of this region.

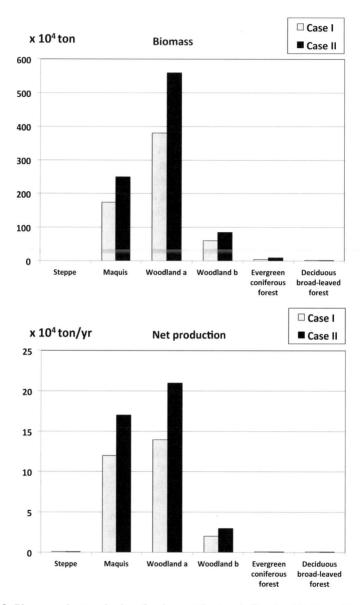

Fig. 10.9 Biomass and net production of each vegetation type in Case I and II. *Source* The authors

10.7 Conclusions

As the vegetation of the Eastern Mediterranean region of Turkey was severely disturbed by anthropogenic pressures over many centuries, it was difficult to estimate the vegetation in the past. However, we could estimate the impact of the anthropogenic pressure by the analysis of the difference between the actual and potential vegetation, especially in the grassland area, meadow and wheat field. To this extent, we should collect additional biomass data to estimate the productivity of vegetation such as the maquis, *Pinus brutia, Quercus* spp. and conifers. Moreover, the analysis of the numerous environmental factors and vegetation functioning related to the distribution and productivity of vegetation such as the soil conditions, physiological factors and the human impacts would be required in this context. Using these data, we estimated the actual vegetation distribution in 2070 and the impact of climate change on the patterns of vegetation in this region.

The actual vegetation was remarkably changed from the potential by the anthropogenic pressure, especially in the lower areas. The climate changes were not strongly affected in the area occupied by each of the vegetation types except the evergreen coniferous forests. All vegetation types in the future will shift from their present area to more northern and/or higher altitudes. The area of steppe will increase and the evergreen coniferous forests will decrease. The biomass in this area will also decrease compared with that of the present, but the net primary production will increase by the 2070s.

As mentioned by Henne et al. (2015), the Mediterranean region is vulnerable to negative climate change impacts, including forest declines, increased wildfires, and biodiversity loss. Thus, reviving the original Mediterranean forest communities may improve the ecosystem resiliency in a warmer future. Ultimately, in order to conserve the appropriate vegetation distribution and in turn to mitigate climate change, it would be indispensable to preserve the natural forests and enhance reforestation activities to mitigate the climate change of this region in the future.

Acknowledgements The authors sincerely thank Prof. Dr. Tsugihiro Watanabe (the leader of the ICCAP: the Research Project on the Impact of Climate Changes on Agricultural Production Systems in Arid Areas) and Prof. Dr. Rıza Kanber (the coordinator of the Turkish Team of the ICCAP) for their great efforts in organising and completing the ICCAP Project; Drs. Türker Altan and Mustafa Artar at the Çukurova University; Drs. Ekrem Aktoklu, Mustafa Atmaca and Kayhan Kaplan at the Mustafa Kemal University; Dr. Meryem Atik at the Akdeniz University; Mr. Ramazan Gökdemiroğlu at the Forest Office in Aladağ; and many people in Turkey for their arrangements and help with our field research and useful discussions. We also greatly thank Dr. Hakan Alphan and Yüksel İzcankurtaran for their help with our preliminary research; Drs. Takanori Nagano, Yoichi Fujihara, Ms. Noriko Sasaki, and many of the staff of RIHN, Dr. Fujio Kimura at Tsukuba University and Dr. Kenji Tanaka at Kyoto University, and all ICCAP members for their great help and support with our research, and Tottori University students, Mr. Keisuke Kato and Ms. Yuki Kishibe for their data analysis, especially of the land-use pattern using Landsat images and the potential vegetation patterns in the present and the future. We are deeply grateful to Prof. Dr. Mehmet Aydın at Mustafa Kemal University and Prof. Dr. Selim Kapur at Çukurova University for their insightful comments and useful suggestions on our manuscript. We had many valuable and precious experiences during the study of the vegetation and climate change in Turkey. Therefore, we gratefully appreciate all of them once again. Finally we thank the editors and reviewers of this book, and Prof. Dr. Marc D. Abrams at the Pennsylvania State University, who made informative comments and corrections.

References

Abrams MD (1990) Adaptations and responses to drought in *Quercus* species of North America. *Tree Physiology* 7(1–4):227–238.

Allen CD, Macalady AK, Chenchouni H, Bachelet D, McDowell N, Vennetier M, Kitzberger T, Rigling A, Breshears DD, (Ted) Hogg EH, Gonzalez P, Fensham R, Zhang Z, Castro J, Demidova N, Lim JH, Allard G, Running SW, Semerci A, Cobb N (2010) A global overview of drought and heat-induce tree mortality reveals emerging climate change risks for forests. *Forest Ecology and Management* 259:660–684.

Altan T (2000) *Natural Vegetation*. University of Çukurova, Agricultural Faculty Publication: 235, A-76.

Atik M, Tamai S, Altan T, Ando M, Sano J, Atmaca M, Aktoklu E, Kaplan K, Artar M, Guzelmansur A, Cincinoglu A, Buyukasik Y (2007) Possible scenario for the vegetation change in Seyhan river basin and role of land uses "anthropogenic pressures". *The Final Report of ICCAP* (Impact of Climate Changes on Agricultural Production System in Arid Areas), Research Institute for Humanity and Nature, Kyoto, Japan, pp 111–118.

Bachelet D, Neilson RP, Lenihan JM, Drapek RJ (2001) Climate change effects on vegetation distribution and carbon budget in the United States. *Ecosystems* 4:164–185.

Baytop T (1994) *A Dictionary of Vernacular Names of Wild Plants of Turkey*. Turkish History Institution Pub, Ankara.

Berberoğlu S, Dönmez C, Evrendilek F (2015) Coupling of remote sensing, field campaign, and mechanistic and empirical modelling to monitor spatiotemporal carbon dynamics of a Mediterranean watershed in a changing regional climate. *Environmental Monitoring and Assessment* 187:179.

Cannel MGR (1982) *World Forest Biomass and Primary Production Data*. Academic Press, London.

Dönmez C, Berberoğlu S, Curran PJ (2011) Modelling the current and future spatial distribution of NPP in a Mediterranean watershed. *International Journal of Applied Earth Observations* 13:336–345.

Evrendilek F, Wali MK (2004) Changing global climate: historical carbon and nitrogen budgets and projected responses of Ohio's Cropland ecosystems. *Ecosystems* 7:381–392.

Henne PD, Elkin C, Franke J, Colombaroli D, Calo C, La Mantia T, Pasta S, Conedera M, Dermody O, Tinner W (2015) Reviving extinct Mediterranean forest communities may improve ecosystem potential in a warmer future. *Frontiers in Ecology and Environment* 13:356–362.

IPCC (2013) Summary for policymakers. In: Climate change 2013: the physical science basis. In Stocker TF, Qin D, Plattner GK, Tignor M, Allen SK, Boschung J, Nauels A, Xia Y, Bex V, Midgley PM (eds) *Contribution of Working Group I to the Fifth Assessment Report of the Intergovernmental Panel on Climate Change*. Cambridge University Press, Cambridge.

Kılıç S, Evrendilek F, Berberoğlu A, Demirkesen C (2006) Environmental monitoring of land-use and land-cover changes in a Mediterranean region of Turkey. *Environmental Monitoring and Assessment* 114:157–168.

Kimura F, Kitoh A, Sumi A, Asanuma J, Yatagai A (2007) Downscaling of the global warming projections to Turkey. *The Final Report of ICCAP* (Impact of Climate Changes on Agricultural Production System in Arid Areas), Research Institute for Humanity and Nature, Kyoto, Japan, pp 21–32.

Kira T (1976) *Terrestrial Ecosystems – A General Description*. Kyoritsu, Tokyo.

Kramer K, Leinonen I, Loustau D (2000) The importance of phenology for the evaluation of impact of climate change on growth of boreal, temperate and Mediterranean forests ecosystems: an overview. *International Journal of Biometeorology* 44:67–75.

Matthews HD, Weaver AJ, Meissner KJ, Gillett NP, Eby M (2004) Natural and anthropogenic climate change: incorporating historical land cover change, vegetation dynamics and the global carbon cycle. *Climate Dynamics* 22:461–479.

Osborne CP, Mitchell PL, Sheehy JE, Woodward FI (2000) Modelling the recent historical impacts of atmospheric CO_2 and climate change on Mediterranean vegetation. *Global Change Biology* 6:445–458.

Polunin O, Huxley A (1990) *Flowers of the Mediterranean*, 3rd ed. London, Chatto and Windus.

Roda F, Retana J, Gracia CA, Bellot J (1999) *Ecology of Mediterranean Evergreen Oak Forests*. Springer, Berlin.

Sabate S, Gracia CA, Sanchez A (2002) Likely effects of climate change on growth of Quercus ilex, Pinus halepensis, Pinus pinaster, Pinus sylvestris and Fagus sylvatica forests in the Mediterranean region. *Forest Ecology and Management* 162:23–37.

Thornthwaite CW (1931) The climates of North America – according to a new classification. *Geography Review* 21:633–655.

Thornthwaite CW (1948) An approach toward a rational classification of climate. *Geography Review* 38:55–94.

Vera FWM (2000) *Grazing Ecology and Forest History*, p 506, CABI Publishing, Wallingford, New York.

Yamamoto K (2000) Estimation of the canopy gap size using two photographs at different heights. *Ecological Research* 15:203–208.

Yılmaz KT (1993) *A Research on the Anthropogenic Impacts of Some Upland Settlements on Natural Vegetation in Dörtyol District of the Amanos Mountains*. Ph.D. thesis, University of Çukurova, Institute of Natural and Applied Sciences, Department of Landscape Architecture, Adana.

Yılmaz KT (1998) Ecological diversity of the Eastern Mediterranean region of Turkey and its conservation. *Biodiversity Conservation* 7:87–96.

Yılmaz KT (2001) *Mediterranean Vegetation*. University of Çukurova, Agricultural Faculty Publication No: 141, B-13.

Chapter 11
Climate Change and Animal Farming

Nazan Koluman (Darcan), Hasan Rüştü Kutlu and İnanç Güney

Abstract In recent years, extreme climate change (CC) and atmospheric events have become a global issue. It is now well known that livestock production contributes to global warming to a certain extent. The impacts of climate change on animal production can be analysed as the direct or indirect effects of the climatic factors. The direct impacts of climate change on animals are caused by climate factors with direct impacts on physiology, such as atmosphere temperature, relative humidity and wind speed. The animals' reactions to changing climate conditions differ, depending on being a ruminant or non-ruminant and their climate comfort zones. Animal farming impacts on climate change, as well. Global warming and climate change are mainly caused by three gases: carbon dioxide (CO_2), methane (CH_4) and dinitrous oxide (N_2O). Moreover, the increased humidity level in the atmosphere is another factor contributing to global warming. In addition to these gases, chlorofluorocarbon, which has commonly been used in industry in the past, has also made a considerable contribution to global warming. The gases which are naturally produced have no harmful effect; on the contrary, their presence in the atmosphere within normal limits contributes to preventing some heat loss from the Earth and establishing the atmospheric conditions which ensure the sustainability of life in the earth. But the high-level release of these gases causes an increase in the rational shares of this layer and thus keeps the long- and short-wave infrared rays from the sun at a higher level. Climate change is likely to have indirect impacts on the quality and the amount of animal feeds, feeding strategies, seasonal usability of grasslands, genetic studies (hybridisation, etc.), performance and the number of animals. The other element of pressure on animal production is political, social and economic sanctions which are aimed at decreasing greenhouse gas emissions. Climate experts have been made to adjust livestock production systems to forecast future climate changes based on climate modelling. From this point of view, it is

N. Koluman (Darcan), Professor, Çukurova University, Faculty of Agriculture, Department of Animal Science, Adana, 01330, Turkey; e-mail: ndarcan@gmail.com.

H. R. Kutlu, Professor, Çukurova University, Faculty of Agriculture, Department of Animal Science, Adana, Turkey; e-mail: hrkutlu@gmail.com.

İ. Güney, researcher, Çukurova University, Vocational School, Adana, Turkey; e-mail: iguney@cu.edu.tr.

© Springer Nature Switzerland AG 2019   223
T. Watanabe et al. (eds.), *Climate Change Impacts on Basin Agro-ecosystems*,
The Anthropocene: Politik—Economics—Society—Science 18,
https://doi.org/10.1007/978-3-030-01036-2_11

very important to make correct estimations with regard to questions such as which feed, or which animal species or breeds will be the most appropriate for different regions.

Findings obtained from the brief analyses which have been done in this chapter should be urgently implemented in national livestock farming in order to overcome environmental degradation and the possible effects of climate change. Some recommendations are provided in order to alleviate the negative influences of climate change on livestock production and, vice versa, the negative influences of animal production on climate change.

Keywords Animal farming · Climate change · Direct and indirect effects Implementation

11.1   Introduction

Weather patterns on a global basis have changed noticeably over the past few decades (IPCC 2007). These climatic changes are not restricted to tropical, arid or Mediterranean zones. Weather patterns in central and northern European (EEA-JRC-WHO 2008) and US (Adams et al. 1990) regions have also changed noticeably over the past few decades, and are characterised by more extreme events and seasonal changes, such as warmer summers and wetter and longer rainy seasons (Silanikove/Koluman 2015). It is widely predicted that warming will continue for centuries and will be associated in central-northern North America and Europe with more frequent heatwaves during the summer and heavy rainfall during the winter. This trajectory is a foreseeable future, even under the most modest warming scenarios, and thus most likely will have a significant impact on livestock farming (Silanikove/Koluman 2015). The effect of climate change (CC) on dairy production is both direct, through effects on the animals themselves, and indirect, through effects on the production of crops and increased exposure to pests and pathogens (Gauly et al. 2013). These negative impacts occur in the face of increasing demands for food, which are related to the increase of the population on Earth (Godfray/ Garnett 2014). The demand for animal products relates to rapid increase in income in some countries (Haq/Ishaq 2011), particularly in China (Qian et al. 2011), and the perception of dairy products as high quality and gourmet food (Silanikove et al. 2010). On the other hand, there is an increased awareness of the contribution of livestock to the greenhouse effect (Steinfeld et al. 2006; EPA 2011), and hence to global warming. The interaction between animal production systems and climate change has recently become a popular subject on the agenda (Silanikove/Koluman 2015).

The impacts of climate change on animal production could be analysed as the direct or indirect effects of the climatic factors. Within this chapter, an animal's interaction with the environment, stemming from its physiological stage, as well as issues such as the use of natural resources by animals and waste management of crop production, become prominent. The other element of pressure on animal production is political, social and economic sanctions which are aimed at decreasing greenhouse gas emission. The total greenhouse gas emission is closely related to the number of animals. In this regard, the use of highly productive animals should become an important strategy in the future in terms of animal production. To that end, certain applications particularly aimed at improving the environment and genotype interactions will become prominent. An improvement in climate conditions is closely related to an animal's biological capacity and the economic level of production. Accordingly, these two factors have an impact on the level of productivity to be gained per animal. Hence, efforts should be made to improve the genetic capacity with regard to the type of animal to be studied and to increase the production level per animal. The biological environmental conditions should be taken into consideration in terms of pollution by preferring conventional methods in the use of natural resources in order to increase production in the unit area. In this respect, issues of protecting the natural life and organic production become prominent. Furthermore, the negative effects on biology deriving from the uncontrolled use of substances cause the emergence of new diseases. The polluted environmental conditions are caused by the greenhouse gas which is released in animal production, as well as ineffective waste management. Thus, certain negative conditions emerge in production and human health.

Other than water, food is naturally the only indispensable input of life for humanity. Accordingly, agricultural activities involving food production and the nature-agriculture relationship have always been on the agenda. As mentioned above, the intensive use of conventional inputs in the nature-agriculture cycle and the pressure on natural resources have caused agriculture to be put under scrutiny and questioned on the process of climate change. Animal and crop production, as well as the methods which are applied in the course of these productions, should be firstly considered with regard to amounts of greenhouse gas emission. Problems such as the inability to come up with an alternative to stubble burning and similar activities, the use of an appropriate diet and high-quality feed in animal breeding, and the pollution which emerges during the production and animal transfer processes should be taken into consideration when analysing the process of climate change. At this point, conventional or traditional production systems will become another issue to be discussed.

A better understanding of the advantages and disadvantages of native resources of goat breeds and crop production within a regional projection is highly significant with regard to the success of adaptation programs to be implemented in the long run. Applying genetic and biotechnological approaches is another means of improving adaptation to CC. However, in the present review we would like to emphasise the local gene resources, which offer resistance and endurance to local extreme environmental conditions. Methane emission, in which the live weight and productivity level of local goats is considered, should also be taken into account in such an approach. A pilot study which was performed on this subject is described below. The calculations that we have made on the methane emission from livestock in Turkey took into account methane released from the gastrointestinal tract, or

Table 11.1 The annual methane emission of the cattle, sheep and goats in Turkey of enteric and manure origin. *Source* Based on Görgülü et al. (2009: 26)

Species	Enteric (ton)	Manure (ton)	Total (ton)	Enteric (%)	Species (%)
Cattle	675,394	108,457	783,850	86.16	76.53
Sheep	203,800	6,114	209,914	97.09	20.49
Goat	29,600	888	30,488	97.09	2.98
Total	**908,794**	**115,459**	**1,024,252**		

through the manure of cattle, sheep and goats per animal of different ages and physiology according to the IPCC (2007), as indicated in the table above (Table 11.1; based on Görgülü et al. 2009).

The majority of greenhouse gas produced by livestock originates mainly from the digestion process in the rumen, while the minority part comes from manure and feed production (Table 11.1). It is estimated that the methane emission by ruminants in Turkey is approximately 1 million tons and more than 85% of that is of enteric origin (Table 11.1). Furthermore, it is considered that the cattle population causes 76% of the total emission. This estimation represents an increase in methane emission in comparison to that made earlier by the State Institute of Statistics' (DIS) Branch Office of Environmental Statistics in the year 2000. According to this source, the methane release caused by enteric fermentation totalled 692,000 tons, and the emission caused by fertilisers totalled 37.6 tons (TÜİK 2009).

The relationship of the methane emission to the animals' dry matter consumption is a fact. Hence, it could be stated that the insufficient intake of feeds might limit methane production, whereas the reason for the high methane production is caused by the imbalance in the feed ratio of Turkey. This could be relatively compensated by low feed intake, though such an approach is not favoured.

Furthermore, higher amounts of methane gas are released from the unit animal, as the animal population is increased in favour of a highly productive animal in an intensive system. The native breeds with less live weight and less productivity also have less methane emission, which will become an advantage in the future with regard to the use of these animals for production purposes.

Greenhouse gasses (CH_4, CO_2 and nitrous oxide) have the capacity to raise the earth's temperature and are indirectly responsible for climate change and global warming. After CO_2, CH_4 is the second most important gas in global warming. Animal husbandry is responsible for approximately 2.5–4% of the total greenhouse effect (Crutzen et al. 1986) where different species take a role in this problem and their effects can be determined by different methods. Goat production is also linked to the methane emission which contributes to global warming. Approximately 700 g/kg of methane production originate from anthropogenic sources. Agriculture accounts for about two-thirds of this, with enteric fermentation being responsible for a third of methane from agriculture (Moss et al. 2000). According to Johnson/Johnson (1995), 80 million Mt of CH_4 is produced by livestock enteric fermentation annually, 730 kg of which is attributed to cattle (Johnson/Johnson 1995).

The emission from enteric fermentation of 1,057 million sheep (each responsible for 9 kg of CH_4 emission) and 677 million goats accounts for 200 g/kg (Mbanzamihigo et al. 2001). CH_4 is not the only important greenhouse effect caused by goats and sheep, but also indicates loss of energy intake (Johnson/Johnson 1995).

To calculate the methane emission rate from sheep and goats, different regression equations can be used, such as the Blaxter/Clapperton (1965), and different parameters can be considered depending on the species and breeding conditions. The Blaxter/Clapperton (1965) equation stands for

$$E_{CH4} = 1.3 + 0.112D + L(2.0 - 0.05D)$$

where E_{CH4} is the CH_4 energy of gross intake (%), D is digestibility of gross energy intake (%), L is the level of feeding.

Some parameters, such as the gross energy intake, emission per head day, dry matter, intake of crude nutrient and protein and digestibility.

Some researchers have reported that the condensed tannins, *acacia mearnsii*, *calliandra calothyrsus* and *lespedeza cuneata*, reduced CH_4 emissions when added to goat feed material at different rates (Hess et al. 2013; Carulla et al. 2005; Puchala et al. 2005).

11.2   Impacts of Global Warming on Animal Production

Climate change is likely to create indirect impacts on the quality and amount of animal feeds, feeding strategies, seasonal usability of grasslands, genetic studies (hybridisation, etc.) and the number of animals. The quality and quantity of animal feed is important with regard to the rations prepared. The grain (wheat, barley, etc.) which is used as animal feed and oily seed residues (cottonseed residues, sunflower seed residues, etc.) ensure that the rations are prepared at the lowest cost without losing any nutritional value (Koluman Darcan et al. 2009).

Climate change will have impacts on animal production in some ways; these namely are the physical and biologic environments and the climate and chemical environments as given in Fig. 11.1.

The impacts of global warming on farm animals emerge as the physical environment, biological environment, chemical environment or direct climate impacts. The physical environmental conditions appear with possible changes to feeding conditions. Shelter becomes more costly due to extreme climate conditions (too warm or too cold), and certain qualities aimed at performance, such as milk and meat productivity, are likely to lag behind. It was put forth by the studies conducted that extreme climate conditions have an impact on cattle in terms of the milk quality and quantity and cause a decrease in the lactation period of milkers (Chase/Sniffen 1988; Bucklin et al. 1991; Alnaimy et al. 1992). The studies conducted on reproductive performance versus climate change report that fertility decreases (Alnamier et al. 2002; De Rensis et al. 2002) and in turn the oestrus cannot be clearly

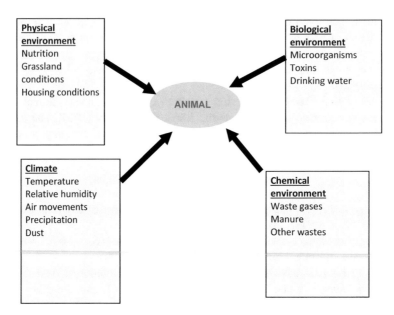

Fig. 11.1 Classification of expected effects of climate change on animal production. *Source* Koluman Darcan et al. (2009: 322)

determined in goats. Subsequently, the first insemination interval (days) increases together with the decreasing pregnancy rate (Alnamier et al. 2002; De Rensis et al. 2002), and the blood flow into the uterus decreases in parallel due to the high atmospheric temperature which in turn affects the temperature of animals. Increase in body temperature leads to a rise in the uterus temperature, which decreases the fertilisation rate limits and embryonic development, aggravating early embryonic fatalities (De Rensis et al. 2002) in farm animals. It has also been reported that the growth and development of young kids and lambs are also negatively affected by these conditions (Darcan 2005; Çoban et al. 2008; Koluman Darcan et al. 2009). Furthermore, it was reported that a high environmental temperature reduces voluntary feed intake, feed efficiency and the extended fattening period of farm animals (Linn 1997; Harner et al. 1999; Silanikove 2000; Davis et al. 2001a, b; Göncü/ Özkütük 2003).

The grassland or crop production and its quality will slump due to the decrease in water resources and the land available for crop production. There will be a tendency to use available land to produce food primarily for human consumption, as the lands appropriate for crop production will shrink due to the increasing seawater level, drought or salinity. The competitive power of animal feed production will decrease due to economic reasons and the priorities of general agricultural and industrial activities. Accordingly, a decrease in this competitive power will primarily emerge as difficulties in feeding animals (IPCC 2007; Koluman Darcan et al. 2009; Silanikove et al. 2010; Steinfeld et al. 2006). As the

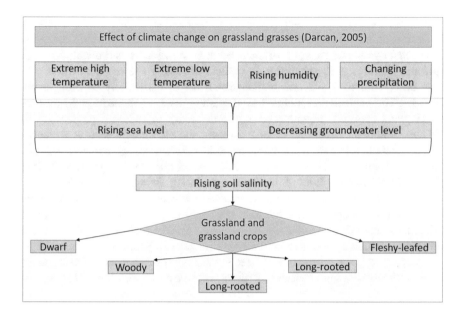

Fig. 11.2 Effects of climate change on grassland grasses. *Source* Darcan (2005: 28)

precipitation regime changes, some cultivated lands will suffer drought, whereas some others will experience salinity. Thus, there will be competition with regard to the feed crop and alternative plants on lands consisting of agricultural production, and the cost will be a prominent issue at this point. Furthermore, short, woody and leafy plant species with long roots which are resistant to salinity and drought and capable of storing water will become prominent in vegetation. These plant species are among the special species which could be efficiently utilised by goats (IPCC 2007; Koluman Darcan et al. 2009; Silanikove et al. 2010; Steinfeld et al. 2006) (Fig. 11.2).

The studies on the genetic improvement of farm animals in terms of production qualities and productivity have always been on the agenda. But local animals' capacity for maintaining their productivity in all circumstances and their advantageous situation with regard to negative environmental impacts should be taken into consideration. Therefore, the work on genotype should be based on selection. The qualities of local gene resources such as length of life, animal health and productivity have an impact on greenhouse gas emissions, thus they will be able to maintain their advantages over hybrid stocks even under climate changes. Furthermore, they are less affected by short heatwaves due to their anatomic and physiological structures and they easily excrete the extra heat with the help of their mechanisms and thus maintain their body temperature (IPCC 2007; Koluman Darcan et al. 2009; Silanikove et al. 2010; Steinfeld et al. 2006).

The excellent quality of their resistance to heat and cold has been especially put forth by certain studies. Besides the productivity level, the length of life could also be affected by climate changes. An increase in such incidents as drought, flood and epidemic diseases is expected due to climate change. Therefore it's important to start using animals which are particularly resistant particularly to drought and diseases. The characteristics aimed at adaptation, such as the skin and hair type, sweat gland capacity, reproduction rate and ability to maintain productivity in difficult conditions, resistance to local diseases and parasitic infestation, metabolic heat production, tolerance to draught, anatomic and morphologic structure are some important impacts which are likely to derive from climate changes (IPCC 2007; Koluman Darcan et al. 2009; Silanikove et al. 2010; Steinfeld et al. 2006).

Certain animal diseases in terms of treatment expenses, productivity losses and immunity (blue tongue, gastroenteritis, etc.) will pose a serious threat in stock-breeding. Many factors, such as easier spread of disease vectors and an increase in atmosphere temperature due to shorter cold seasons with temperatures above 15 °C, will play a role in the emergence of these diseases with subsequent problems. Goats are the animals with high resistance to diseases and external parasites. They are also likely to be more resistant than other animals to disease vectors, which will change under the drought and extreme climate conditions in the future. The social and economic impacts of global warming are likely to be serious, such as the lack of food under limited natural resources. Accordingly, conventional resources will become widespread in production and conventional methods will be developed and used in order to gain higher productivity from the unit area. Thus timbalanced and unhealthy nutrition is likely to occur. It's known that animal protein is of considerable importance, especially with regard to the body and mental health of the human organism. Therefore, the provision of continuity in animal production should be considered as part of the healthy society perspective (Koluman Darcan et al. 2009).

The interaction between food, animal feed and fuel is another factor expected to have negative impacts on the animal feed sector in the near future, which will most probably cease to be a profitable production field for stockbreeding, as feed preparation costs increase. The possibility of using plant residues in bio-diesel production in order to meet energy needs will cause these by-products to be used less in animal production in the future. The low atmospheric temperature is one of the most important factors creating negative impacts on the dry matter content of grassland feed crops, particularly in spring. The stored water of the soil will fall with decreasing precipitation, which will lead to a decrease in the dry matter content in plants. Additionally, the biomass of by-products which are used in animal production will be greatly influenced by climate change, and the nutritional value of the rations to be prepared will decrease. Therefore, silage and fodder production will lag behind (Koluman Darcan et al. 2009).

Besides the productivity level, livestock's length of life could also be affected by climate changes. An increase in such incidents as drought, flood and epidemic diseases is expected due to climate change. This makes it even more important to start using animals which are particularly resistant to drought and diseases. The local gene resources are more advantageous than exotic breeds, particularly in terms of resistance to diseases, ability to utilise poor vegetation and endurance to drought. Native animals are able to use the plants rich in lignin effectively by consuming less feed and water.

Genetic diversity in farm animals is important with regard to food safety and rural development. The applications that are carried out towards these ends will provide stockbreeders with the advantage of animal selection and new genotype development with regard to climate change, varying market opportunities and turning towards new fields which are needed by people. FAO (2007) reported that a great number of animal stocks in the world were under threat of perishing. The majority of these animals are raised by small enterprises which are run with low inputs, where the plant and animal production based on grasslands is performed simultaneously.

The direct impacts of climate change on animals are caused by climate factors with direct impacts on physiology, such as atmospheric temperature, relative humidity and wind speed. Animals' reactions to changing climate conditions differ, depending on being a ruminant or non-ruminant and their climate comfort zones. The values related to temperature limits in view of species for optimum production are shown in Fig. 11.3 (Koluman Darcan et al. 2009).

As can be seen in Fig. 11.3, climate impacts differ, depending upon such factors as an animal's age, sex, productivity level and products. The farm animals adapting to and living in climate conditions which are available in a region face certain problems related to adaptation, where younger animals even perish if a change occurs in these conditions. For example, rainy and mild climates decrease the mortality rate in young animals and thus become an advantage, whereas extremely high heat creates negative impacts particularly on dairy cows in terms of feed consumption, productivity and milk quality. Therefore the sheltering conditions gain importance in animal production (Koluman Darcan et al. 2009). The climate biological capacities of goat species are more flexible than those of other species.

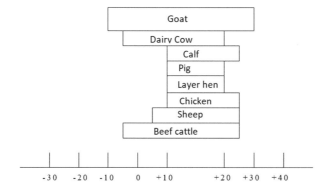

Fig. 11.3 Temperature limits in view of species for optimum production. *Source* Johnson (1987)

Goats can use some part of their rumens as water storage, and thus they can adapt to dry and appropriately limited conditions. Furthermore, based on these data, it could be indicated that goat species are likely to be advantageous in extreme climate conditions which might occur under global climate changes.

11.3   Impacts of Animal Production on Global Warming

Global warming and climate change are mainly caused by three gases; carbon dioxide (CO_2), methane (CH_4) and nitrous oxide (N_2O). Moreover, the increased humidity level in the atmosphere is another factor contributing to global warming. In addition to these gases, chlorofluorocarbon, which has been commonly used in the industry in the past, has also made a considerable contribution to global warming. The gases which are naturally produced have no harmful effect; on the contrary, their presence in the atmosphere within normal limits contributes to preventing some heat loss from the Earth and establishing the atmospheric conditions which ensure the sustainability of life on Earth. But the high-level release of these gases causes an increase in the rational shares of this layer and thus keeps the long- and short-wave infrared rays from the sun at a higher level.

Steinfeld et al. (2006) reported that, considering the effects of global warming and assuming the CO_2 effect, 1.1 ton of methane corresponds to the effect of 23 tons of CO_2, whereas 1 ton of N_2O equals the effect of 296 tons of CO_2. The intensive emergence of these gases, particularly methane, as a result of agricultural activities, caused people's attention to focus on this issue in the course of the Kyoto Protocol.

11.3.1   Carbon Dioxide Emission

The CO_2 in the global atmosphere in the last 250 years is the outcome of the rapid decrease in forest lands and intensive use of fossil fuels. The CO_2 emission mainly results from the tendency towards rapid industrialisation, which was induced in the 1970s (IPCC 2007). The IPCC (2007) reported that the CO_2 concentration, which was 280 ppm in 1750, had reached 379 ppm in 2005 with an increase by approximately 30%. However, the current CO_2 emission that stems from animal production equals only 9% of the total global CO_2 emission (Clarke 2007). The sources of the animal-production-based CO_2 are the energy used in feed production, manure production, product processing and transportation. For example, 1 kg of compound feed is likely to cause 0.57–2.21 kg CO_2 production, whereas rough feed produces slightly lower amounts of CO_2 (O'Mara 2004). Similarly, a huge amount of fossil fuel is utilised in order to produce fertilizers that are used for obtaining animal feed, adding up to 41 million tons of CO_2 production annually (Steinfeld et al. 2006). Moreover, the nitrogen fertilizers incorporated into the soil in order to produce animal feed increases the di-nitrous oxide emission. Additionally, the overgrazed and degraded grasslands and deforestation also have an impact on the increasing CO_2 emission (Görgülü et al. 2009).

11.3.2   Nitrous Oxide (N_2O) Emission

The crop products which are used as feed are not sufficient alone for a balanced ruminant diet. Approximately 50% of the corn produced in the world and 80% of the soybeans are used in animal production. Corn production involves considerable use of nitrogenous manures (HSUS 2008). Therefore the amount of nitrogen incorporated into the soil via manures and fertilizers in order to produce feeds reaches serious levels. This has deteriorated the global nitrogen cycle to a great extent. These fertilizers and manure used for herbal production also increase the level of substances with nitrogen in water resources, which, it should be emphasised, is millions of tons. Nitrogen pollution dramatically decreases the water, air and thus the life quality of the societies living near animal production areas. The nitrogen originating from the excessive use of nitrogen fertilizers in lagoons for large-scale aquaculture is likely to leak and mix into floods and rivers via the dead fish in the manure lagoons. There are two examples for this event, namely the 1995 episode concerning the discharges from the manure lagoon at a pig-raising enterprise in the US to the New River and the discharge of a manure lagoon at a dairy cow enterprise into the Black River in 2005 (NYSDEC 2006), which had drastic consequences.

The nitrogen is also likely to have serious impacts on global warming. The global warming potential (GWP) of this gas is important, as it shrinks the ozone layer. It is known that the di-nitrous oxide concentration in the atmosphere has increased in the last 150 years in an unprecedented wa,y as stated by Steinfield et al. (2006).

Once the manures of ruminants such as cattle, sheep, goat, buffalo and camel are stored, serious amounts of di-nitrous oxide gas are released. Approximately 70% of global N_2O emission derives from herbal and animal production, 65% of which results from animal production (Steinfeld et al. 2006). Similarly, once the pig and poultry manures are stored in liquid forms under unventilated conditions, they become serious methane and N_2O emission resources. Bos/Wit (1996) reported that the manure pit management in two-thirds of poultry and pig manures in the world was based on liquid manures.

In order to decrease di-nitrous oxide emission (O'Mara 2004; Gworgwor et al. 2006; Clarke 2007), the following suggestions are made:

- The amount of nitrogen which is applied via manures should be decreased;
- Animal productivity should be increased, whereas the number of animals should be decreased (but this situation poses an economic problem for farmers);

- The amount of nitrogen applied to soil for crop and roughage production should be decreased;
- The amount of feed sources rich in nitrogen used in animal feeding should be decreased.

11.3.3 Methane (CH_4) Emission

It is now well known that the global warming potential of methane is 23 times greater than that of CO_2 and its concentration in the atmosphere has increased by 150%, compared to the level measured in 1750. Animal production is responsible for 16% of the methane production which is caused by human activities (Steinfeld et al. 2006). Watson et al. (1992) calculated that 39% of the annual methane emission originating from agriculture is caused by enteric fermentation, 29% from rice production areas, 19% from burning herbal wastes and 12% from animal wastes. Bolle et al. (1986) calculated the contribution made by non-biological factors as 32%, rice production areas as 18%, ruminants as 18%, marshes as 13% and other factors as 6%.

The amount of methane produced by individual animals is quite low. For example, a mature cow produces approximately 80–110 kg methane annually (O'Mara 2004; USEPA 2008). The main problem related to ruminants is that their presence in the world has reached nearly 1 billion (FAO 2008) and that this number makes serious contributions to methane emissions. Enteric methane production is likely to reach approximately 15–18% of digestible energy taken from cattle which are fed with bad-quality rough feeds. This situation results from the lack of food substances and inadequate dietary balance for the cultivation of optimal rumen microorganisms, and thus insufficient and inefficient microbial development. The amount of methane released from beef and milk cattle to the atmosphere in the US constitutes 71% of animal methane and 19% of the total methane emission (Steinfeld et al. 2006; EPA 2008).

The contribution made by animal production to the total methane emission is low by enteric fermentation, but high by manure fermentation. It is now well known that the methane-production capacity of stored manures from poultry and pig production constitutes serious methane resources. The methane emission caused by pig manure constitutes approximately half of the methane emission derived from livestock manure in the world (Steinfeld et al. 2006).

If global warming and climate change continue their courses at the estimated speed, they might lead to drought, famine and social conflicts. Desertification-linked social conflicts and settlement shifts in the world are responsible for the decrease in food production and inability to meet water needs. In such cases, the sectors of the community who will be most affected by this situation are farmers. The number of people dealing with animal production in the world is approximately

2 billion (IPCC 2007). Animal production is their only source of income and the number of grazing animals they own is nearly 200 million.

11.3.4   Ways of Decreasing the Greenhouse Gas Production Caused by Animals

As generally discussed before, the direct animal contribution in the greenhouse gas emission originating from animals is caused by enteric fermentation. The others are caused by indirect ways, such as feed production, product processing and transportation, manure storage and its use in plant production. The most important greenhouse gas which is released during these processes is methane. The methane emission directly originating from animals is caused by inefficient fermentation, namely the food substance of feeds in rumens, and inability to meet the food substance needs of microbial flora in a sufficient and balanced way.

The activation of microbial fermentation activity in both the digestive system and manure lagoons plays an important role in terms of decreasing the greenhouse gas emission to be produced.

As indicated before, the basic principle with regard to decreases in greenhouse gas production directly originating from animals will be realised by increasing the efficiency of feed substances in the digestive system. Methane is derived from fermentation under unventilated conditions. After food substances are decomposed in the rumen, they are fermented to volatile oil acids, hydrogen, CO_2 and NH_3, incorporated into CO_2 hydrogen and finally reduced to methane ($CO_2 + 4H_2 \rightarrow CH_4 + 2H_2O$) under unventilated conditions. Furthermore, the hydrogen which is produced under unventilated conditions could be used to produce volatile oils (*Hydrogen production*: Pyruvate $\rightarrow$ Acetate (C2) + CO_2 + 2H; *Hydrogen usage*: Pyruvate + 4H $\rightarrow$ propionate (C3) + H_2O or $2C_2$ + 4H $\rightarrow$ butyrate (C4) + $2H_2O$). At this point, the aim is to direct the fermentation from acetate towards propionate. In this case, methane production decreases and energy utilisation is improved.

The previous studies primarily deal with increasing the digestibility of feed substances in rations. McGinn et al. (2004a) reported in a study that a rise in the rate of roughage in the diet increased methane production. Moreover, different starch resources were determined to be effective in methane production where corn was likely to increase methane production in diets based mainly on roughage, and barley could do the same in diets based on concentrated feed. Similarly, in the case of fat/oil used in the diet as a source of energy, the microbial flora and energy efficiency change in the rumen and thus the methane production decreases (McGinn et al. 2004a; Beauchemin/McGinn 2006). This situation emerges as a result of a decrease in ruminal fermentation due to dietary fat/oil, as these inhibit microbial growth. Furthermore, different species of a particular cereal could produce different levels of methane. Ovenell-Roy et al. (1998) reported in a study on lambs that four different varieties of barley produced methane in different ways. Lovett et al. (2003) reported similar indications with regard to roughage (ryegrass varieties). In line with these assessments, given that such manipulations are used in animal diets (roughage/concentrated ratio, cereal resource, dietary fat/oil) as energy sources, it is considered necessary to suggest ways of limiting methane production.

Reynolds et al. (1991) studied the impacts of concentrate feed rate on the methane production of heifers by comparing a diet of 75% clover fodder with one of 75% concentrate feed, and finally determined that the energy loss caused by methane greatly decreased in high concentrate feed. Similarly, Oliveira et al. (2007) reported that the methane emission was low in beef cattle receiving a diet based on concentrate feed. Christophersen et al. (2008) observed that methane production decreases with an increase in cereal rate in the diet of sheep. They considered that this decrease derived from the rumen pH, which is altered by an increase in the concentrate feed and a fall in methanogens which are sensitive to low pH. Moreover, a decline in the number of protozoa and the duration of concentrate feed present in the digestive system (the duration of fermentation gets shorter) and the increase in propionate production are among the most important incidents in the course of removing the hydrogen formed in the rumen. Consequently, these biochemical reactions are followed by a decrease in the acetate/propionate ratio.

The oil/fat ratio, which is among the main energy resources used in food, is also capable of decreasing methane production to a great extent. Martin et al. (2008) used raw flaxseed, extruded flaxseed or flaxseed oil in diets prepared for dairy cattle and concluded that the amount of methane greatly declined in the cattle which consumed the rations containing 5.5% oil. McGinn et al. (2004b) used monensin, sunflower oil, enzymes, ferment and fumaric acid to decrease the methane emission in beef cattle and determined that the use of oil in the diet decreases methane production to a great extent. Additionally, Sauer et al. (1998) conducted a study on Black Spotted cattle and reported that when soybean oil is used at a concentration of 3.5% in rations, which are arranged in the form of TMR, the methane and CO_2 production remained unchanged. The reasons for the decreases which were observed in the use of oil in rations and methane production are indicated as follows:

1. Some hydrogen in the rumen is used to feed unsaturated fatty acids and thus the amount of hydrogen is decreased for methane production,
2. Both the cellulose fermentation and the acetate/propionate rate decrease due to the negative impacts of fatty acids on microorganisms,
3. The number of protozoa where the methanogen bacteria live symbiotically decreased.

Furthermore, Newbold/Rode (2006) reported that the limitation in methane production is among the issues which have been researched most recently, and that probiotics, organic acids and plant extracts could be used for this purpose. As indicated before, the aim is to manipulate the rumen fermentation in a desired way by dietary means. But it could be said that *in vitro* and *in vivo* studies conducted to control methane production in the rumen have not totally promoted each other.

Wallace (2004) reported that a considerable improvement has been obtained from certain studies which were conducted in order to decrease the methane emission in ruminants with the use of natural herbal feed additives. Kamra et al. (2006) stated that plant secondary components were likely to decrease methane production and plants containing secondary components of saponin, tannin and volatile oils played an important role. Researchers stated that plants consisting of both saponin and tannin decrease the number of protozoa in rumens and thus the methane production. Patra et al. (2006) determined that fennel, clove, garlic, onion and ginger volatile oil inhibited *in vitro* methane production. Beauchemin/McGinn (2006) reported that the total methane emission of the essential oil used in the diet was bound to the methane production per dry matter intake in kg, and the energy which is lost due to methane as the percentage of gross energy intake remained unchanged.

The amount and efficiency of secondary plant components differ, depending on the region, vegetation/harvesting time and location. The efficiency of plant extracts depends on their antimicrobial content, anti-protozoal and antioxidant substances. It is of the utmost importance to determine the enteric methane production decreasing capacities of the plant volatile oils available in each country. The botanic composition of Turkey provides great advantages with regard to producing plant essential oils and diversity. However, while the antimicrobial, antiprotozoal and antioxidant qualities of plant volatile oils from Turkey have been commonly researched, their potential impacts on methane emission remains unknown.

The productivity in ruminants directly depends on improving the efficiency of feed utilisation in the rumen. Choice feeding studies revealed significant improvements in the efficiency of feed utilisation in the rumen and provide a way to manipulate and optimise rumen fermentation (Görgülü et al. 1996, 2003, 2008; Yurtseven/Görgülü 2004). The animals receiving feeds as choice could have a chance to optimise rumen conditions by synchronising rumen microorganisms, as well as to meet nutrient requirements by selecting the appropriate amount of feed ingredients offered as a free choice. Yurtseven and Öztürk (2008) determined that methane and CO_2 emission could be decreased by choice feeding in sheep. In this respect, it's considered that the impacts of choice feeding, which are carried out under different physiological conditions and in diverse species of ruminant animals, on the production of greenhouse gases could be of value.

Wallace et al. (2006) reported that the addition of fumaric acid by 0.1% into the diet of sheep caused declines in the methane gas emission by 0.4–0.75% (O'Mara et al. 2008). Some researchers stated that, in terms of methane emission, there was no difference in Angus beef cattle which are fed with malate. The studies on this issue dealt with the economic quality of organic acid addition to rations and they concluded that this application would be costly (McGinn et al. 2004a; Foley et al. 2007).

11.4 Concluding Remarks

Sustainable animal production systems should be urgently implemented in national agricultural development programmes in order to overcome environmental degradation and the possible effects of climate change.

The following recommendations could be made in order to alleviate the negative influences of climate change on animal production or, vice versa, the negative influences of animal production on climate change.

- It will be important to preserve the local gene breed resources, which are naturally resistant to heat and cold stresses and are capable of utilising all sorts of feed resources. In this respect, all local breeds, mainly goat, sheep and cattle, should be studied for their resistance to possible climate changes.
- Priority should be given to selection of the goat breeds which are resistant to diseases and possess high tolerance of warm and drought conditions, particularly in the course of conducting regional projections. It will be important to implement such studies in combination with bio-technological studies in order to ensure sustained production of animal protein.
- Herd compositions should be improved by selecting and crossbreeding to increase productivity and sustain income. Priority should be given to resistance to diseases and minimisation of the release of greenhouse gases by ruminants.
- Regional markets of ruminant products should be established to meet the enhanced future milk and meat demands due to the increasing population.
- Farmer awareness should be raised towards the conservation of natural resources and possible impacts of climate change.
- The support to new scientific and technologic research on goat-farming versus climate change should be enhanced
- The composition of grassland food crops for the future should be considered, with preference given to growing crops which are likely to minimise methane and nitrous oxide emission and capture more CO_2. Goat-farming should be adjusted for the optimal utilisation of these resources.
- Social and economic problems, which are likely to emerge, particularly in rural areas, should be considered. The appropriate locations for ruminant farming in these areas should be determined and organic breeding should be predominantly supported.

References

Adams RM, Rosenzweig C, Peart RM, Ritchie JT, Mccarl BA, Glyer JD, Curry RB, Jones JW, Boote KJ, Allen LH (1990) Global climate change and United-States agriculture. *Nature* 345:219–224.

Alnaimy AM, Habeeb I, Fayaz I, Marai M, Kamal TH (1992) Heat Stress, Farm Animals and the Environment. In Philips C, Piggins D (eds) *International Conference on Farm Animals and the Environment, University College of North Wales*, CAB International England, Cambridge.

Alnamier M, De Rosa G, Grasso F, Napolitana F, Bordi A (2002) Effect of climate on the response of three oestrus synchronisation techniques in lactating dairy cows. *Animal Reproduction Science* 71:157–168.

Beauchemin KA, McGinn SM (2006) Methane emissions from beef cattle: effects of fumaric acid, essential oil and canola oil. *Journal of Animal Science* 84:1489–1496.

Blaxter KL, Clapperton JL (1965) Prediction of the amount of methane produced by ruminants. *British Journal of Nutrition* 19:511–522.

Bolle HJ, Seiler W, Bolin B (1986) Other green house gases and aerosols. Trace gases in the atmospheres. In Bolin B, Doos BOR, Jager J, Warrick RA (eds) *The Green House Effect, Climate change and Ecosystems (SCOPE 29)* Wiley and Sons, Chichester 157–203.

Bos J, Wit J (1996) Environmental Impact Assessment of Landless Monogastric Livestock Production systems. In: *Livestock and the Environment: Finding a Balance*. FAO/World Bank/ USAID, Rome.

Bucklin RA, Turner LW, Beede DK, Bray DR, Hemken RW (1991) Methods to relieve heat stress for dairy cows in hot, humid climates. *Applied Engineering Agriculture* 7(2):241–247.

Carulla JE, Kreuzer M, Machmuller A, Hess HD (2005) Supplementation of *Acacia mearnsii tannins* decreases methanogenesis and urinary nitrogen in forage-fed sheep. *Australian Journal of Agricultural Research* 56:961–970.

Chase LE, Sniffen CJ (1988) Feeding and managing dairy cows during hot weather. Feeding and Nutrition.

Clarke J (2007) Climate change pushes diseases north: expert. Reuters, March 9; at: www.reuters. com/article/healthNews/idUSL0920787420070309?sp=true (April 23, 2008).

Cristophersen CT, Wright ADG, Vercoe PE (2008) In vitro methane emission and acetate: propionate ratio are decreased when artificial stimulation of the rumen wall is combined with increasing grain diets in sheep. *Journal of Animal Science* 86:384–389.

Crutzen PJ, Aselmann I, Seiler W (1986) Methane production by domestic animals, wild ruminants, other herbivorous fauna, and humans. *Tellus* 38B:271–284.

Çoban Y, Darcan N, Aslan N, Karakök SG (2008) Global warming and its effects on animal husbandry. 4. National Zootechnics Student Congress, May 2008, Samsun, Proceedings.

Darcan N (2005) Global warming effects on animal husbandry and alleviation to heat stress. *Hasad Animal Journal* 21(243):27–29.

Davis S, Mader T, Holt S, Cerkoney W (2001a) Effects of feeding regimen on performance, behaviour and body temperature of feedlot steers. Beef cattle report.

Davis S, Mader T, Cerkoney W (2001b) Managing heat stress in feedlot cattle using sprinklers. Beef cattle report; at: www.Liru.asft.ttu.edu/pdf/mp76pg77-81.pdf.

De Rensis F, Marconi P, Capelli T, Gatti F, Facciolongo F, Franzini S, Scaramuzzi RJ (2002) Fertility in postpartum dairy cows in winter or summer following estrous synchronization and fixed time AI after the induction of an LH surge with GnRH or hCG. *Theriogenology* 58:1675–1687.

EEA-JRC-WHO (2008) Impacts of Europe's changing climate – 2008 indicator-based assessment. Joint EEA-JRC-WHO report, EEA Report No 4/2008, *JRC Reference Report* No JRC47756, EEA, Copenhagen.

EPA (Environmental Protection Agency) (2011) DRAFT: *Global Anthropogenic Non-CO$_2$ Greenhouse Gas Emissions: 1990–2030*. Publication 430-D-11-003. EPA, Washington, DC.

Foley PA, Callan J, Kenny DA, Johnson KA, O'Mara FP (2007) *Proceedings of the Agricultural Research Forum*, 112 p.

FAO (2007) Global Plan of Action for Animal Genetic Resources and the Interlaken Declaration. Rome; at: http://www.fao.org/ag/againfo/programmes/en/genetics/documents/Interlaken/GPA_en.pdf.

FAO (2008) Food and Agriculture Organization of the United Nations.2008. FAO Statistical Database, FAOSTAT.

Gauly M, Bollwein H, Breves G, Brugemann K, Danicke S, Das G, Demeler J, Hansen H, Isselstein J, Konig S, Loholter M, Martinsohn M, Meyer U, Potthoff M, Sanker C, Schroder B, Wrage N, Meibaum B, Von Samson-Himmelstjerna G, Stinshoff H, Wrenzycki C (2013) Future consequences and challenges for dairy cow production systems arising from climate change in Central Europe – a review. *Animal* 7:843–859.

Godfray HCJ, Garnett T (2014) Food security and sustainable intensification. *Philosophical Transactions of the Royal Society B* 369(1):23–38.

Göncü S, Özkütük K (2003) Shower Effect at Summer Time on Fattening Performances of Black and White Bullocks. *Journal of Applied Animal Research* 23:123–127.

Görgülü M, Kutlu HR, Demir E, Öztürkcan O, Forbes JM (1996) Nutritional Consequances of Free-Choice Among Feed Ingredients by Awassi Lambs. *Small Ruminant Research* 20:23–29.

Görgülü M, Güney O, Torun O, Özuyanik O, Kutlu HR (2003) An Alternative Feeding System for Dairy Goats: Effects of Free-Choice Feeding on Milk Yield and Milk Composition of Lactating Suckling Damascus Goats. *Journal of Animal Feed Science* 12:33–44.

Görgülü M, Boğa M, Şahin A, Serbester U, Kutlu HR, Şahinler S (2008) Diet Selection and Eating Behaviour of Lactating Goats Subjected to Time Restricted Feeding in Choice and Single Feeding System. *Small Ruminant Research* 78:41–47.

Görgülü M, Koluman Darcan N, Göncü Karakök S (2009) Animal husbandry and global warming. 5. Ulusal Hayvan Besleme Kongresi, 30 Eylül- 3 Ekim 2009, Çorlu.

Gworgwor ZA, Mbahi TF, Yakubu B (2006) Environmental Implications of Methane Production by Ruminants: A Review. *Journal of Sustainable Development in Agriculture and Environment* 2(1). ISSN 0794-8867.

Haq Z, Ishaq M (2011) Economic growth and agrifood import performance of emerging economies and Next-11. *African Journal of Business Management* 5:10338–10344.

Harner JP, Smith J, Brook M, Murphy JP (1999) Sprinkler systems for cooling dairy cows at a feed line. Kansas State University Agricultural Experiment Station and Cooperative Extension service.

Hess HD, Monsalve LM, Lascano CE, Carulla JE, Diaz TE, Kreuzer M (2013) Supplementation of a tropical grass diet with forage legumes and Sapindus saponaria fruits: effect on in vitro ruminal nitrogen turnover and methanogenesis. *Australian Journal of Agricultural Research* 54:703–713.

HSUS (2008) HSUS Fact Sheet: Animal Agriculture and Climate Change.

IPCC (2007) *Fourth Assessment Report. Climate Change 2007: Synthesis Report. Summary for Policymakers*, 2–5.

Johnson HD (1987) *Bioclimates and Livestock, World Animal Science*, Elsevier, Amsterdam, Publication No: B5.

Johnson KA, Johnson DE (1995) Methane emissions from cattle. *Journal of Animal Science* 73:2483–2492.

Kamra DN, Agarwal N, Chaudhary LC (2006) Inhibition of ruminal methanogenesis by tropical plants containing secondary compounds. *International Congress Series* 1293:156–163.

Koluman Darcan N, Karakök Göncü S, Daşkıran İ (2009) Strategy of adaptation animal production to global warming in Turkey. *Ulusal Kuraklık ve Çölleşme Sempozyumu*, 16–18 June 2009, Konya.

Linn JG (1997) Nutritional management of lactating dairy cows during periods of heat stress.

Lovett D, Lovell S, Stack J, Callan J, Finlay M, Conolly J, O'mara FP (2003) Effect of forage/concentrate ratio and dietary coconut oil level on methane output and performance of finishing beef heifers. *Livestock Production Science* 84:135–146.

Martin C, Rouel J, Jouany JP, Doreau M, Chilliard Y (2008) Methane output and diet digestibility in response to feeding dairy cows crude linseed, extruded linseed, or linseed oil. *Journal of Animal Science* 86:2642–2650. https://doi.org/10.2527/jas.2007-0774.

Mbanzamihigo L, Fievez V, Da Costa Gomez C, Piattoni F, Carlier L, Demeyer DD (2001) Methane emission from the rumen of sheep fed a mixed grass-clover pasture at two fertilisation rates in early and late season. *Canadian Journal of Animal Science* 82:69–77.

McGinn SM, Beauchemin KA, Coates TW (2004a) Measurement of methane emissions from cattle using Chambers and micrometeorological techniques; at: http://ams.confex.com/ams/pdfpapers/79664.pdf.

McGinn SM, Beauchemin KA, Coates T, Colombatto D (2004b) Methane emissions from beef cattle: Effects of monensin, sunflower oil, enzymes, yeast, and fumaric acid. *Journal of Animal Science* 82:3346–3356.

Moss AR, Jouany J, Newbold J (2000) Methane Production by Ruminants: Its Contribution to Global Warming. *Annales De Zootechnie* 49:231–253.

Newbold CJ, Rode LM (2006) Dietary additives to control methanogenesis in the rumen. *International Congress Series* 1293:138–147.

NYSDEC (2006) New York State Department of Environmental Conservation. 2006. DEC announces consent order with Lewis County farm for manure spill; at: www.dec.ny.gov/environmentdec/18863.html (23 April 2008).

O'Mara F (2004) Greenhouse gas production from dairying: reducing methane production. *Advances in Dairy Technology* 16:295–309.

O'Mara FP, Beauchemin KA, Kreuzer M, McAllister TA (2008) Reduction of greenhouse gas emissions of ruminants through nutritional strategies. International Congress on Livestock and Global Climate Change, 17–20 Mayıs 2008, Hamamet, Tunus, *Proceedings* 40–43

Oliveira SG, Berchielli TT, Pedreira MS, Primavesi O, Frighetto R, Lima MA (2007) Effect of tannin levels in sorghum silage and concentrate supplementation on apparent digestibility and methane emission in beef cattle. *Animal Feed Science and Technology* 135:236–248.

Ovenell-Roy KH, Nelson ML, Westberg HH, Froseth JA (1998) Effects of barley cultivar on energy and nitrogen metabolism of lambs. *Canadian Journal of Animal Science* 8:389–397.

Patra AK, Kamra DN, Agarwal N (2006) Effect of plant extract on in vitro methanogenesis, enzyme activities and fermentation of feed in rumen liquor of buffalo. *Animal Feed Science and Technology* 128:276–291.

Puchala R, Min BR, Goetsch AL, Sahlu T (2005) The effect of a condensed tannin-containing forage on methane emission by goats. *Journal of Animal Science* 83(1):182–186.

Qian G, Guo X, Guo JJ, Wu JG (2011) China's dairy crisis: impacts, causes and policy implications for a sustainable dairy industry. *International Journal of Sustainable Development and World Ecology* 18:434–441.

Reynolds CK, Tyrrell HF, Reynolds PJ (1991) Effects of diet forage-to-concentrate ration and intake on energy metabolism in growing beef heifers: whole body energy and nitrogen balance and visceral heat production. *Journal of Nutrition* 121:994–1,003.

Sauer FD, Fellner V, Kinsman R, Kramer JK, Jackson HA, Lee AJ, Chen S (1998) Methane output and lactation response in Holstein cattle with monensin or unsaturated fat added to the diet. *Journal of Animal Science* 76(3):906–914.

Silanikove N (2000) Effects of heat stress on the welfare of extensively managed domestic ruminants. *Livestock Production Science* 67:1–18.

Silanikove N, Koluman N (2015) Impact of climate change on the dairy industry in temperate zones: Predications on the overall negative impact and on the positive role of dairy goats in adaptation to earth warming. *Small Ruminant Research* 123(1):27–34.

Silanikove N, Leitner G, Merin U, Prosser CG (2010) Recent advances in exploiting goat's milk: Quality, safety and production aspects. *Small Ruminant Research* 89:110–124.

Steinfeld H, Gerber P, Wassenaar T, Castel V, Rosales M, De Haan C (2006) *Livestock's long shadow: environmental issues and options.* Food and Agriculture Organization of the United Nations, 82–114.

TUIK (2009) Environmental Statistics; at: http://www.tuik.gov.tr (9 May 2009).

USEPA (2008) Environmental Protection Agency, 2008. Ruminant livestock: frequent questions.
Wallace RJ (2004) Antimicrobial properties of plant secondary metabolites. *Proceedings of the Nutrition Society* 63:621–629.
Wallace RJ, Wood TA, Rowe A, Price J, Yanez DR, Williams SP, Newbold CJ (2006) In Soliva CR, Takahashi J, Kreuzer M (eds) *Greenhouse Gases and Animal Agriculture : An Update*; Elsevier International Congress Series 1293, Amsterdam, The Netherlands 148–151.
Watson RT, Meira Filho LG, Sanhueza E, Janetos T (1992) Sources and Sinks. In Houghton JT, Callander BA, Varney SK (eds) *Climate Change*, Cambridge University Press, Cambridge 25–46.
Yurtseven S, Görgülü M (2004) Effects of Grain Sources and Feeding Methods, Free-Choice vs Total Mixed Ration, on Milk Yield and Composition of German Fawn X Hair Crossbred Goats in Mid Lactation. *Journal of Animal Feed Science* 13(3):417–428.
Yurtseven S, Öztürk İ (2008) Influence of Two Sources of Cereals (Corn or Barley), in Free Choice Feeding on Diet Selection, Milk Production Indices and Gaseous Products (CH_4 and CO_2) in Lactating Sheep. *Asian Journal of Animal Veterinary Advances* 4(2):76–85.

Part IV
Climate Change Impacts
on Soil-Water and Crop Interactions

Chapter 12
Interactive Effects of Elevated CO_2 and Climate Change on Wheat Production in the Mediterranean Region

Burçak Kapur, Mehmet Aydın, Tomohisa Yano, Müjde Koç and Celaleddin Barutçular

Abstract Global climate change could be harmful to agriculture. In particular, water availability and irrigation development under changed climatic conditions already pose a growing problem for crop production in the Mediterranean region. Wheat is the major significant crop in terms of food security. Therefore, in relation to these issues, this review gives an overview of climate change effects on wheat production in the Mediterranean environment of Turkey. Future climate data generated by a general circulation model (e.g., CGCM2) and regional climate models (e.g., RCM/MRI, CCSR-NIES and TERCH-RAMS) have been used to quantify the wheat growth and the soil-water-balance around the Eastern Mediterranean region of Turkey. The effects of climate change on the water demand and yield of wheat were predicted using the detailed crop growth subroutine of the SWAP (Soil-Water-Atmosphere-Plant). The Soil evaporation was estimated using the E-DiGOR (Evaporation and Drainage investigations at Ground of Ordinary Rainfed-areas) model. This review revealed that the changes in climatic conditions and CO_2 concentration have caused parallel changes in the wheat yield. A close correspondence between measured and simulated yield data was obtained. The grain yield increased by about 24.7% (measured) and 21.9% (modelled) under a two-fold CO_2 concentration and the current climatic conditions. However, this increase in the yield was counteracted by a temperature rise of 3 °C. Wheat biomass decreases under the future climatic conditions and the enhanced CO_2 concentration,

B. Kapur, Asst. Prof. Dr., Çukurova University, Department of Agricultural Structures and Irrigation Engineering, Adana 01330, Turkey; e-mail: bkapur@cu.edu.tr.

M. Aydın, Retired Professor, Mustafa Kemal University, Department of Soil Science, Faculty of Agriculture, Antakya; e-mail: maydin08@yahoo.com.

T. Yano, Emeritus Professor, Tottori University, Arid Land Research Center, Tottori, Japan; e-mail: yano@ant.bbiq.jp.

M. Koç, Retired Professor, Çukurova University, Department of Field Crops, Faculty of Agriculture, Adana; e-mail: mkoc@cu.edu.tr.

C. Barutçular, Professor, Çukurova University, Department of Field Crops, Faculty of Agriculture, Adana; e-mail: cebar@cu.edu.tr.

© Springer Nature Switzerland AG 2019
T. Watanabe et al. (eds.), *Climate Change Impacts on Basin Agro-ecosystems*,
The Anthropocene: Politik—Economics—Society—Science 18,
https://doi.org/10.1007/978-3-030-01036-2_12

regardless of the model used. Without CO_2 effects, grain yield also decreases for all the models. By contrast, the combined impact of elevated CO_2 and increased temperature on grain yield of wheat was positive, but varied with the climatic models. Among the models, the CCSR-NIES and TERCH-RAMS denote the highest (24.9%) and lowest (6.3%) increases in grain yield respectively. The duration of the regular crop-growing season for wheat was 24, 21, and 27 days shorter as calculated for the future, mainly caused by the projected air temperature rise of 2.2, 2.4, and 3 °C for a growing period by the 2070s for CGCM2, CCSR-NIES and TERCH-RAMS respectively. The experimental results show large increases in the water use efficiency of wheat, due to the increases in CO_2 concentration and air temperature. Despite the increased evaporative demand of the atmosphere, the increases in water use efficiency can be attributed to the shorter growing days and a reduction in the transpiration due to stomata closure. Unlike reference evapotranspiration and potential soil evaporation, actual evaporation from bare soils was estimated to reduce by 16.5% in response to a decrease in rainfall and consequently soil wetness in the future, regardless of the increases in the evaporative demand. It can be concluded that to maintain wheat production in the future, the water stress must be managed by proper irrigation management techniques.

Keywords CO_2 · Crop growth · Crop water use · Growth chamber Soil evaporation

12.1 Introduction

In recent decades, the issue of climate change has been at the centre of many scientific studies at global level (Southworth et al. 2000; IPCC 2001a; Ibrahim 2014; Alpert et al. 2008). Global climate change may have serious impacts on water resources and agriculture in the future. The concerns about these issues have stimulated interdisciplinary research (Rosenzweig 1985; Rosenberg 1992; Maytin et al. 1995; Wolf/Van Diepen 1995; El Maayar et al. 1997; Alexandrov/ Hoogenboom 2000; Olesen/Bindi 2002; Aggarwal 2003; Jones/Thornton 2003; Izaurralde et al. 2003; Quinn et al. 2004; Yano et al. 2007a; Guo et al. 2017; Ibrahim 2014). In the past, researchers began to use crop and agro-ecosystem models to assess how agricultural production could be affected according to the projections of climate models. This approach has been proven useful because it also yields insights into some of the extended impacts of climate change on ecosystem processes (Mearns et al. 1992; Brown/Rosenberg 1999; Guerena et al. 2001; Izaurralde et al. 2003; Steduto et al. 2009; Asseng et al. 2013). Based on a range of several climate models, the mean annual global surface temperature is projected to increase by 1.1–6.4 °C over the period of 1990–2100 (IPCC 2007), with changes in the spatial and temporal patterns of precipitation.

Climate change is a long-term issue with short-term risks for the production systems in each Mediterranean country. The updated report of IPCC (2013) indicates that increased temperature and droughts, and decreased precipitation, as well as a decline in soil moisture, would dramatically influence agricultural production. Projections from the General Circulation Model-Regional Climate Model (GCM-RCM) reveal the changes in mean and extreme temperature and precipitation for Europe and the Mediterranean (van der Linden/Mitchell 2009). In terms of mean annual temperature, projections show a warming of 0.5–1.5 °C (2046–2065) and 1–5 °C (2081–2100), over the land portion of the Mediterranean. Moreover, decreased precipitation in the Mediterranean of 10–20% in 2046–2065 and 10–30% in 2081–2100, is likely under the "Representative Concentration Pathways" scenario (IPCC 2013). Based on the multi-model projections of the 21st century, the mean annual soil moisture is estimated to decrease in the Mediterranean region. In other words, the risk of agricultural drought will increase in the future. A group of regional climate models (RCMs) driven by several general circulation models (GCMs) using the A1B scenario foresees a significant decrease in runoff emerging only after 2050 (Sanchez-Gomez et al. 2009) for the Mediterranean basin. Thus, considerable increases in meteorological drought are also projected (IPCC 2013). Furthermore, Giorgi/Lionello (2008) presented a review of climate change projections over the Mediterranean region, based on the most recent and comprehensive group of global and regional climate change simulations. These simulations represent substantial drying and warming of the Mediterranean region between 2071 and 2100, especially in the warm season (precipitation decrease exceeding −25 to –30% and warming exceeding 4–5 °C). The only exception to this picture is an increase in precipitation during the winter over some areas of the northern Mediterranean basin and most noticeably in the Alps. Inter-annual variability is projected to experience a general increase, as is the occurrence of extreme heat and drought events.

Despite intensive research efforts, many aspects of climate change are still unpredictable, particularly at site scale and regional level (Zhang/Liu 2005). Therefore, numerous simulation studies have been carried out by some interdisciplinary teams under many scenarios (Evrendilek et al. 2005, 2008; Yano et al. 2005, 2007a, b, c; Aydın et al. 2008; Önder et al. 2009a; Hu et al. 2015) projected by the general circulation models (GCMs) and regional climate models (RCMs) to predict the impacts of climate change on the soil-water balance and crop growth in the Mediterranean environment of Turkey. In this chapter, the changes in biomass and grain yield of wheat under future climatic conditions and a two-fold CO$_2$ increase scenario are discussed.

12.2 Climate Change Scenarios

The Special Report on Emission Scenarios (SRES) produced forty future emission scenarios for greenhouse gases and sulfate aerosols. These scenarios cover a wide range of the main driving forces of future emissions, from demographic to technological and economic developments. The GCMs are the best available tools to study the possible change in the future climate globally. Many climate change studies use downscaled results from GCMs to estimate the future climate (Alexandrov/Hoogenboom 2000; Schulze 2000; Eckersten et al. 2001; Van Ittersum et al. 2003; Kimura et al. 2007; Kitoh 2007; Yano et al. 2007a; Hu et al. 2015), and some of these projections are available on the IPCC website (http://ipcc-ddc.cru.uea.ac.uk). However, GCMs sometimes do not provide accurate data for the specific site due to downscaling from data in the surrounding grid points and due to the inherent model performance. Thus, it is sometimes assumed that while data for the present climate differs between the control simulation and the actual observation, the change between the present and the future climate remains accurate. On the other hand, the RCMs are developed to evaluate climate change on a regional scale (Sato et al. 2007). In order to predict the impacts of climate change on the soil-water balance components and crop production, Yano et al. (2007a, b, c) have obtained the climate change data for a Mediterranean environment of Turkey from the outputs of the three GCMs [the second version of the Canadian Global Coupled Model – CGCM2 – (Flato/Boer 2001), the model developed from the atmospheric model of the European Center for Medium-Range Weather Forecasts, and parameterized at Hamburg – ECHAM4 – (Roeckner et al. 1996), and the general circulation model developed at the Meteorological Research Institute of Japan – MRI – (Yukimoto et al. 2001; Kitoh et al. 2005)] for the A2 scenario in the SRES. The data of the two RCMs [RCM/MRI and RCM/CCSR-NIES] were additionally used for complementarity. The A2 scenario describes a very heterogeneous world of high population growth, slow economic development and strong regional cultural identities. Scenario A2 is one of the emission scenarios with the highest projected CO_2 increase (up to 800 ppm) by the end of the 21st century (Nakicenovic/Swart 2000).

   In the present chapter, we compared the climate change data of CGCM2, RCM/MRI, RCM/CCSR-NIES and TERCH-RAMS for the period of 2070–2079. The forcing data for the boundary conditions of the RCM/MRI are given by MRI. Hereafter, the RCM/CCSR-NIES is referred to as CCSR-NIES, which represents the Center for the Climate Systems Research and the National Institute for Environmental Studies of Japan. The acronym TERCH-RAMS stands for the Terrestrial Environmental Research Center-Regional Atmospheric Modelling System. The projection years for wheat production were 1994–2003 (baseline) and 2070–2079. Using the daily data for the baseline period, the future climate changes have been applied to the available baseline climate series in a straightforward way. Temperature changes have been added as absolute changes to the baseline series; the other climate parameters have been adapted according to their relative changes

(Tao et al. 2003; Dhungana et al. 2006; Yano et al. 2007a). In other words, the future climate data have been created by superimposing the observed values on the change between the present and the future estimated values.

All the models projected a temperature rise by the 2070s in the Eastern Mediterranean region of Turkey. Rises in air temperature projected by different climatic models for the future (2070–2079) compared with the baseline (1994–2003) are depicted in Fig. 12.1. Among the four models, the TERCH-RAMS and RCM/MRI denote the highest and lowest increases in temperature respectively. The averaged air temperature is estimated to increase by 2.2, 2.4, 1.6, and 3.0 °C based on the projections of CGCM2, CCSR-NIES, RCM/MRI and TERCH-RAMS, respectively, for a growing period of wheat. Önol/Unal (2012) have projected climate change over the Mediterranean region of Turkey using the RegCM3 regional climate model for A2 IPCC scenario and reported that a 2–4 °C temperature increase was detected which is similar to the findings of the models mentioned in this review.

Although the annual precipitation denoted noticeable variations year-by-year, it was not likely that it will have increased in the future. Percentage changes in the mean annual precipitation projected by different climatic models for the future are shown in Fig. 12.2. The mean annual precipitation during 2070–2079 compared with 1994–2003 is projected to decrease by 11%, 42%, 46%, and 25% according to the CGCM2, CCSR-NIES, RCM/MRI and TERCH-RAMS models respectively. On average, regarding the models, future precipitation is predicted to decrease by 31%. This average is similar to the results obtained by some of the other researchers. For example, according to Kimura et al. (2007), precipitation is expected to decrease by 27% over the basin. Önder et al. (2009a) have reported that the projected precipitation is 29.6% less compared to the present for the southern region of Turkey, especially along the coast of the Mediterranean by the 2070s.

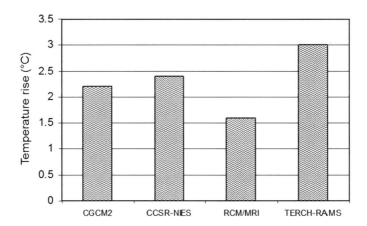

Fig. 12.1 Rises in air temperature projected by different climatic models for the future (2070–2079) compared with the baseline (1994–2003) during the wheat-growing period. *Source* The authors

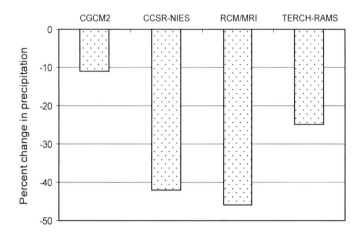

Fig. 12.2 Changes in precipitation projected by different climatic models for the future (2070–2079) compared with the baseline (1994–2003). *Source* The authors

Moreover, Lionello/Scarascia (2017) indicated that the precipitation decrease would affect most of the Mediterranean region, where a 7% decrease in annual precipitation for each degree of global warming is expected for the southern parts of Mediterranean, according to CMIP5 estimates.

For a specific region, if the climate models show consistent results, then they can be used to predict potential impacts (Van Ittersum et al. 2003). From this point of view, it seems that using the RCM/MRI data would not correctly reflect the effect of climate change on crop growth. Therefore, further comparisons have been conducted with CGCM2, CCSR-NIES, and TERCH-RAMS data due to the consistency of their results. In other words, the climate change data projected by these models have been used for quantifying potential impacts of climate change on wheat growth. Although it is generally difficult to downscale GCMs data with large grid point distances to the study area accurately, using the CGCM2 data to predict the future change in crop growth is more appropriate under the present situation than using the less reliable RCM/MRI data (Yano et al. 2007c).

12.3 Crop Growth

Crop yields are affected by variations in climatic elements such as air temperature and precipitation and CO_2 concentration. Moreover, the changes in both the total seasonal precipitation and its pattern are important (Alexandrov/Hoogenboom 2000; Olesen/Bindi 2002; Asseng et al. 2015) in crop yield studies concerning climate change. Thus, sites which are already at the limit with respect to water supply under current conditions are likely to be most sensitive to climate change, leading to an increase in the need for irrigation in dry areas, while more humid areas may be less affected (Brumbelow/Georgakakos 2001; Fuhrer 2003; Hatfield/Prueger 2015). In agro-ecosystems under a changing climate, water deficit stress could cause decreasing yield or require irrigation to maintain crop production. This negative effect of climate warming may be counteracted by effects of elevated CO_2 on the crop tolerance to water stress (Lawlor/Mitchell 2000; Young/Long 2000; Robredo et al. 2011; Bencze et al. 2014; Pazzagli et al. 2016). Changes in atmospheric CO_2 levels have important physiological effects on plants. Elevation in CO_2 affects photosynthesis by increasing the photosynthetic rate (Heinemann et al. 2006; Bencze et al. 2014). In Australia, the impact of elevated CO_2 and increased temperature on grain yield of wheat was on average positive, but varied with seasonal rainfall distribution (Asseng et al. 2004). According to Richter/Semenov (2005), simulated scenarios for the 2020s and 2050s show that in comparison to the baseline, wheat yields in England are likely to increase more until the 2020s than in the following 30 years in spite of CO_2 and temperature increasing linearly. Erice et al. (2014) concluded that the stimulation of plant growth by elevated CO_2 was found in durum wheat genotypes with high harvest indices and optimal water supply.

Future cereal production will depend not only on climate change effects, but also on further developments in technology and crop management, including changes in the sowing dates, plant densities, and irrigation applications (Olesen/Bindi 2002; Dhungana et al. 2006; Godfray et al. 2010). Technological innovations, including the development of new crop hybrids and cultivars that may be developed to match the changing climate, are considered to be a promising adaptation strategy (Alexandrov/Hoogenboom 2000; Jones/Thornton 2003; Tester/Langridge 2010). The semi-arid regions differ in socio-economic development, technological possibilities and climatic regime, but all have relatively ample water supplies for agriculture in the current climate (Krol/Bronstert 2007). However, the continued population growth will cause an increasing demand for food and water alongside the rising temperatures and CO_2, and uncertainties in rainfall associated with global climatic change (Aggarwal 2003; Godfray et al. 2010). Thus, the estimates of future food production and demand are associated with high uncertainties (Olesen/Bindi 2002). Major agricultural regions may be developed under changing climate conditions, since they may become even more important as food-producing centres relative to agricultural areas in more marginal, semi-arid regions that have been found to be vulnerable to climate change (IPCC 2001b). In this regard, Olesen/Bindi (2002) reported that developing countries would be more severely affected by climate change than the developed countries that are generally located in temperate regions. At conditions yielding higher climate change impacts that were tested in the A1 and A2 SRES scenarios, the disparities in cereal yields between the developed and the developing countries are likely to increase (Parry et al. 2004). Consequently, based on some scenarios, the yield impact studies using crop response models have been conducted for many countries (Table 12.1). For example, the effects of the projected climate change on the wheat yield in a Mediterranean environment of Turkey, have been studied by Yano et al. (2007a, b, c). The results of these studies have revealed a wide variation concerning the climatic scenarios, the methods used

Table 12.1 The impact of the two-fold increased CO_2 and raised temperature on the yield of wheat in different parts of the world. *Source* IPCC (1995: 33), Kapur (2010: 15)

Region	Percentage change in yield	Countries
Europe	−10 to +10	France, England, and Northern Europe
Latin America	−61 to +5	Argentina, Brazil, Chile, and Mexico
Former Soviet Union	−19 to +41	–
North America	−100 to +234	USA and Canada
Africa	−65 to +6	Egypt, Kenya, South Africa, and Zimbabwe
South Asia	−61 to +67	Bangladesh, India, Philippines, Thailand, and Indonesia
China	−78 to +28	–
Australia and Far East Asia	−41 to +65	Australia, Japan

for crop response, and the site-specific conditions (Peiris et al. 1996; Wolf et al. 2002).

Crop growth was simulated using the detailed crop growth sub-model of the SWAP (Soil-Water-Atmosphere-Plant) incorporated from WOFOST (WOrld FOod STudies) with a daily time step from sowing to maturity (van Dam et al. 1997; Boogaard et al. 1998). As a rule, experimentation is needed to obtain specific parameters and to calibrate and verify the model results. Therefore, the SWAP model was parameterised for wheat and calibrated with the crop growth measurements (Yano et al. 2006, 2007a) in addition to photosynthesis and transpiration rates of the plant leaves measured under Mediterranean field conditions, using the Photosynthesis Monitor system (PM-48M). As reported by Evrendilek et al. (2008), the net photosynthetic rate of wheat, especially, exhibited a peak at mid-morning, and a photosynthetic midday depression under the limiting effects of high evaporative demand. On the other hand, to evaluate the positive effect of the elevated CO_2 concentration and the negative effect of the risen air temperature on the wheat yield, the confidence level in the model used can be obtained through the simulation of the measured data. Therefore, the measured data under the controlled conditions and estimated data by the SWAP model are compared in Fig. 12.3, for a two-fold CO_2 increase scenario, as well as for an increased temperature of 3 °C. Similar results were obtained in this study between the measured and the modelled wheat yield data for the enhanced CO_2. For example, the grain yield of wheat increases were about 24.7% (measured) and 21.9% (modelled) under a two-fold CO_2 increase scenario and the current climatic conditions. By contrast, the increase in temperature without the effects of CO_2 caused decreases in the wheat yield.

A further simulation was done by a two-fold CO_2 scenario and by increasing the maximum and minimum temperatures by 1 and 3 °C relative to the current conditions. The yield of wheat estimated by the SWAP model is shown in Fig. 12.4. The increases in both the biomass and grain yield obtained via the use of the two-fold CO_2 scenario experimentation under the conditions of current climate are

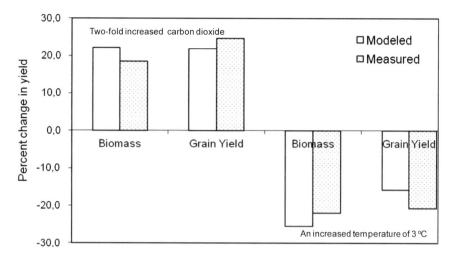

Fig. 12.3 A comparison of estimated versus measured wheat yield, for a two-fold CO$_2$ increase scenario, as well as for an increased temperature of 3 °C. *Source* The authors

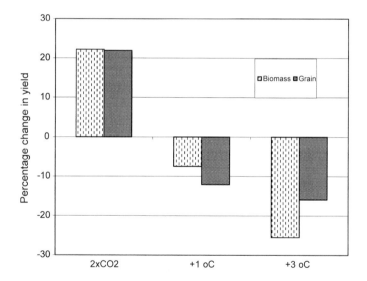

Fig. 12.4 Effects of the two-fold CO$_2$ increase scenario and temperature rise separately on wheat growth. *Source* The authors

about 22% for wheat as a C3 crop. The increases in biomass and grain yield are counteracted by a temperature rise of 3 °C. Although several variations of temperature rise are recognised among the models, more than 1 °C of temperature rise is estimated in the 2070s when the CO$_2$ concentration is assumed to increase two-fold (See Fig. 12.1). Yields were also simulated using climate inputs generated

Table 12.2 Percentage changes in biomass and grain yield of wheat under future (2070–2079) climatic conditions and a two-fold CO_2 increase scenario compared with the baseline (1994–2003). *Source* The authors

Climate model	Percentage change in biomass		Percentage change in grain yield		Shortness in growth duration (days)
	Without CO_2 effects	With CO_2 effects	Without CO_2 effects	With CO_2 effects	With CO_2 effects
CGCM2	–24.1	–3.3	–11.9	+15.3	–24
CCSR-NIES	–9.2	–	–6.0	+24.9	–21
TERCH-RAMS	–22.0	–2.3	–20.8	+6.3	–27

from GCM and RCM climate data. Changes in yield were evaluated by comparing the future yields with the current ones of the same crop varieties, and then stating the change as difference in per cent. Biomass and grain yield predictions for wheat during the period of 2070–2079 denoted the complicated results in accordance with the different photosynthesis rates and air temperature rises in the future. Wheat biomass decreases under the future climatic conditions and the elevated CO_2 concentration, regardless of the model used. Without CO_2 effects, grain yield also decreases for all the models. However, the combined impact of elevated CO_2 and increased temperature on grain yield of wheat was positive, but varied with climatic models (Table 12.2). Since the rise in temperature for TERCH-RAMS is higher than the other two models, the lower increase in grain yield is expected to occur. It is assumed that the use of climatic data alone would cause erroneous results, because the CO_2 is the fundamental parameter for crop growth prediction. Thus, the effects of increased CO_2 on the plants should be imposed on the model simulations. Although wheat is a C3 plant and raised CO_2 concentration promotes photosynthesis, the risen temperature may partially compensate with stomatal closure. That is to say, stimulation of photosynthesis does not directly translate to increased biomass, or yield. The reason why grain yields increase for all the models cannot be fully explained. However, in determinate crops such as cereals, the grain yield not only depends on photosynthesis but also on the length of the active phase of leaf photosynthesis and the sink capacity of the grains (Fuhrer 2003). Kapur (2010) reported that the grain yield was dependent on differences in the tillers rather than the main spike. The positive effects of elevated atmospheric carbon dioxide on wheat were lower at the main spike. Consequently, the substances produced during the process of photosynthesis were stored in the vegetative organs and especially in the roots (Parry et al. 2011). Thus, the yield increase depends on the sink capacity of the wheat variety, rather than the amount of assimilated resources.

Climate change data used in the SWAP model have projected shorter growing seasons for wheat during the period 2070–2079 compared to 1994–2003 (Table 12.2). These changes are caused by the predicted temperature rises during the growing periods of the 2070s. The duration of the regular crop-growing season for wheat was 24, 21, and 27 days shorter in the future, mainly caused by the projected air temperature rise of 2.2, 2.4, and 3 °C for a growing period by the 2070s, according to CGCM2, CCSR-NIES, and TERCH-RAMS respectively. In other words, high temperatures accelerate the phenological development of plants, resulting in quicker maturation. This may cause a shift in harvest maturity dates for wheat from May to April in nearly three-quarters of this century. Without the positive effects of the elevated CO_2 the shortened growth cycle may reduce crop yield (Yano et al. 2007a; Hatfield/Prueger 2015). Similarly, Alexandrov/ Hoogenboom (2000) reported that the GCM scenarios for the 21st century projected a decrease in yield of winter wheat in Bulgaria, caused by a shorter crop-growing season due to higher temperatures and precipitation deficit. When the direct effects of CO_2 were included in the study above, all GCM scenarios resulted in an increase in winter wheat yield. They emphasised that the selection of a suitable sowing date will probably be the appropriate response to offset the negative effect of a potential increase in temperature. This change in the planting date will allow for the crop to develop during a period of the year with lower temperatures, thereby decreasing developmental rates and increasing the growth duration, and especially increasing the grain-filling period (Yano et al. 2007a). On the other hand, Brown/Rosenberg (1999) emphasised that regardless of the GCM scenario used, an enhancement in atmospheric CO_2 increased the yield of winter wheat. In their study, the crop was negatively influenced by the increasing temperature, and yields fell dramatically when global mean temperatures increased by 5 °C. Moreover, Barlow et al. (2015) revealed that increased temperature caused reduction in grain number and reduced duration of the grain-filling period of wheat.

Unfortunately, there are several limitations in such simulations. The crop models assume that nutrients are not limited. Also, pests (insects, diseases, weeds) are assumed to pose no limitation to crop growth and yield under both current and future climate scenarios (El Maayar et al. 1997; Scherm et al. 2000; Tubiello/Ewert 2002; Asseng et al. 2015). Yano et al. (2007a) have stated that simulated quantitative effects with the models should be interpreted cautiously based on the decadal time studies, limited field experimentation, and the large possible range of factorial interactions that are not tested. As an alternative to assessing climate change impacts over decadal timescales, climate variability impacts may be more useful for more immediate decision-making at inter-annual time scales (Schulze 2000). Present results are what we would expect if farmers continue to grow the same varieties in the same way in the same locations. However, the research and plant breeding studies may mitigate many of the detrimental effects. Similarly, Kapur (2010) indicated that irrigation must be considered an important application for eliminating the yield losses in the future. New varieties are also expected to benefit more from the high assimilate due to the increasing carbon dioxide. Though the results are clearly not conclusive, they certainly suggest the potential impacts of climate change on crop production in the region. As the relationships between climate, crop growth and soil are complicated, this cannot be described in terms of simple and average relationships. Despite the mentioned drawbacks, the results from such studies not only improve our understanding of the possible impacts of climate change, but may also provide valuable information to adapt a strategy for mitigating the detrimental effects.

12.4 Crop Water Use

Agriculture is the largest consumer of water resources in the semi-arid to arid regions. The demand for irrigation water is projected to rise in a warmer climate (Olesen/Bindi 2002). According to Mehta et al. (2013), the warmer and drier A2 climate scenario ultimately encourages higher irrigation demand than the B1 scenario. This indicates the need to assess the effects of global warming on irrigation water requirements. In Mediterranean countries, cereal yields are limited by water availability, heat stress and the short duration of the grain-filling period. Irrigation is important for crop production in Mediterranean countries (Alexandrov/ Hoogenboom 2000; Olesen/Bindi 2002; Fader et al. 2015), including Turkey, due to high evapotranspiration and restricted rainfall (Yano et al. 2007a). Soil and climatic conditions in Turkey are suitable for the cultivation of cereals. However, sometimes a precipitation deficit and high air temperatures occur during the critical period of crop development. Both factors are the ones that most limit the growth and final yield of crops. Agricultural production in the Mediterranean region of Turkey is vulnerable to changes in precipitation; and agriculture is dependent on irrigation for its viability. Projected future changes foreseen in water use for irrigation in this region, based on the predicted climate change, may have serious consequences for the economic situation of the country. Thus, the studies with global and regional climate change projection models are necessary to assess the full range of potential climate change impacts and adaptation strategies. The selection of this area has been based on its significance in current or potential crop production in Turkey, and on its sensitivity to current and future climate regimes. On the specific site or regional scale, the variability of climate, soils and management need to be superimposed and tested for their relative impacts. Climate data needs to be generated, which includes changes in climatic variability, as predicted by GCMs. Introducing changes in variability may have profound impacts on the predictions of crop yield in areas with an inherent water shortage (Richter/Semenov 2005). Simulation models are useful tools to analyse the potential effects of climate change on crop growth, but testing model performance against measured data under such scenarios is essential for such an analysis to be meaningful (Asseng et al. 2015). Some preliminary studies have indicated that decreased precipitation and increased temperature will lead to increased irrigation water requirements. On the other hand, some other studies have suggested that increased CO_2 concentrations will increase the water use efficiency of the crops and in turn decrease the growing days due to the increased temperature ultimately decreasing the irrigation water requirements. Yano et al. (2007a) and Fujihara et al. (2008) have also stated that the effects of global warming on irrigation water requirements should be offset by increased CO_2 concentrations. In addition, to obtain some realistic estimation for the future, all the water balance components, including precipitation, irrigation, losses by runoff, soil evaporation, crop transpiration, and percolation from the root zone, should be taken into account in further studies. In other words, the models should take climate-crop-soil interactions into consideration, thereby facilitating a differential crop development and yield under the varying environmental conditions.

Experimental data indicating considerable decreases in the actual evapotranspiration (*ETa*) due to stomatal closure under elevated CO_2 concentration have received wide recognition (Ainsworth/Long 2005; Asseng et al. 2015). Cure/Acock (1986) reported a decrease of 17% in the transpiration of wheat. In fact, elevated atmospheric CO_2 concentration increases photosynthesis in C3 plants and reduces evapotranspiration in both C3 and C4 plants due to reduced stomatal conductance, thereby improving water use efficiency (Thomson et al. 2006). Kapur (2010) observed that actual evapotranspiration of wheat decreased about 11% (from 304.7 to 269.9 mm) because of a shorter growing period under an increased temperature of 3 °C and CO_2 concentration of 700 ppm. Evrendilek et al. (2008) concluded that diurnal variations in water use efficiency and light use efficiency of wheat showed a bimodal behaviour with the maximum values in the early morning and late afternoon. As the impacts of global climate change become increasingly felt, continuous measurements of climate-crop-soil-management interactions under natural conditions play a pivotal role not only in exploring changes in eco-physiological properties of strategic crops for food security such as wheat but also in devising preventive and mitigative management practices to ensure sustained agricultural productivity. The experimental data obtained by Kapur (2010) indicated large increases in the water use efficiency of wheat, due to an increase in the CO_2 concentration, as well as due to a rise in temperature (Fig. 12.5). The comparisons of the variations in water use efficiency were made between the current (about 400 ppm) and future (700 ppm) CO_2 concentrations based on the SRES A2 scenario for the 2070s. It is well-known that the *ETa* decreases considerably due to stomatal closure under elevated CO_2 concentration and risen temperature (Ainsworth/Long 2005; Bernacchi/VanLoocke 2015; Gao et al. 2015), in spite of the increased evaporative demand of the atmosphere. However, in this study, increases in water use efficiency can be attributed to the shorter growing days and reduction of transpiration due to stomatal closure, regardless of the increase in evaporative demand. In brief, plant water stress would cause the irrigation requirement to maintain wheat production in the future, since precipitation is projected to dramatically decrease.

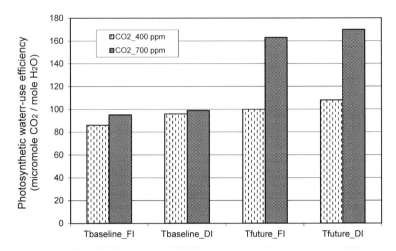

Fig. 12.5 The combined effects of the increased air temperature, elevated CO_2 concentration, and applied irrigation on the water use efficiency of wheat [modified from Kapur (2010). (Tbaseline: Baseline Temperature, Tfuture: Future Temperature, FI: Full Irrigation, DI: Deficit Irrigation)]. *Source* The authors

12.5 Soil Evaporation

Arid and semi-arid regions comprise almost 40% of the world's land surface. The low and erratic precipitation pattern is the single most significant contributor for limiting crop production in such lands (Aydın 1995). Evaporation from bare soils has a direct impact on crop yield in rainfed agriculture of the semi-arid to arid regions. The soil surface remains bare under field crops for many weeks during the periods of seed germination, seedling establishment and subsequent growth of the young crop when the moisture content of the upper soil layer can be of critical importance. In the orchards, the soil surface between the trees is kept bare by frequent tillage and continuously subjected to evaporation (Mellouli et al. 2000; Aydın et al. 2008). Some earlier results indicated that estimates of soil evaporation in semi-arid environments ranged from 30% to more than 60% of the seasonal rainfall (Jackson/Wallace 1999). Similarly, some other results have demonstrated that in regions where summer fallow is practised, direct evaporation from the soil surface accounted for about 50% or more of total precipitation (Hillel 1980; Hanks 1992). Önder et al. (2009b) reported that the actual soil evaporation in different parts of Turkey accounted for 34–83% of the incoming precipitation. In most parts of Turkey, the present precipitation is hardly adequate for good crop yields, and a further decrease in precipitation may seriously damage agriculture and the ecosystems (Kimura et al. 2007), although these areas have relatively ample water supplies for agriculture under the conditions of the current climate.

Evaporation from a bare soil surface is a complex process. The most important transport processes are characterised by a simultaneous change in the amount of energy or material with time and place (Aydın/Huwe 1993; Aydın 1994). In general, soil evaporation is modelled by limiting potential evaporation (e.g., from Penman-Monteith) with soil and/or aerodynamic resistance, although newer approaches (e.g., the Aydın model) derive soil evaporation successfully from the soil water potential (Aydın et al. 2005; Falge et al. 2005). However, many researchers distinguish two different stages of evaporation, related to the soil water content as: (1) when the actual water content is high, evaporation is controlled by the atmospheric evaporative demand; (2) when the amount of water is low, the evaporation is limited by the actual soil water content and, therefore, driven by the hydrodynamic characteristics of the soil. In this context, the Aydın model developed to address this contemporary issue is based on energy fluxes and soil properties, and experimental data are used to define a threshold separating the two stages of evaporation (Quevedo/Frances 2007; Romano/Giudici 2007; Bellot/Chirino 2013).

The quantification of the components of the water balance is still questionable in the assessment of soil water management under fallow conditions. In the present study, evaporation from bare soils was estimated using the E-DiGOR [Evaporation and Drainage investigations at Ground of Ordinary Rainfed-areas] model developed by Aydın (2008). The E-DiGOR model, as a helpful tool, incorporates the quantification of runoff, drainage, actual soil evaporation, and soil water storage (Aydın 2008; Aydın et al. 2014, 2015). The model also takes into account the important physical processes to quantify these components (Önder et al. 2009b; Aydın/Kececioglu 2010; Aydın 2012; Aydın et al. 2012a, b). The model part for actual evaporation had been previously validated by Aydın et al. (2005, 2008) using measured data. The model has been additionally adapted to provide a method of assessing drainage losses from the soil profile using field capacity concepts. The theory of the processes simulated by the E-DiGOR program has been extensively described by Aydın (2008). The model requires the input of daily climate data and information on soil properties (Aydın/Polat 2010). In principle, the E-DiGOR model can simulate the components of the soil-water balance on the scale of a plot. The drainage rates, actual soil evaporation, and soil water storage for a ten-year period in the future (2070–2079) were estimated by Aydın et al. (2008) using the E-DiGOR model with a daily time step.

The actual soil evaporation (Ea) is jointly dominated by the atmospheric evaporative demand and soil water availability, as well as the size of rainfall events. In this section, therefore, the future climate data of the RCM/MRI model have been used to calculate Ea rates, because this model projected a dramatic decrease in precipitation by the period of 2070–2079. Percentage changes in the mean annual evaporation for the present and future are compared in Fig. 12.6. Both potential evapotranspiration from a reference crop (ETr) and potential evaporation from bare soils (Ep) are projected to increase by 8.0% and 7.3% (equivalent to 92 mm and 69 mm increase), based on calculations from RCM/MRI data respectively. This change would be caused by the predicted temperature increase, as well as changes in the other climatic variables (data not shown). Both reference evapotranspiration and potential soil evaporation can serve the purpose of comparison of the climatic

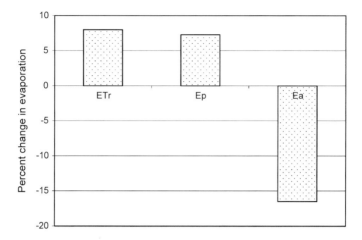

Fig. 12.6 Percentage changes in reference evapotranspiration (*ETr*), potential (*E*p) and actual (*Ea*) soil evaporation by 2070s. *Source* Derived from Aydın et al. (2008: 128)

conditions. Unlike both, however, the actual soil evaporation (*Ea*) is largely influenced by the soil wetness. In other words, the magnitude of *Ea* is strongly related to the temporal rainfall pattern, because the contribution of rainfall to soil water content is considerably dependent on the size of the rainfall events. The size of the rainfall events may play a key role in the magnitude of the actual soil evaporation, whereas, on the other hand, the variations in actual soil evaporation may reflect the effects of the rainfall pattern (Aydın et al. 2015). The projected annual actual soil evaporation (*Ea*) during 2070–2079 would decrease by 16.5% (equivalent to 50 mm decrease) relative to the baseline period of 1994–2003. This result, without any validation, may not allow such a clear-cut conclusion; however, the credibility of the E-DiGOR model has been demonstrated by different researchers in Japan and Turkey, using field-based measurements in a wide range of environments (Aydın 2008; Aydın et al. 2005, 2008; Kurt 2011). Since the projected mean annual precipitation during 2070–2079 compared with the baseline would decrease by 46% according to the RCM/MRI data, in a warmer climate and under lower precipitation, increased evaporative demand of the atmosphere favours soil dryness.

In fact, the simulation of the soil evaporation rates from bare soils in a Mediterranean region under a changing climate has revealed a reduction in *Ea* during the hot and dry summer season and an increase in *Ea* during the mild and rainy spring season, depending mainly on soil wetness. During the wet period of November to May, when *Ea* is close to *E*p, the fields should be kept cropped to increase beneficial use of soil water by crops and consequently prevent water loss through evaporation from soils. Alternatively, adoption of such agronomic practices as retention of crop residues or formation of a natural layer of mulch on the soil surface by proper and timely tillage when the evaporation rate is most rapid plays a significant role in the magnitude of reducing the soil water loss. Summer fallow is not practised under the current climatic conditions of the study area but may be adopted in the case of a decrease in rainfall. In that case, beneficial effects of fallowing in terms of the amount of water stored in the soil before planting depend on the timing, length and type of the fallow period. Fallowing with crop residue cover is most likely to be more beneficial in terms of increased soil water storage during the period of highest evaporation rate than lowest (Aydın et al. 2008).

As mentioned above, in the future, potential evapotranspiration from the reference crop (Allen et al. 1994) and potential evaporation from bare soils in the Mediterranean environment was found to increase, whereas actual soil evaporation was found to decrease, mainly due to lesser and erratic rainfall pattern, presumably soil wetness.

12.6   Conclusions

The climate change effects would be harmful, particularly for the water and agriculture sectors. The effects vary from ordinary to scary, as in the case of water limitations and increased temperature, thus hitches in the context of global food security via wheat production. In a changing global climate, the regions under the Mediterranean climate would increasingly face the challenges of water scarcity and of devising innovative and adaptive ecosystem management practices to make efficient use of the water available in the soil profile for plant growth. Moreover, expected temperature rises are likely to affect the maturation of wheat, thereby reducing total yield potential, with high temperatures causing more severe losses. Thus, temperature effects are increased by water stress, so water will be needed to develop more strong adaptation options to counterbalance the impacts of the raised temperatures associated with a changing climate. The foremost adaptation strategies are the conservation of water, the improvement of sustainable water projects and proper cultivation practices. The effects of climate change on water resources are part of a more comprehensive environmental problem. Adaptations to climate change will thus have to be part of a more extensive strategy to surpass the environmental problems of the present, as well as the future. The strategy could be clarified as the conservation and active use of water in a conscious and forceful approach. In addition, the response of individual producers to changes in the climate regime will need to include the selection of stress-tolerant wheat species, and innovation and investments in environmentally sound technologies for cultivation and irrigation techniques. Consequently, further studies are needed to assess the full range of climate change impacts and mitigative and adaptation strategies under the projections of climate change models.

Acknowledgements This research was conducted as part of the ICCAP (Impact of Climate Change on Agricultural Production System in Arid Areas) Project, a collaboration between the Research Institute of Humanity and Nature (RIHN) of Japan and the Scientific and Technical Research Council of Turkey (TÜBİTAK). The authors would like to extend their thanks to the editors, S. Kapur, T. Watanabe, M. Aydın, E. Akça and R. Kanber.

References

Aggarwal PK (2003) Impact of climate change on Indian agriculture. *Journal of Plant Biology* 30:189–198.

Ainsworth EA, Long SP (2005) What have we learned from 15 years of free-air CO_2 enrichment (FACE, Free Air CO_2 Enrichment) A meta-analytic review of the responses of photosynthesis, canopy properties and plant production to rising CO_2. *New Phytologist* 165:351–372.

Alexandrov VA, Hoogenboom G (2000) The impact of climate variability and change on crop yield in Bulgaria. *Agricultural and Forest Meteorology* 104:315–327.

Allen RG, Smith M, Perrier A, Pereira LS (1994) An update for the definition of reference evapotranspiration. *ICID Bulletin* 43:92.

Alpert P, Krichak KO, Shafir H, Haim D, Osetinsky I (2008) Climatic trends to extremes employing regional modeling and statistical interpretation over the E. Mediterranean. *Global and Planetary Change* 63:163–170. Elsevier, Amsterdam.

Asseng S, Jamieson PD, Kimball B, Pinter P, Sayre K, Bowden JW, Howden SM (2004) Simulated wheat growth affected by rising temperature, increased water deficit and elevated atmospheric CO_2. *Field Crops Research* 85:85–102.

Asseng S, Ewert F, Rosenzweig C et al (2013) Uncertainty in simulating wheat yields under climate change. *Nature Climate Change* 3:827–832.

Asseng S, Zhu Y, Wang E, Zhang W (2015) Crop modelling for climate change impact and adaptation. *Crop Physiology: Applications for Genetic Improvement and Agronomy Chapter 20*, 2nd ed. 505–546; at: https://doi.org/10.1016/B978-0-12-417104-6.00020-0.

Aydın M (1994) Hydraulic properties and water balance of a clay soil cropped with cotton. *Irrigation Science* 15:17–23.

Aydın M (1995) Water: key ingredient in Turkish farming. *Forum for Applied Research and Public Policy* (A Quarterly Journal of the University of Tennessee, USA) 10:68–70.

Aydın M (2008) A model for evaporation and drainage investigations at ground of ordinary rainfed-areas. *Ecological Modelling* 217(1–2):148–156.

Aydın M (2012) Improvements in E-DiGOR model: quantifying the water balance components of bare soils. *Journal of Agricultural, Life and Environmental Sciences* 24(2):68–71.

Aydın M, Huwe B (1993) Test of a combined soil moisture/soil heat simulation model on a bare field soil in Southern Turkey. *Zeitschrift fur Pflanzenernahrung und Bodenkunde* 156:441–446.

Aydın M, Kececioglu SF (2010) Sensitivity analysis of evaporation module of E-DiGOR model. *Turkish Journal of Agriculture and Forestry* 34(6):497–507.

Aydın M, Polat V (2010) A computer program for E-DiGOR model. International Soil Science Congress on Management of Natural Resources to Sustain Soil Health and Quality, 26–28 May 2010, Samsun, Turkey.

Aydın M, Yang SL, Kurt N, Yano T (2005) Test of a simple model for estimating evaporation from bare soils in different environments. *Ecological Modelling* 182:91–105.

Aydın M, Yano T, Evrendilek F, Uygur V (2008) Implications of climate change for evaporation from bare soils in a Mediterranean environment. *Environmental Monitoring and Assessment* 140:123–130.

Aydın M, Jung YS, Yang JE, Lee HI, Kim KD (2012a) Simulation of soil hydrological components in Chuncheon over 30 years using E-DiGOR model. *Korean Journal of Soil Science and Fertilizer* 45:484–491.

Aydın M, Vithanage M, Mowjood MIM, Jung YS, Yang JE, Ok YS, Kim SC, Dissanayake CB (2012b) Estimation of evaporation and drainage losses from two bare soils in Sri Lanka. *Eurasian Journal of Soil Science* 1:1–9.

Aydın M, Jung YS, Yang JE, Lee HI (2014) Long-term water balance of a bare soil with slope in Chuncheon, South Korea. *Turkish Journal of Agriculture and Forestry* 38:80–90.

Aydın M, Jung YS, Yang JE, Kim SJ, Kim KD (2015) Sensitivity of soil evaporation and reference evapotranspiration to climatic variables in South Korea. *Turkish Journal of Agriculture and Forestry* 39:652–662.

Barlow KM, Barlow BP, Christy GJ, O'Leary PA, Riffkin JG et al (2015) Nuttall Simulating the impact of extreme heat and frost events on wheat crop production: a review. *Field Crops Research* 171:109–119.

Bellot J, Chirino E (2013) Hydrobal: an eco-hydrological modelling approach for assessing water balances in different vegetation types in semi-arid areas. *Ecological Model* 266:30–41.

Bencze S, Bamberger Z, Janda T, Balla K, Varga B, Bedö Z et al (2014) Physiological response of wheat varieties to elevated atmospheric CO2 and low water supply levels. *Photosynthetica* 52:71–82; at: https://doi.org/10.1007/s11099-014-0008-y.

Bernacchi CJ, Van Loocke A (2015) Terrestrial ecosystems in a changing environment: a dominant role for water. *Annual Review of Plant Biology* 66:599–622; at: https://doi.org/10. 1146/annurev-arplant-043014-114834.

Boogaard HL, Van Diepen CA, Rötter RP, Cabrera JMCA, Van Laar HH (1998) WOFOST 7.1. User's guide for the WOFOST 7.1 crop growth simulation model and WOFOST Control Center 1.5. Technical Document 52, DLO Winand Staring Centre, Wageningen.

Brown RA, Rosenberg NJ (1999) Climate change impacts on the potential productivity of corn and winter wheat in their primary United States growing regions. *Climate Change* 41:73–107.

Brumbelow K, Georgakakos A (2001) An assessment of irrigation needs and crop yield for the United States under potential climate changes. *Journal of Geophysical Research: Atmospheres* 106:27383–27405.

Cure JD, Acock B (1986) Crop responses to carbon dioxide doubling: a literature survey. *Agricultural and Forest Meteorology* 38:127–145.

Dhungana P, Eskridge KM, Weiss A, Baenziger PS (2006) Designing crop technology for a future climate: an example using response surface methodology and the CERES-Wheat model. *Agricultural Systems* 87:63–79.

Eckersten H, Blombaeck K, Kaetterer T, Nyman P (2001) Modelling C, N, water and heat dynamics in winter wheat under climate change in southern Sweden. *Agriculture Ecosystem Environment* 86:221–235.

El Maayar M, Singh B, Andre P, Bryant CR, Thouez JP (1997) The effects of climate change and CO_2 fertilisation on agriculture in Quebec. *Agricultural and Forest Meteorology* 85(3–4):193–208.

Erice G, Sanz-Sáez A, Urdiain A, Araus JL, Irigoyen JJ, Aranjuelo I (2014) Harvest index combined with impaired N availability constrains the responsiveness of durum wheat to elevated CO2 concentration and terminal water stress. *Functional Plant Biology* 41:1138–1147; at: https://doi.org/10.1071/fp14045.

Evrendilek F, Ben-Asher J, Aydın M, Celik I (2005) Spatial and temporal variations in diurnal CO_2 fluxes of different Mediterranean ecosystems in Turkey. *Journal of Environmental Monitoring* 7:151–157.

Evrendilek F, Ben-Asher J, Aydın M (2008) Diurnal photosynthesis, water use efficiency and light use efficiency of wheat under Mediterranean field conditions, Turkey. *Journal of Environmental Biology* 29(3):397–406.

Fader M, Shi S, Bloh W, Bondeau A, Cramer W (2015) Mediterranean agriculture: more efficient irrigation needed to compensate increases in future irrigation water requirements. *Geophysical Research Abstracts* 18(HESSD 12):8459–8504.

Falge E, Reth S, Bruggemann N, Butterbach-Bahl K, Goldberg V, Oltchev A, Schaaf S, Spindler G, Stiller B, Queck R, Kostner B, Bernhofer C (2005) Comparison of surface energy exchange models with eddy flux data in forest and grassland ecosystems of Germany. *Ecological Modelling* 188:174–216.

Flato GM, Boer GJ (2001) Warming asymmetry in climate change simulations. *Geophysical Research Letters* 28:195–198.

Fuhrer J (2003) Agroecosystem responses to combinations of elevated CO_2, ozone, and global climate change. *Agriculture, Ecosystems and Environment* 97:1–20.

Fujihara Y, Tanaka K, Watanabe T, Nagano T, Kojiri T (2008) Assessing the impacts of climate change on the water resources of the Seyhan River Basin in Turkey: use of dynamically downscaled data for hydrologic simulations. *Journal of Hydrology* 353:33–48.

Gao J, Han X, Seneweera S, Li P, Zong YZ, Dong Q et al (2015) Leaf photosynthesis and yield components of mung bean under fully open-air elevated [CO_2]. *Journal of Integrative Agriculture* 14:977–983; at: https://doi.org/10.1016/s2095-3119(14)60941-2.

Giorgi F, Lionello P (2008) Climate change projections for the mediterranean region. *Global Planet Change* 63:90–104.

Godfray HCJ, Beddington JR, Crute IR, Haddad L, Lawrence D, Muir FJ, Pretty J, Robinson S, Thomas MS, Toulmin C (2010) Food security: the challenge of feeding 9 billion people. *Science* 327(5967):812–818; at: https://doi.org/10.1126/science.1185383.

Guerena A, Ruiz-Ramos M, Diaz-Ambrona CH, Conde JR, Minguez MI (2001) Assessment of climate change and agriculture in Spain using climate models. *Agronomy Journal* 93:237–249.

Guo MX, Shia HU, Zhong-Hui L, Su-Xiaa L, Jun X (2017) Impacts of climate change on agricultural water resources and adaptation on the North China Plain. *Advances in Climate Change Research* 8:93–98; at: https://doi.org/10.1016/j.accre.2017.05.007.

Hanks RJ (1992) *Applied Soil Physics: Soil water and temperature applications*, 2nd. New York, Springer.

Hatfield JL, Prueger JH (2015) Temperature extremes: effect on plant growth and development. *Weather and Climate Extremes* 10:4–10; at: https://doi.org/10.1016/j.wace.2015.08.001.

Heinemann AB, Maia A de HN, Dourado-Neto D, Ingram KT, Hoogenboom G (2006) Soybean (Glycine max (L.) Merr.) growth and development response to CO_2 enrichment under different temperature regimes. *European Journal of Agronomy* 24:52–61.

Hillel D (1980) *Applications of soil physics*. New York, Academic Press.

Hu S, Mo XG, Lin ZH (2015) Evaluating the response of yield and evapotranspiration of winter wheat and the adaptation by adjusting crop variety to climate change in Huang-Huai-Hai Plain. *Chinese Journal of Applied Ecology* 26(40):1153–1161 (in Chinese).

Ibrahim B (2014) Climate Change Effects on Agriculture and Water Resources Availability in Syria. In: Albrecht E, Schmidt M, Mißler-Behr M, Spyra S (eds) *Implementing Adaptation Strategies by Legal, Economic and Planning Instruments on Climate Change. Environmental Protection in the European Union*, vol 4. Berlin, Heidelberg, Springer.

IPCC (1995) Climate Change 1994: Radiative Forcing of Climate Change and Evaluation of the IPCC IS92 Emission Scenarios. In Houghton JT, Meira Filho LG, Bruce J, Lee H, Callender BA, Haites E, Harris N, Maskell K (eds) *Climate Change 1995: IPCC Second Assessment Report*. Cambridge, Cambridge University Press.

IPCC (2001a) Climate Change 2001: *The Scientific Basis*. Cambridge, Cambridge University Press.

IPCC (2001b) Climate Change 2001: *Impacts, Adaptation, and Vulnerability*. Cambridge, Cambridge University Press.

IPCC (2007) Summary for Policymakers. In Solomon S, Qin D, Manning M, Chen Z, Marquis M, Averyt KB, Tignor M, Miller HL (eds) Climate change 2007: *The Physical Science Basis. Contribution of Working Group I to the Fourth Assessment Report of the Intergovernmental Panel on Climate Change*, Cambridge and New York: Cambridge University Press.

IPCC (2013) Summary for Policymakers. In Stocker TF, Qin D, Plattner GK, Tignor M, Allen SK, Boschung J, Nauels A, Xia Y, Bex V, Midgley PM (eds) *Climate Change 2013: The Physical Science Basis. Contribution of Working Group I to the Fifth Assessment Report of the*

Intergovernmental Panel on Climate Change, Cambridge, New York, Cambridge University Press.

Izaurralde RC, Rosenberg NJ, Brown RA, Thomson AM (2003) Integrated assessment of Hadley Center (HadCM2) climate-change impacts on agricultural productivity and irrigation water supply in the conterminous United States: part II. Regional agricultural production in 2030 and 2095. *Agricultural and Forest Meteorology* 117:97–122.

Jackson NA, Wallace JS (1999) Soil evaporation measurements in an agroforestry system in Kenya. *Agricultural and Forest Meteorology* 94:203–215.

Jones PG, Thornton PK (2003) The potential impacts of climate change on maize production in Africa and Latin American in 2055. *Global Environmental Change* 13:51–59.

Kapur B (2010) Enhanced CO$_2$ and Global Climate Change Effects on Wheat Yield in Çukurova Region. Ph.D. Thesis, Çukurova University, Adana, Turkey (in Turkish).

Kimura F, Kitoh A, Sumi A, Asanuma J, Yatagai A (2007) Downscaling of the global warming projections to Turkey. *The Final Report of ICCAP*: The Research Project on the Impact of Climate Changes on Agricultural Production System in Arid Areas (ICCAP). Research Institute for Humanity and Nature (RIHN) of Japan, and The Scientific and Technological Research Council of Turkey (TÜBİTAK), ICCAP Publication 10 (ISBN 4-902325-09-8), 21–37.

Kitoh A (2007) Future Climate Projections around Turkey by Global Climate Models. *The Final Report of ICCAP*: The Research Project on the Impact of Climate Changes on Agricultural Production System in Arid Areas (ICCAP). Research Institute for Humanity and Nature (RIHN) of Japan, and The Scientific and Technological Research Council of Turkey (TÜBİTAK), ICCAP Publication 10 (ISBN 4-902325-09-8), 39–42.

Kitoh A, Hosaka M, Adachi Y, Kamiguchi K (2005) Future projections of precipitation characteristics in East Asia simulated by the MRI CGCM2. *Advances in Atmospheric Sciences* 22:467–478.

Krol MS, Bronstert A (2007) Regional integrated modelling of climate change impacts on natural resources and resource usage in semi-arid Northeast Brazil. *Environmental Modelling Software* 22(2):259–268.

Kurt N (2011) Monitoring of Soil Water Budget Using E-DiGOR Model in Olive Producing Area. Ph.D. Thesis (Code: 27), Mustafa Kemal University, Hatay, Turkey (in Turkish).

Lawlor DW, Mitchell RAC (2000) Crop Ecosystem Responses to Climatic Change: Wheat. In: Reddy KR, Hodges HF (eds), *Climate Change and Global Crop Productivity*. New York, CABI Publishing, pp 57–80.

Lionello P, Scarascia L (2017) Linking the Mediterranean regional and the global climate. *EGU General Assembly*, vol 19, EGU2017-3954.

Maytin CE, Acevedo MF, Jaimez R, Anderson R, Harwell MA, Robock A, Azocar A (1995) Potential effects of global climatic change on the phenology and yield of maize in Venezuela. *Climate Change* 29:189–211.

Mearns LO, Rosenzweig C, Goldberg R (1992) Effects of changes in interannual variability on CERES-wheat yields: sensitivity and 2 × CO$_2$ general circulation model studies. *Agricultural and Forest Meteorology* 62:159–189.

Mehtaa VK, Haden VR, Joycea BA, Purkeya DR, Jackson LE (2013) Irrigation demand and supply, given projections of climate and land-use change, in Yolo County, California. *Agricultural Water Management* 117:70–82; at: https://doi.org/10.1016/j.agwat.2012.10.021.

Mellouli HJ, van Wesemael B, Poesen J, Hartmann R (2000) Evaporation losses from bare soils as influenced by cultivation techniques in semi-arid regions. *Agricultural Water Management* 42:355–369.

Nakicenovic N, Swart R (2000) *Special Report on Emissions Scenarios: A Special Report of Working group III of the Intergovernmental Panel on Climate, Change.* Cambridge, Cambridge University Press.

Olesen JE, Bindi M (2002) Consequences of climate change for European agricultural productivity, land use and policy. *European Journal of Agronomy* 16:239–262.

Önder D, Aydın M, Berberoğlu S, Önder S, Yano T (2009a) The use of aridity index to assess implications of climatic change for land cover in Turkey. *Turkish Journal of Agriculture and Forestry* 33:305–314.

Önder D, Aydın M, Önder S (2009b) Estimation of actual soil evaporation using E-DiGOR model in different parts of Turkey. *African Journal of Agricultural Research* 4(5):505–510.

Önol B, Unal YS (2012) Assessment of climate change simulations over climate zones of Turkey. *Regional Environmental Change*; at: https://doi.org/10.1007/s10113-012-0335-0.

Parry ML, Rosenzweig C, Igleasias A, Livermore M, Fischer G (2004) Effects of climate change on global food production under SRES emissions and socio-economic scenarios. *Global Environmental Change* 14:53–67.

Parry MAJ, Reynolds M, Salvucci ME, Raines C, Andralojc PJ, Zhu X, Price GD, Condon AG, Furbank RT (2011) Raising yield potential of wheat. II. Increasing photosynthetic capacity and efficiency. *Journal of Experimental Botany* 62(2):453–467; at: https://doi.org/10.1093/jxb/erq304.

Pazzagli PT, Weiner J, Liu F (2016) Effects of CO_2 elevation and irrigation regimes on leaf gas exchange, plant water relations, and water use efficiency of two tomato cultivars. *Agricultural Water Management* 169:26–33.

Peiris DR, Crawford JW, Grashoff C, Jefferies RA, Porter JR, Marshall B (1996) A simulation study of crop growth and development under climate change. *Agricultural and Forest Meteorology* 79:271–287.

Quevedo DI, Frances F (2007) A conceptual dynamic vegetation-soil model for arid and semiarid zones. *Hydrology and Earth System Sciences Discussions* 4:3469–3499.

Quinn NWT, Brekke LD, Miller NL, Heinzer T, Hidalgo H, Dracup JA (2004) Model integration for assessing future hydroclimate impacts on water resources, agricultural production and environmental quality in the San Joaquin Basin, California. *Environmental Modelling Software* 19:305–316.

Richter GM, Semenov MA (2005) Modelling impacts of climate change on wheat yields in England and Wales: assessing drought risks. *Agricultural Systems* 84:77–97.

Robredo A, Pérez-López U, Miranda-Apodaca J, Lacuesta M, Mena-Petite A, Muñoz-Rueda A (2011) Elevated CO_2 reduces the drought effect on nitrogen metabolism in barley plants during drought and subsequent recovery. *Environmental and Experimental Botany* 71:399–408; at: https://doi.org/10.1016/j.envexpbot.2011.02.011.

Roeckner E, Arpe K, Bengtsson L, Christoph M, Claussen M, Dümenil L, Esch M, Gioretta M, Schlese U, Schulzweida U (1996) *The Atmospheric General Circulation Model ECHAM4: Model Description and Simulation of Present-day Climate* (Report No. 218). Max-Planck Institute for Meteorology (MPI), Hamburg, Germany.

Romano E, Giudici M (2007) Experimental and modeling study of the soil-atmosphere interaction and unsaturated water flow to estimate the recharge of a phreatic aquifer. *Journal of Hydrologic Engineering* 12(6):573–584.

Rosenberg NJ (1992) Adaptation of agriculture to climate change. *Climate Change* 21:385–405.

Rosenzweig C (1985) Potential CO_2-induced climate effects on North American wheat-producing regions. *Climate Change* 7:367–389.

Sanchez-Gomez E, Somot S, Mariotti A (2009) Future changes in the Mediterranean water budget projected by an ensemble of regional climate models. *Geophysical Researcg Letters*; at: https://doi.org/10.1029/2009gl040120.

Sato T, Kimura F, Kitoh A (2007) Projection of global warming onto regional precipitation over Mongolia using a regional climate model. *Journal of Hydrology* 333:144–154.

Scherm H, Sutherst RW, Harrington R, Ingram JSI (2000) Global networking for assessment of impacts of global change on plant pests. *Environmental Pollution* 108:333–341.

Schulze R (2000) Transcending scales of space and time in impact studies of climate and climate change on agrohydrological responses. *Agriculture, Ecosystems and Environment* 82:185–212.

Southworth J, Randolph JC, Habeck M, Doering OC, Pfeifer RA, Rao DG, Johnston JJ (2000) Consequences of future climate change and changing climate variability on maize yields in the midwestern United States. *Agriculture, Ecosystems and Environment* 82:139–158.

Steduto P, Hsiao TC, Raes D, Fereres E (2009) AquaCrop – The FAO crop model to simulate yield response to water: I. Concepts and underlying principles. *Agronomy Journal* 101: 426–437.

Tao F, Yokozawa M, Hayashi Y, Lin E (2003) Terrestrial water cycle and the impact of climate change. *Agriculture, Ecosystems and Environment* 95:203–215.

Tester M, Langridge P (2010) Breeding technologies to increase crop production in a changing world. *Science* 327(5967):818–822; at: https://doi.org/10.1126/science.1183700.

Thomson AM, Izaurralde RC, Rosenberg NJ, He X (2006) Climate change impacts on agriculture and soil carbon sequestration potential in the Huang-Hai Plain of China. *Agriculture, Ecosystems and Environment* 114(2–4):195–209.

Tubiello FN, Ewert F (2002) Simulating the effects of elevated CO$_2$ on crops: approaches and applications for climate change. *European Journal of Agronomy* 18:57–74.

Van Dam JC, Huygen J, Wesseling JG, Feddes RA, Kabat P, van Walsum PEV, Groendijk P, van Diepen CA (1997) Theory of SWAP version 2.0. *Simulation of water flow, solute transport and plant growth in the Soil-Water-Atmosphere-Plant environment.* Technical Document 45, DLO Winand Staring Centre, Report 71, Dept. Water Resources, Agricultural University, Wageningen.

Van der Linden P, Mitchell JFB (2009) ENSEMBLES: *Climate Change and its Impacts: Summary of research and results from the ENSEMBLES project*) European Environment Agency. Office Hadley Centre, FitzRoy Road, Exeter EX1 3 PB, UK.

Van Ittersum MK, Howden SM, Asseng S (2003) Sensitivity of productivity and deep drainage of wheat cropping systems in a Mediterranean environment to changes in CO$_2$, temperature and precipitation. *Agriculture, Ecosystems and Environment* 97:255–273.

Wolf J, Van Diepen CA (1995) Effects of climate change on grain maize yield potential in the European Community. *Climate Change* 29(3):299–331.

Wolf J, Van Oijen M, Kempenaar C (2002) Analysis of the experimental variability in wheat responses to elevated CO$_2$ and temperature. *Agriculture, Ecosystems and Environment* 93:227–247.

Yano T, Koriyama M, Haraguchi T, Aydın M (2005) Prediction of future change of water demand following global warming in the Çukurova region of Turkey. International Conference on Water, Land and Food Security in Arid and Semi-Arid Regions (in CD-ROM), Mediterranean Agronomic Institute Valenzano (Bari), CIHEAM-MAIB, Italy, 6–11 September 2005.

Yano T, Koriyama M, Haraguchi T, Aydın M (2006) Implications of future climate change for irrigation water demand in the Çukurova region, Turkey. The International Workshop for the Research Project on the Impact of Climate Change on Agricultural Production System in Arid Areas (ICCAP), Research Institute for Humanity and Nature (RIHN), Kyoto, Japan, 51–55.

Yano T, Aydın M, Haraguchi T (2007a) Impact of climate change on irrigation demand and crop growth in a Mediterranean environment of Turkey. *Sensors* 7(10):2297–2315.

Yano T, Koriyama M, Haraguchi T, Aydın M (2007b) Simulation of Crop Productivity for Evaluating Climate Change Effects. The Final Report of ICCAP: The Research Project on the Impact of Climate Changes on Agricultural Production System in Arid Areas (ICCAP). Research Institute for Humanity and Nature (RIHN) of Japan, and the Scientific and Technological Research Council of Turkey (TÜBİTAK), ICCAP Publication 10 (ISBN 4-902325-09-8), 181–184.

Yano T, Haraguchi T, Koriyama M, Aydın M (2007c) Prediction of future change of water demand following global warming in the Çukurova region, Turkey. The Final Report of ICCAP: The Research Project on the Impact of Climate Changes on Agricultural Production System in Arid Areas (ICCAP). Research Institute for Humanity and Nature (RIHN) of Japan, and the Scientific and Technological Research Council of Turkey (TÜBİTAK), ICCAP Publication 10 (ISBN 4-902325-09-8), 185–190.

Young KJ, Long SP (2000) Crop ecosystem responses to climatic change: maize and sorghum. In Reddy KR, Hodges HF (eds) *Climate Change and Global Crop Productivity* 107–131. CABI Publication, New York.

Yukimoto S, Noda A, Kitoh A, Sugi M, Kitamura Y, Hosaka M, Shibata K, Maeda S, Uchiyama T (2001) A new Meteorological Research Institute coupled GCM (MRI-CGCM2) – its climate and variability. *Papers in Meteorology and Geophysics* 51:47–88.

Zhang XC, Liu WZ (2005) Simulating potential response of hydrology, soil erosion, and crop productivity to climate change in Changwu tableland region on the Loess Plateau of China. *Agricultural and Forest Meteorology* 131:127–142.

Chapter 13
Enhanced Growth Rate and Reduced Water Demand of Crop Due to Climate Change in the Eastern Mediterranean Region

Jiftah Ben-Asher, Tomohisa Yano, Mehmet Aydın and Axel Garcia y Garcia

Abstract The specific objectives of this study were to: (a) test the reliability of a regional climate model (RCM) as a tool for climate change projection in the Eastern Mediterranean, (b) compare the observed yield variables of maize and wheat in the region with results of two crop models, (c) compare the models DSSAT and SWAP and (d) use DSSAT and SWAP to generate future productivity of wheat and maize under the A2 global warming scenario. Reference evapotranspiration was highly correlated with the models with average $r^2 = 0.98$ and a unit slope. The two models accurately predicted observed dry mass production (DMP) and leaf area index (LAI) of wheat and maize. The correlations strengthen the legitimacy of DSSAT, SWAP and RCM to serve as predicting models for future climate change on a regional scale.

A simulation was carried out to describe the effects of climate change on crop growth and irrigation water requirements for a wheat-maize-wheat cropping sequence. Climate change scenarios were projected using data of three general circulation models (CGCM2, ECHAM4 and MRI) for the period of 1990–2100 and one RCM for the period of 2070–2079. Daily RCM data were consistent with actual meteorological data in the region and therefore were used for computations of

---

The original version of this chapter was revised: Incorrect author name tagging has been corrected. The correction to this chapter is available at https://doi.org/10.1007/978-3-030-01036-2_19

J. Ben-Asher, Professor Emeritus, Ben Gurion University Agroecology Group, The Katif R&D Center, Ministry of Science and Technology, Sedot Negev Regional Council 85200, Israel; e-mail: benasher@bgu.ac.il.

T. Yano, Professor Emeritus, Tottori University, Arid Land Research Center, Tottori, Japan; e-mail: yano@ant.bbiq.jp.

M. Aydın, Retired Professor, Mustafa Kemal University, Department of Soil Science, Antakya, Turkey; e-mail: maydin08@yahoo.com.

A. Garcia y Garcia, Assistant Professor, Department of Agronomy and Plant Genetics, University of Minnesota, Southwest Research and Outreach Center, Lamberton, MN, United States; e-mail: axel@umn.edu.

© Springer Nature Switzerland AG 2019
T. Watanabe et al. (eds.), *Climate Change Impacts on Basin Agro-ecosystems*,
The Anthropocene: Politik—Economics—Society—Science 18,
https://doi.org/10.1007/978-3-030-01036-2_13

269

present and future water balance and crop development. Predictions derived from the models about changes in irrigation and crop growth covered the period of 2070–2079 relative to a baseline period of 1994–2003. The effects of climate change on wheat and maize water requirements and yields were predicted using the detailed crop growth subroutine of the DSSAT (Decision Support System for Agrotechnology Transfer) and SWAP (Soil-Water-Atmosphere-Plant) models. Precipitation was projected to decrease by about 163, 163 and 105 mm during the period of 1990–2100 under the A2 scenario of the CGCM2, ECHAM4 and MRI models respectively (an average of about 1.3 mm/year). The models projected a temperature rise of 4.3, 5.3 and 3.1 °C, by the year 2100. An increase in temperature may result in a higher evaporative demand of the atmosphere under combined doubling CO_2 concentration and temperature rise by about 2 °C for the period of 2070–2079. The temperature rise accelerated crop development and shortened the growing period by a maximum of thirteen days for wheat and nine days for maize during the period 2070–2079. When yield and available water (rain + applied irrigation) were normalised by extension of the growing period with respect to the baselines years, DMP of maize increased by 1–3 ton ha^{-1} and that of wheat by 3–4 ton ha^{-1}. Consequently, water use efficiency (WUE) increased for both crops. It was concluded, therefore, that the effect of increased temperature and doubling CO_2 on agro-productivity may be positive. This positive effect can be explained if elevated temperature meets the optimal level of a crop response to temperature. Effects of elevated CO_2 on crop tolerance to water stress may counteract the expected negative effects of rising temperature. Increased atmospheric CO_2 levels have important physiological effects on crops such as the increase in photosynthetic rate, which is associated with higher yield and WUE, at least for some cereal crops in the Eastern Mediterranean.

Keywords Climate change DSSAT · CSM-CERES-Wheat · CSM-CERES-Maise SWAP · Atmospheric CO_2 enrichment

13.1 Introduction

13.1.1 The Effects of Climate Change on Crop Production

Agriculture will be affected by changes in climate. The increase in greenhouse gases such as carbon dioxide will affect the physiology of crop growth. Increases in temperature will affect the growing seasons, rate of growth and transpiration. In general, temperature increases have been found to reduce yields and quality of many crops, most importantly cereal and feed grains (Adams et al. 1990, 1998).

In the effort to assess the impacts of global warming on agriculture, many models have been used for crop simulations. As many different crop models are available, it is important to compare them in order to assess uncertainties in the predicted crop response to climate change scenarios. Contradicting results and philosophies can be found in the literature but quantification is still needed. Rosenzweig/Hillel (1998) emphasised the negative role of agriculture that contributes to the increase of greenhouse gases in the atmosphere. When forest lands are cleared in order to develop agricultural fields, carbon dioxide in the air increases. Stored soil organic carbon is released when soil is ploughed. According to the literature, higher temperatures result in higher respiration rates, shorter periods of seed formation and, consequently, lower biomass production.

Francesco et al. (2000) in Italy suggested that the combined effects of elevated atmospheric CO_2 and warmer air would decrease crop yields by 10–40%. By contrast, Grant et al. (1999) found increased crop growth under elevated atmospheric CO_2 concentration by 10–20% especially when water availability was limited. These results were in agreement with simulated results by the ECOSYS model. Although the experiment was carried out on wheat in Arizona, USA, the theoretical approach provided a means to evaluate the effects of elevated CO_2 on ecosystem behaviour across a wide range of hydrologic conditions and added some confidence in crop simulation models. Tubiello et al. (2002) conducted an aggregated study over a wide range of US climate conditions using DSSAT (Decision Support System for Agrotechnology Transfer) (Jones et al. 2003) crop simulation models. The climate data in this study were based on two scenarios developed with two General Circulation Models (GCMs). Their results emphasised the importance of locality in predicting the effects of climate change. They found that positive results largely depend on precipitation increases projected by the climate scenarios. Where precipitation was projected to decrease, climate change resulted in 30–40% reductions in grain yield. When compared with actual data, the DSSAT simulations under current climate conditions were agreeable within the range of 20–25% with reported yields across the United States.

Confidence in a crop model can be gained through simulations of observed data. A crop model is a simplification of the real world. However, it is a very complex system, with many components. The model must be tested at the system level; that is, by running the full model and comparing the results with observations (Grant et al. 1999; Tubiello/Ewert 2002; Jones et al. 2003; Eitzinger et al. 2004). Jones et al. (2003) and Dhungana et al. (2006) provided an overview based on hundreds of published studies in which DSSAT simulations were compared with observations. This evaluation included observations of wheat, maize and soybean yields. The correlation between the observed and the simulated data was acceptable for field data with a slope larger than 0.8 and $r^2 > 0.8$. A comparison between DSSAT and the SWAP (Soil Water Atmosphere Plant; van Dam et al. 1997) models was made by Eitzinger et al. (2004). They compared observed and simulated barley and wheat grain yields where prediction of yields indicated that the slope was far from 1 (approximately 0.6 and $r^2 \geq 0.6$ and 0.4 and $r^2 < 0.8$ for DSSAT and SWAP respectively). Both DSSAT and SWAP models integrate soil-water balance and crop growth as a function of weather conditions, soil characteristics, and management practices.

13.1.2 The Role of Climate and Crop Models to Advance Our Understanding of the Effects of Global Warming on Crops Production

In order to provide the world with a clear scientific view on the current state of knowledge in climate change, the WMO (World Meteorological Organization) established the IPCC (Intergovernmental Panel on Climate Change), which compares the output of about 23 climate models (Randall et al. 2007). The models generally agree in predicting that the whole Earth will be somewhat warmer in the future. However, concerning the rate of change, the extent of overall change, and the effects in particular regions of the globe, the predictions of some GCMs differ from the predictions of others. Global warming is usually estimated by GCMs; however, horizontal resolution of the ordinary GCMs is quite low, i.e., grid interval is about 100–300 km and the resolution is still not enough to estimate the climate change on a basin scale (Kimura/Kitoh 2008). Yet the use of GCM projections has been at the core of climate change assessments for agriculture and water resources over the past few decades (Mearns et al. 1992; Izaurralde et al. 2003). Although, with the increase in computer power,GCMs are being improved day by day, the resolution is still not high enough to estimate the climate change in a specific basin, such as the Seyhan river basin in Turkey. To avoid this difficulty, Kimura/Kitoh (2007) applied a Pseudo Global Warming Method to downscale a GCM to a regional climate model (RCM) by gradual nesting of a coarse grid that covers Europe and the Mediterranean region with the interval of 100 km to a second grid that covers the whole of Turkey with a finer grid of 25 km, which in turn covers the Seyhan basin with an even finer grid of 8.3 km. Comparative runs of the current climate with the various climate models were acceptable and are summarised in the Appendix. The assessments of climate change impacts on agricultural production often use outputs of more than one GCM to drive agro-ecosystem models. Therefore, adequate downscaling of GCM projections is necessary to study most impacts of climate change at local to regional scales. Although many aspects of climate change are still uncertain, particularly at the local and regional scales (Christensen et al. 2002), there is a need to make extensive use of regional climate models (RCMs). For example, Kimura/Kitoh (2008) report that although climate change is usually estimated by GCMs, their low horizontal resolution of 100–300 km is a limitation to applications at the basin scale. To overcome this issue, the authors developed a GCM downscaling procedure using RCMs (Kimura/Kitoh 2007). Here, their data were used to estimate the climate change and provide the scenarios of the likely climate change in the Seyhan basin, Turkey.

Although GCMs and the methods of downscaling still have many problems with issuing a reliable projection, to our knowledge the uniqueness of our study is based on (a) the information on the future changes of the Eastern Mediterranean climate (especially the projection of agro-meteorological changes due to global warming) and, (b) the RCM-based projections of the impact of global warming on agricultural productivity of two major crops. So far most projections are based on GCMs, while in this study the resolution is higher because RCM was used on a relatively small Eastern Mediterranean region.

Like the IPCC approach in terms of testing climate models, the Agricultural Model Intercomparison and Improvement Project (AgMIP) was established by several organisations (Rosenzweig et al. 2013) in order to capture the complexity of the multiple impacts of climate change on crop production. AgMIP has compared several models; one of those studies (Zhao et al. 2017) tested the impact of global warming on agricultural production by multi-crop models for wheat, rice, maize, and soybean under a scenario of constant CO_2 concentration and irrigation to avoid covariance of $CO_2 \times$ precipitation. Their study reports negative temperature effects on yield of crops at the global scale; without CO_2 fertilization, each degree-Celsius increase in global mean temperature would, on average, reduce global yields of wheat by 6.0%, rice by 3.2%, maize by 7.4%, and soybean by 3.1%. This methodology has a higher potential than using a single model to assess the effects of global warming on agriculture.

The specific objectives of this study were to: (a) test the reliability of RCM as a tool for climate change projection in the Eastern Mediterranean, (b) compare the observed yield variables of maize and wheat in the region with results of two crop models, (c) compare the models of DSSAT and SWAP, and, (d) use DSSAT and SWAP to generate future productivity of wheat and maize under the A2 global warming scenario.

13.2 Materials and Methods

13.2.1 The Study Region

The study site (Adana) (36°59′ N, 35°18′ E) is located in the Çukurova plain of ca. 38,500 km^2 on the Eastern Mediterranean coast of Turkey (Fig. 13.1).

The region selected for this study is the major growing belt of field crops (cotton, maize, wheat, and soybean) and the permanent crops (citrus, fruit trees, and grapes) in Turkey. A typical Mediterranean climate prevails in the study region with the long-term (1975–2006) mean annual temperature, precipitation and potential evapotranspiration of 19.0 °C, 650 mm, and 1320 mm respectively. The temperature extremes are −6.4 °C in February and 44.0 °C in July. About 87% of the precipitation occurs from mid-autumn to mid-spring (November–May). The region's soils, with different proportions of sand, silt, and clay fractions, are predominantly fine-textured. The soil at the study location is a Vertic Luvisol (Aydın 1994).

13.2.2 Climate Change Scenarios

The climate change data were obtained from the outputs of the Meteorological Research Institute of Japan (MRI) (Yukimoto et al. 2001; Kitoh et al. 2005).

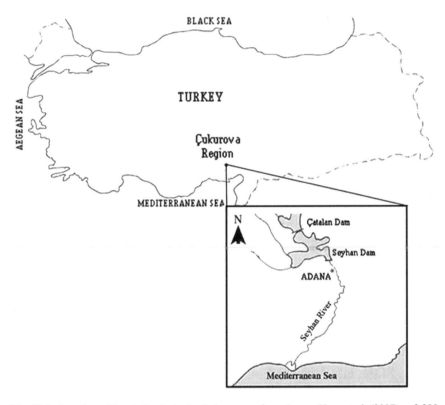

Fig. 13.1 Location of the study site in the Çukurova region. *Source* Yano et al. (2007): p 2,300

The MRI control run simulates the current climate condition, while the MRI global warming run is based on the A2 scenario of the SRES (Special Report on Emission Scenarios). Scenario A2 is one of the emission scenarios with the highest projected CO_2 increase (up to 750–800 ppm) by the end of the 21st century (Nakicenovic/ Swart 2000). The MRI data were computed for Adana from the projected values at the four nearest GCM grid points. We moved from the coarse grid scale of the GCMs to a regional climate model which was obtained by downscaling from the MRI model, bearing in mind that it is unreasonable to expect that the resolution of an ordinary GCM grid (100–300 km) would properly represent climate change in a relatively small basin such as the Seyhan river basin.

The impacts of generated climate data on evapotranspiration (ET) and irrigation water requirements for wheat and second crop maize were simulated using the DSSAT and SWAP models for the periods of 1994–2003 and 2070–2079. The two models can simulate water transport in relation to plant growth at the field scale for the entire growing seasons. The models integrate soil-water balance and crop growth, and require input data for soil, crop-specific coefficients, management practices and weather for calibration and validation in different environments. Daily weather data, soil hydraulic functions and crop management data (e.g., crop calendar, growth information and irrigation) were obtained for the study location.

The daily soil-water balance was calculated during the growing period of the studied crops. The amount of irrigation was calculated for optimal irrigation conditions. When soil moisture was depleted to 100% of the readily available water, the calculated amount of water was added by irrigation to bring the soil back up to field capacity. The available amount of soil water in the root zone was calculated from the water balance. This balance included precipitation, irrigation, losses by runoff, soil evaporation, crop transpiration, and percolation from the root zone. The model simulates evaporation from soil water and plant transpiration separately, thus facilitating a differential crop development and yield under varying temperatures as well as rainfall conditions. Partitioning of potential evapotranspiration into potential rates of soil evaporation and crop transpiration is calculated based on the leaf area index (LAI). Evapotranspiration from grass as a reference crop (ETr) was calculated to represent evaporative demand of the atmosphere.

13.2.3 Crop Growth Models Parameterisation and Validation

The parameterisation and validation of the models was based on experiments with wheat-maize rotation during the years 2003 and 2004. Fertilisation and water application was applied at a local routine. Details are given in Table 13.1.

Wheat was grown with two agronomic variables during the study; one was irrigated wheat and the second was rain-fed wheat. The sowing density of wheat was 500 seeds per m^2 with an 82% germination rate, whereas that of maize was 8.5

Table 13.1 Sowing and harvest dates of the maize-wheat rotation. *Source* The authors

Activity	Crop	Day	Month	Year
Sowing	Maize	10	4	2003
Harvest	Maize	10	9	2003
Sowing	Wheat [for both, irrigated and rain-fed]	17	11	2003
Harvest 1	Irrigated Wheat	25	5	2004
Harvest 2	Rain-fed Wheat	17	5	2004
Sowing	Maize	19	6	2004
Harvest	Maize	9	11	2004
Sowing	Wheat [for both, irrigated and rain fed]	4	3	2004
Harvest 1	Irrigated Wheat	9	6	2004
Harvest 2	Rain-fed Wheat	14	6	2004

Table 13.2 Climate conditions used as input for all DSSAT and SWAP simulations runs. *Source* The authors

Present baseline simulations for calibration and validation			
Crop/ Years	2003 and 2004	1994–2003	1994–2003
Maize and Wheat	Actual measured data	Actual climate data for averaging the baseline year	Simulated RCM data
Future projections			
Crop/ years	2070–2079	2070–2079	2070–2079
Maize and Wheat	Simulated RCM data	Case 1: Only temperature rise	RCM Case 2: Temperature rise + doubling CO_2 $2 \times CO_2$

seeds per m^2 with a 95% germination rate. Other measured and observed parameters were LAI, total dry mass production (DMP), and phenology (Table 13.2).

Crop growth was simulated using DSSAT and the detailed crop growth sub-model of the SWAP, both with a daily time step from sowing to maturity, based on eco-physiological processes that describe daily growth and development in response to environmental factors such as soil and climate, and crop management. For DSSAT the Cropping System Model (CSM)-CERES-Maize (Jones/Kiniry 1986) and -Wheat (Ritchie 1991; Ritchie et al. 1998) models were used. The sub-model of SWAP is a version of WOFOST (WOrld FOod Studies) (Supit et al. 1994; Boogaard et al. 1998), which requires the input of daily climate data as well as information on soil properties and crop-specific growth parameters. In principle, the CERES and WOFOST models simulate the growth, development, and yield of a single plant at a point which is then upscaled to a single farm field, therefore assuming spatially uniform response of crops to limiting production factors. The major processes for crop growth are phenological development, CO_2-assimilation, transpiration, respiration, partitioning of assimilates to the various organs, and dry matter accumulation. The models simulate grain yield from biomass accumulation until anthesis and during grain filling. The two models are particularly suited to quantifying the combined effect of changes in CO_2 and temperature. Both models were parameterised for maize and wheat and calibrated with the crop growth variables, including LAI and biomass. Measurements were taken during the wheat-maize-wheat rotation cycle via two growing seasons for wheat and one growing season for maize. The growth of the crops was simulated in single stands of each crop. Changes in yield were evaluated by comparing the future crop yields to the current yields, as also described by Southworth et al. (2000).

13.2.4   Simulations Analysis

In order to compare the simulated crop development variables with the present experimental results, the following approaches were taken: (a) the graphical course of simulated LAI, DMP and atmospheric reference evapotranspiration were compared to experimental and measured results, (b) the best fit correlation curves (slope intercept and r^2) were reported as text data, (c) the detailed totals of all simulations are presented as Tables 13.B1 and 13.B2 in the appendix, (d) the simulation of DSSAT and SWAP based on the present weather data were averaged to set the baseline conditions. Projected DSSAT and SWAP simulations without doubling CO_2 were averaged to represent the effect of global warming on cereal productivity. Projected DSSAT and SWAP simulations with doubling CO_2 were averaged to represent the combined effect of global warming and CO_2 elevation on the cereals' productivity, (e) normalised values based on the totals (rather than on the daily changes) were calculated by averaging the data simulated by DSSAT and SWAP and are given in Table 13.B1 for maize and 13.B2 for wheat in the appendix. The normalised procedure was as follows:

1. Actual scenarios: Two simulation sources based on the present climate conditions. The two simulations were made by DSSAT and SWAP.
2. Projected scenarios: Climate conditions were projected by RCM for the years 2070–2079 and used for simulations by DSSAT and SWAP for two cases: (i) temperature rise only, and (ii) temperature rise and doubling CO_2 enrichment. The data presented in Table 13.3 are the averages of the projected results of the models.
3. Normalising procedure: Bearing in mind that shorter growing duration of seed formation, consequently may result in lower biomass production we normalised the future productivity with respect to the growing duration of the baseline years. The projected DMP and available water (rain + irrigation) for the years 2070–2079 were multiplied by the ratio between the longer baseline duration and the future shorter growing duration. Water use efficiency (WUE) was then determined by the normalised seasonal DMP divided by normalized seasonal available water.

13.3   Results

13.3.1   Validation of RCM Projection Properties of Air Temperature

In order to study the effect of climate change in the Eastern Mediterranean on crop production a regional downscaling of the GCM models to RCM is necessary. However the method used here for downscaling to RCM needs some validation.

Table 13.3 Predicted averages of actual and A2 scenarios by DSSAT and SWAP are based on detailed calculations in the appendix Tables 13.B1 and 13.B2. *Source* The authors

Averages of scenarios from DSSAT and SWAP	Rain (mm)	ETa (mm)	Irrig (mm)	DMP (t ha^{-1})	Grain (t ha^{-1})	GD (day)	DMP Rate (kg d^{-1} ha^{-1})	DMP* (t ha^{-1})	Available H$_2$O* (mm)	WUE§ (kg m^{-3})
Maize normalised values										
Present actual 1994–2003	22	416	341	27	14	99	273	27	363	7.4
Temperature rise 2070–2079	6	435	372	27	13	95	279	28	394	7.1
Temperature rise 2070–2079 + 2 × CO$_2^{£}$	6	389	327	27	13	90	300	30	366	8.2
Wheat										
Present actual 1994–2003	525	381	47	18	7	173	104	18	572	3.2
Temperature rise 2070–2079	274	355	126	20	7	160	125	22	400	5.5
Temperature rise 2070–2079 + 2 × CO$_2^{£}$	272	377	138	19	7	158	120	21	449	4.7

Normalised maize DMP = daily production × duration from actual sowing to actual maturity (99 days)
Normalised wheat DMP = daily production × duration from actual sowing to actual maturity (173 days)
§WUE = Normalised yield/normalised water availability (see materials and methods section C)
$^{£}$Doublings CO$_2$ marked as 2 × CO$_2$

Variations of the mean annual temperature from 1990 to 2100 in Adana are shown in Fig. 13.2.

The data in Fig. 13.2 were computed for Adana with the projected values of CGCM2, (Canadian Global Coupled Model – CGCM2) (Flato/Boer 2001), ECHAM4 (European Centre for Medium Range Weather Forecasting, parameterised at HAMburg) (Roeckner et al. 1996) and MRI models (Yukimoto et al. 2001; Kitoh et al. 2005). The averaged surface temperature is estimated to increase by 4.3, 5.3 and 3.1 °C by 2100 for the CGCM2, ECHAM4 and MRI models respectively.

The RCM model compares well with the observed temperature and precipitation (Fig. 13.3) and hence can be used reliably as a tool for the projection of future changes in the region.

13.3.2   Reference Evapotranspiration (ETr)

Monthly variations of ETr were calculated by DSSAT and SWAP based on the same data that were generated by RCM and compared to observed data. The calculations used climate data that were generated by the RCM model for the current conditions. Results are shown in Fig. 13.4.

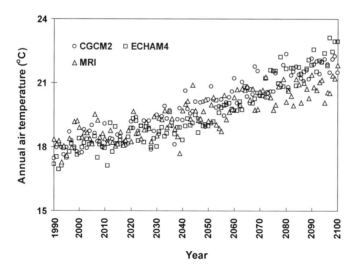

Fig. 13.2 The course of temperature increase predicted by two widely used GCMs and the MRI downscaling for the in Adana region. *Source* Yano et al. (2007): p 2,303

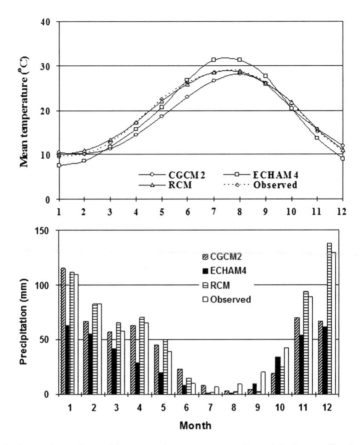

Fig. 13.3 Comparison of monthly mean air temperature and precipitation predicted with the CGCM2, ECHAM4 and RCM models, and observed data for a period of ten years from 1994. *Source* Yano et al. (2007). p 2,304

The total annual ETr (calculated and observed) for the periods of 1994–2003 is about 1.100 ± 76 mm with a daily average of about 3 mm. The correlation between observed and simulated results demonstrates good agreement between ETr calculated by DSSAT and SWAP based on RCM climate data and the ETr calculations based on observed data. The slope of both models with respect to the observed ETr is very close to a unit and the correlation coefficient is $r^2 = 0.96$ for DSSAT and 0.99 for SWAP. This high correlation between observations and predictions based on RCM, DSSAT and SWAP strengthens the legitimacy of RCM to serve as a predicting model for future climate change on a regional scale. It also demonstrates the consistency of DSSAT and SWAP and their ability to mimic the baseline conditions.

Bearing in mind the definition of ETr as a hypothetical Penman-Monteith evaporation (Monteith/Szeicz 1962) from well-watered grass, the ETr for a crop depends on its growth duration and the season. For example, the ETr of maize

Fig. 13.4 Monthly reference evapotranspiration (ETr) calculated by DSSAT and SWAP using, RCM data and compared to observed data for average of a period of 10 years from 1994. *Source* The authors

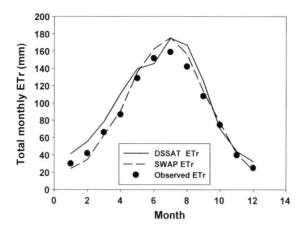

during the baseline growing season was about 750 mm over 100 growing days (ETr = 7.5 mm day^{-1}), such that shortening the growing season by ten days may be associated with 'savings' of 75 mm of water or 10%. On the other hand, the ETr of wheat during the baseline growing season (from November to May) was about 420 mm over 180 growing days (2.3 mm day^{-1}), shortening the growing season by fifteen days, 'saving' about 8% less than the present conditions. In the future, lesser actual evapotranspiration is also expected. This aspect is one of the objectives that will be examined later in the study.

13.3.3 DSSAT and SWAP Simulations and Observed Crop Growth

Calculated and measured LAI and biomass for wheat and maize are compared in Figs. 13.5 and 13.6 respectively.

The calibration of models was based on adjusting the specific-cultivar coefficients of winter wheat and summer maize. This calibration included phenology, DMP and grain yield (Grain). The first step was to calibrate the coefficients related to phenology and LAI and then the coefficients related to DMP characteristics.

The simulated LAI values of maize are in agreement with the measured values. However, for maize, the two model simulations and especially SWAP represented the observation very well qualitatively and quantitatively. LAI estimated by DSSAT was slightly smaller than SWAP and the observed values. Although some discrepancies were recognised, the correlation between the observed and DSSAT simulated data was acceptable with the linear equation: DSSAT = 1.1 * observed data – 0.2 and r^2 = 0.87. The correlation between SWAP and observed data was similar to the correlation with DSSAT, but the correlation coefficient was larger (SWAP = 1.0 * observed data and r^2 = 0.90).

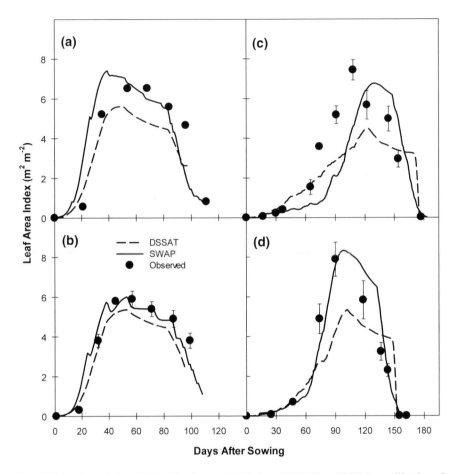

Fig. 13.5 Leaf area index (LAI) calibration and validation of DSSAT and SWAP, **a** calibration of the simulation maize models 2003, **b** validation of maize model 2004, **c** calibration of wheat model, **d** validation of wheat model. *Source* The authors

As can be seen for wheat, in Fig. 13.5c and d, the agreement between DSSAT and the observation was mostly qualitatively acceptable. It agreed with observation on time span because all three tests lasted about 180 days. However, the LAI predicted by DSSAT was only about 5 m^2 m^{-2}, while SWAP and the observed data reached about 8 m^2 m^{-2}. The correlation between SWAP and the observed data was: SWAP = 0.94 * observed data − 0.02 and r^2 = 0.90, while the correlation equation for DSSAT was far from a unit slope in that DSSAT = 0.59 * observed data + 0.05 and r^2 = 0.81. Thus, LAI simulations for maize reflected the measurements sufficiently well. On the other hand, the wheat correlation with measurements was not as good.

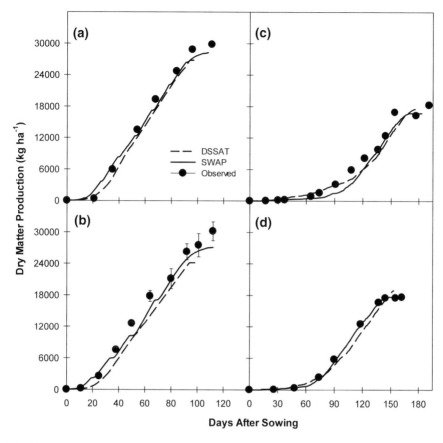

Fig. 13.6 Dry Matter Biomass (DMP) calibration and validation of DSSAT and SWAP, **a** calibration of the simulation maize models 2003, **b** validation of maize model 2004, **c** calibration of wheat model, **d** validation of wheat model. *Source* The authors

In Fig. 13.6 we displayed the DMP of maize and wheat. Crop DMP simulations of maize and wheat with the two models are in good agreement with the observations. In general, the models gave good predictions of crop development and final yields. For wheat the correlation between SWAP and the observed DMP was SWAP = 0.99 * observed DMP – 333 and r^2 = 0.99, and the correlation equation for DSSAT was also very close to a unit slope, in that DSSAT = 0.98 * observed DMP – 111 and r^2 = 0.99. Thus, DMP simulations for wheat reflected the measurements by the two models. It is well known that the DMP is related to LAI and intercepted radiation (Wolf et al. 2002; Asseng et al. 2004); in spite of the weak correlation between the LAI of DSSAT and the observed data, other phenological variables masked it and enable the DSSAT to reasonably predict DMP.

For maize, the prediction was acceptable with the correlation of SWAP versus the observed DMP, which was somewhat smaller than that of wheat. SWAP = 0.90

* observed DMP + 1212 and $r^2 = 0.99$, which is also slightly smaller than that of the correlation with DSSAT: DSSAT = 0.93 * observed DMP – 562 and $r^2 = 0.99$.

This comparison was based on two models and comprised a set of one-year calibration and a second-year validation. With this approach, we achieved greater confidence in our assessment of the impact of future climate change on productivity of field crops.

In summary, three variables with a strong impact on agricultural productivity were thoroughly analysed. They were reference ET, LAI and cumulative dry matter production. The analysis was based on two crops (maize and wheat), two experimental years (2003 and 2004), and two simulation models (DSSAT and SWAP) that were compared to the actual experimental results. All aspects of this analysis strengthen the reliability of the two models as simulation tools for prediction of the future impact of global climate change on agro-productivity, irrigation water requirements, and WUE.

13.3.4   Global Climate Change and Future Agro-productivity

Future cereal crop production will depend not only on climate change effects but also on further developments in technology and crop management (Olesen/Bindi 2002; Jones/Thornton 2003; Evrendilek/Wali 2004). For example, Alexandrov/Hoogenboom (2000) suggested that the sowing dates of spring crops could shift under climate change scenarios in order to reduce yield losses caused by an increase in temperature. Thus, the selection of an earlier sowing date cultivar for maize will probably be the appropriate response to offset the negative effect of a potential increase in temperature.

This study suffers from several limitations, like most studies on climate change (El Maayar et al. 1997; Tubiello/Ewert 2002). For instance, the crop models assume that nutrients are not limited. Also, pests are assumed to pose no limitation on crop growth under the climate scenarios. On the other hand, research and plant breeding studies may mitigate many of the detrimental effects. Due to decadal time studies, limited field experiments, and the large possible range of factorial interactions not tested, the simulated quantitative effects of the model should be interpreted cautiously. However, model results would meet our expectations if farmers could continue to grow the same varieties in the same way in the same locations. Bearing the above in mind, the results of future simulations are summarised in Table 13.3.

In Table 13.3 we analysed the impact of climate change on two crops: summer maize and winter wheat. The average rain during the summer (maize growing season) is predicted to be lower than the actual present conditions under any scenario. The actual ET (ETa) during the growing season is the lowest in the worst scenario of 2070–2079 [2 × CO_2 + °C], with daily ETa = 4.3 mm, which is 6% lower than the current summer ETa. However, due to lesser rainfall, the irrigation demand increased. Due to the shortening of the growing duration by nine days, the projected daily maize production DMP increased from the current 273 to the future 304 kg d^{-1} ha^{-1}. However, the total and daily grain yield decreased from 14 to 13 ton ha^{-1} and from 141 to 137 kg d^{-1} ha^{-1} when the growing duration was 95 days, and increased to 144 kg d^{-1} ha^{-1} when the growing duration was 90 days. It should be noted, however, that the number of growing days was nine days shorter than the growing duration of the actual conditions, such that when the growing season is extended by nine days (by normalising to the 99 growing days) the DMP may enlarge by 1–3 ton ha^{-1} (about 10%). A relevant result is the impact of global warming and doubling CO_2 on water use efficiency of maize as an indicator to other summer cereals. This was calculated as the ratio between normalised DMP and normalised available water, to include the extended irrigation season. This projected impact of climate change (rising temperature and doubling CO_2) resulted in an increase of WUE by 11% from about 7 to 8 kg m^{-3} (DMP per m^{-3} of water). This is a dramatic increase in the average of the ten years from 2070 to 2079.

The analysis of winter crops represented by wheat shows a similar trend of improving agro-productivity under the regional climate change. The annual precipitation, projected to decline from about 530 mm today to 270 mm in the future (2070–2079 under the A2 scenario), is still enough to fulfil crop water requirements when the ETa of 350–375 mm is supplemented by increased irrigation from about 50 mm today to 125–140 mm in the future. According to Table 13.3, the DMP increased by 1–2 ton ha^{-1} (5–10%) and the grain yield production is projected to remain around 7 ton ha^{-1}. Consequently, with less water availability, the WUE can be higher in the future – about 4.7 to 5.2 kg m^{-3} compared to the current situation of 3.2 kg m^{-3} today. As with maize when the growing season is extended by twelve days (normalising to the 170 growing days), the DMP may be enlarged by about 3 ton ha^{-1} ($\sim 17\%$).

One way to explain this positive effect is to attribute it to the effect of CO_2 on photosynthesis and transpiration. Experiments with CO_2 enrichment showed that it could reduce the stomatal conductance and transpiration (Ainsworth et al. 2002; Bernacchi et al. 2007). However, CO_2 influence on crop photosynthesis and transpiration depends on other factors as well. Hence, it is difficult to determine a single crop response to a certain climate-change scenario (Bunce 2004; Bernacchi et al. 2007). Concerning CO_2 effects on reducing crop evapotranspiration, Bunce (2004) pointed out that results obtained in open top chambers might be not be representative due to unnatural ventilation conditions. However, our study, in which both DSSAT and SWAP take into account the relationships between stomatal conductance, photosynthesis and transpiration, shows that combined increase in temperature and CO_2 concentrations could cause significant reductions in irrigation demands during the growing season and consequently an increase in WUE.

13.4   Discussion

The results of the simulations yielded the unexpected effect of projected climate change. Elevated temperature was associated with improved wheat DMP and normalised WUE. Maize DMP was not changed significantly but its grain yield was reduced by 1 ton ha^{-1}. This can result from a positive harmony between the climate conditions and optimal level of a crop response. On the other hand, the effects of elevated CO_2 on crop tolerance to water stress may counteract its negative effect. However, bearing in mind that both models run at a daily time step while stomatal aperture is instantaneous, it is hard to expect complete physiological explanations. Alternatively, the response of the crops to harsher drought conditions was tested in additional runs (not shown here) and revealed that a temperature 2 °C higher did not change the reported observation even with forced stomatal closure.

The resulting maize simulations did not change the previous observations significantly because under integration of stomatal closure, doubling the CO_2 concentration and increasing the temperature by 2 °C yielded the same DMP as the simulation with open stomata. The results for maize were different from the results for wheat possibly because of better harmony between the climate conditions and optimal temperature that is included in the model of maize response.

Unlike the result for maize, the wheat simulations were affected by a 2 °C temperature increase but not by stomatal closure. The addition of 2 °C (also not shown here) reduced the DMP by 20–30% and grain yield by 5–20%. Duration of the growing season lasted some 150 days at high temperatures and about 170 days at the lower temperature simulation with no effect of stomatal closure or aperture.

13.5   Concluding Remarks

The air temperature rise projected an increase in the evaporative demand of the atmosphere, especially in the summer. However, the associated increases in ETa for both wheat and maize are prevented by a decrease in the growing days and LAI due to temperature rise and by a decrease in transpiration also due to the decline in the growing days regardless of the increased evaporative demand (Yano et al. 2005). Experimental findings indicating considerable decreases in the actual ET can be explained by stomatal closure under elevated CO_2 concentration. These explanations have received wide recognition (Bunce 2004; Bernacchi et al. 2007) but the models used here do not include this mechanism. The alternative mechanism (not discussed here) mimicked stomatal closure, thereby showing improved WUE under the projected climate change. Irrigation water requirements resulting from the RCM data for wheat and maize in the future are higher compared to the present due to decreases in precipitation. In the Appendix it can be seen that precipitation is projected to decrease during the studied periods under the A2 scenario of the three GCMs (CGCM2, ECHAM4 and MRI) and the RCM.

All the GCMs projected a temperature increase by 2100 in the Çukurova plain in Turkey, although the models were slightly inconsistent in their predictions of the seasonal cycles of temperature and precipitation (albeit not ETr) for both the control run and scenario simulations. Though the results are clearly not conclusive, they are certainly suggestive of potential impacts of climate change on crop production in the Çukurova plain as a representative case for the Eastern Mediterranean.

In the future, water requirements for the irrigation of wheat in the Mediterranean environment may increase due to the decreasing precipitation. Changes in climatic conditions and CO_2 concentration would result in changes in crop yields. The results suggest that when the effect of the increases in CO_2 concentration and temperature was considered, the RCM data projected an increase in both DMP and grain yield of wheat and maize. It is likely that they would behave tolerantly under climate change.

The detailed crop growth subroutines of both models DSSAT and SWAP are able to simulate a crop-specific growth pattern and its interaction with environmental conditions. The parameterised crop models should be validated under different climatic conditions. Elaborated experimental data sets from wheat and second crop maize under a range of environmental conditions are rare but essential for further studies to refine our findings. The combined effects of the increases in CO_2 concentration and air temperature on crop growth response should also be clarified for the other major crops.

Acknowledgements The research was funded by the project Impact of Climate Change on Agricultural Production in Arid Areas (ICCAP), administered by the Research Institute for Humanity and Nature (RIHN) of Japan, and the Scientific and Technological Research Council of Turkey (TÜBITAK). We are grateful to Drs. M. Koç, M. Ünlü and C. Barutçular for providing crop and meteorological data. The study was partially supported by a grant from the Ministry of Science, Israel, the Bundesministerium für Bildung und Forschung (BMBF), and State and Federal funds allocated under the GLOWA project.

Appendix

Climate Change Scenarios

Based on a range of several current climate models, the mean annual global surface temperature is projected to increase by 1.4 to 5.8 °C over the period of 1990–2100 (IPCC 2001), with changes in the spatial and temporal patterns of precipitation (Southworth et al. 2001; Raisanen 2001).

The objective of this appendix was to support the above study by generating future climate data that can be used in the models of the Çukurova plain using GCMs and a known RCM of this region.

The climate change data were obtained from the outputs of the three GCMs: the second version of the Canadian Global Coupled Model – CGCM2 – (Flato/Boer 2001), the model developed from the atmospheric model of the European Centre for

Medium Range Weather Forecasting, and parameterised at HAMburg—ECHAM4 (Roeckner et al. 1996), and the general circulation model developed at the Meteorological Research Institute of Japan – MRI (Yukimoto et al. 2001; Kitoh et al. 2005). The impacts of climate change based on GCM data were estimated for the A2 scenario in the Special Report on Emission Scenarios (SRES). The A2 scenario describes a very heterogeneous world of high population growth, slow economic development and strong regional cultural identities. Scenario A2 is one of the emission scenarios with the highest projected CO_2 increase (up to 750–800 ppm) by the end of the 21st century (Nakicenovic/Swart 2000). The GCM-based climate change data are available with seven climate models on the IPCC website (http://ipcc-ddc.cru.uea.ac.uk/). Monthly temperature and precipitation values projected by CGCM2 and ECHAM4 were obtained from the IPCC database. The MRI model can be used to explore climate change associated with anthropogenic forcing (Yukimoto et al. 2001); however, its data are not available on the IPCC website. The MRI control run simulates the current climate condition, while the MRI global warming run is performed based on the A2 scenario of the SRES. As was mentioned previously in the materials and methods section, the MRI data were computed for Adana from the projected values at the four nearest grid points. It is unreasonable to expect that a large GCM grid box completely represents climate for any particular point. In order to move from the coarse grid scale of the GCM outputs to the specific location, the following procedure was used: GCM data from the four nearest grid points were used to compute climatic data for the specific site. The actual values were calculated using the inverse distance weighted method between the specific site and the GCM grid points (Schulze 2000).

Monthly mean precipitation predictions for Adana for the future period (2070–2079) based on the CGCM2, ECHAM4 and RCM models are depicted in Fig. 13.A1.

Projected mean annual precipitation would decrease by 133, 56, and 306 mm (equivalent to a 25, 12 and 46% decrease), according to the sums of CGCM2, ECHAM4 and MRI models respectively. The discrepancies between the RCM and GCM results can be attributed to the spatial resolutions of the models (Fig. 13.A1).

Variations in mean annual precipitation for 111 years from 1990 to 2100 in Adana are shown in Fig. 13.A2.

Precipitation is projected to decrease by about 163 mm, 163 mm and 105 mm over the period of 1990–2100 under the A2 scenario of the CGCM2, ECHAM4 and MRI models respectively (Fig. 13.A2).

Decreasing rainfall trends in Turkey have already been observed during the 20th century (Türkeş 1996).

Global Climate Change and Future Agro-productivity

The impact of climate change on two crops – summer maize, which is associated with low rainfall amounts and large irrigation water requirements (Table 13.B1),

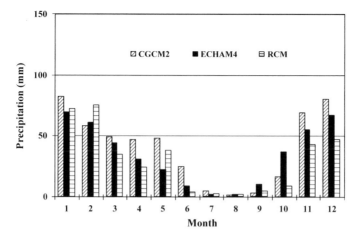

Fig. 13.A1 Comparison of monthly precipitation created from the CGCM2, ECHAM4 and RCM models for a period of ten years from 2070. *Source* Yano et al. (2007): p 2,305

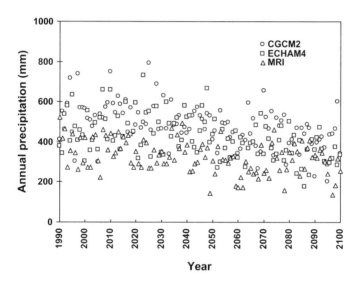

Fig. 13.A2 Variations in annual precipitation from 1990 to 2100 downscaled for Adana with the CGCM2, ECHAM4 and MRI models. *Source* Yano et al. (2007): p 2,303

and winter wheat, with larger rainfall amounts and minimal irrigation water requirement (Table 13.B2) – were simulated using the DSSAT and SWAP models. The water balance components, such as precipitation (Rain), actual crop evapotranspiration (ETa) and irrigation amount (irrig), and crop growth components, such

Table 13.B1 Tabulated results of final values for major production variables of present and future maize. *Source* The authors

Period units	Factor	Rain (mm)	ETa (mm)	Irrig (mm)	DMP (t ha^{-1})	Grain (t ha^{-1})	GD (day)
1994–2003	SWAP measured data	39 ± 20	381 ± 28	291 ± 36	27.2 ± 1.6	14.9 ± 1.3	98 ± 4
1994–2003	DSSAT measured data	35 ± 26	440 ± 24	333 ± 29	24.0 ± 1.7	12.0±1.0	97 ± 3
A2 scenario only rising temperature							
2070–2079	SWAP	6 ± 6	413 ± 14	359 ± 6	28.1 ± 0.9	14.4 ± 0.6	93 ± 1
2070–2079	DSSAT	5 ± 6	457 ± 9	384 ± 18	25.0 ± 0.5	12.0±0.3	97 ± 1
A2 scenario rising temperature and doubling CO_2 (750–800 ppm)							
2070–2079	SWAP + 2 × CO_2	4 ± 5	404 ± 14	355 ± 6	31.0 ± 1.0	14.9 ± 0.6	89 ± 1
2070–2079	DSSAT + 2 × CO_2	4 ± 5	374 ± 7	299 ± 16	23.7 ± 0.5	10.6 ± 0.6	91 ± 2

Table 13.B2 Tabulated results of final values for major production variables of present and future wheat. *Source* The authors

Period units	Factor	Rain (mm)	ETa (mm)	Irrig (mm)	DMP (t ha^{-1})	Grain (t ha^{-1})	GD (day)
1994–2003	SWAP measured data	451 ± 109	354 ± 28	31 ± 50	16.6 ± 1.7	4.5 ± 0.7	165 ± 10
1994–2003	DSSAT measured data	544 ±172	417 ± 15	62 ± 13	19.0 ± 0.9	8.3 ± 0.7	177 ± 5
A2 scenario only rising temperature							
2070–2079	SWAP	256±49	348 ± 24	131 ± 47	17.1 ± 0.9	4.4 ± 0.8	149 ± 5
2070–2079	DSSAT	292±75	367 ± 12	121 ± 18	22.5 ± 0.9	9.7 ± 0.5	170 ± 4
A2 scenario rising temperature and doubling CO_2 (750–800 ppm)							
2070–2079	SWAP + 2 × CO_2	256±49	349 ± 25	141 ± 51	20.0 ± 1.0	5.3 ± 0.9	149 ± 5
2070–2079	DSSAT + 2 × CO_2	288±78	404±12	135 ± 30	18.0 ± 0.7	8.0 ± 0.6	167 ± 4

as dry matter production (DMP), grain yield (Grain) and growing duration, are shown in Tables 13.B1 and 13.B2.

Both tables were obtained from daily runs of the models and used for the average values that are presented in Tables 13.2 and 13.3 in the main part of this article.

Precipitation amount predicted and observed are comparable, especially in the summer, although some considerable deviations are evident.

References

Adams RM, Rosenzweig C, Ritchie J, Peart R, Glyer J, McCarl B, Curry B, Jones J (1990) Global climate change and U.S. agriculture. *Nature* 345:219–224.

Adams RM, Hurd BH, Lenhart S, Leary N (1998) Effects of global climate change on agriculture: an interpretative review. *Climate Research* 11:19–30.

Ainsworth EA, Davey PA, Bernacchi CJ, Dermody OC, Heaton EA, Moore DJ, Morgan PB, Naidu SL, Ra HSY, Zhu XG (2002) A meta analysis of elevated [CO_2] effects on soybean (Glycine max) physiology, growth and yield. *Global Change Biology* 8:695–709.

Alexandrov VA, Hoogenboom G (2000) The impact of climate variability and change on crop yield in Bulgaria. *Agricultural and Forest Meteorology* 104:315–327.

Asseng S, Jamieson PD, Kimball B, Pinter P, Sayre K, Bowden JW, Howden SM (2004) Simulated wheat growth affected by rising temperature, increased water deficit and elevated atmospheric CO_2. *Field Crop Research* 85:85–102.

Aydın M (1994) Hydraulic properties and water balance of a clay soil cropped with cotton. *Irrigation Science* 15:17–23.

Bernacchi CJ, Kimball BA, Quarles DR, Long SP, Ort DR (2007) Decreases in stomatal conductance of soybean under open-air elevation of [CO_2] are closely coupled with decreases in ecosystem evapotranspiration. *Plant Physiology* 143:134–144.

Boogaard HL, van Diepen CA, Rötter RP, Cabrera JMCA, van Laar HH (1998) *User's Guide for the WOFOST 7.1 Crop Growth Simulation Model and WOFOST Control Center 1.5.* DLO-Winand Staring Centre, Wageningen, Technical Document 52.

Bunce JA (2004) Carbon dioxide effects on stomatal responses to the environment and water use by crops under field conditions. *Oecologia* 140:1–10.

Christensen JH, Carter T, Giorgi F (2002) PRUDENCE Employs New Methods to Assess European Climate Change. *EOS* 83:147.

Dhungana P, Eskridge KM, Weiss A, Baenziger PS (2006) Designing crop technology for a future climate: An example using response surface methodology and the CERES-Wheat model. *Agricultural Systems* 87:63–79.

Eitzinger J, Trnka M, Hösch J, Žalud Z, Dubrovský M (2004) Comparison of CERES, WOFOST and SWAP models in simulating soil water content during growing season under different soil conditions. *Ecological Modelling* 171:223–246.

El Maayar M, Singh B, Andre P, Bryant CR, Thouez JP (1997) The effects of climate change and CO_2 fertilisation on agriculture in Quebec. *Agricultural and Forest Meteorology* 85:193–208.

Evrendilek F, Wali MK (2004) Changing global climate: historical carbon and nitrogen budgets and projected responses of Ohio's cropland ecosystems. *Ecosystems* 7(4):381–392

Flato GM, Boer GJ (2001) Warming asymmetry in climate change simulations. *Geophysical Research Letters* 28:195–198.

Francesco NT, Donatelli M, Rosenzweig C, Stockle CO (2000) Effects of climate change and elevated CO_2 on cropping systems: model predictions at two Italian locations. *European Journal of Agronomy* 13:179–189.

Grant RF, Wall GW, Kimball BA, Frumau KFA, Pinter Jr PJ, Hunsakerb DJ, Lamorte RL (1999) Crop water relations under different CO_2 and irrigation: testing of *ecosys* with the free air CO_2 enrichment (FACE) experiment. *Agricultural and Forest Meteorology* 95:27–51.

IPCC (2001) *The Scientific Basis.* Cambridge: Cambridge University Press.

Izaurralde RC, Rosenberg NJ, Brown RA, Thomson AM (2003) Integrated assessment of Hadley Center (HadCM2) climate-change impacts on agricultural productivity and irrigation water supply in the conterminous United States: Part II. Regional agricultural production in 2030 and 2095. *Agricultural and Forest Meteorology* 117:97–122.

Jones JW, Hoogenboom G, Porter CH, Boote KJ, Batchelor WD, Hunt LA, Wilkens PW, Singh U, Gijsman AJ, Ritchie JT (2003) The DSSAT cropping system model. *European Journal of Agronomy* 18:235–265.

Jones CA, Kiniry JR (1986) Ceres-Maize: *A Simulation Model of Maize Growth and Development.* *College Station*: Texas A & M Univ Press.

Jones PG, Thornton PK (2003) The potential impacts of climate change on maize production in Africa and Latin American in 2055. *Global Environmental Change* 13:51–59.

Kimura F, Kitoh A (2007) Downscaling by Pseudo Global Warming Method. In the Final Report of ICCAP, Research Institute for Humanity and Nature and the Scientific and Technological Research Council of Turkey, 43–46.

Kimura F, Kitoh A (2008) Downscaling by Pseudo Global Warning Method. Meteorological Research Institute, Japan Meteorological Agency Tsukuba, Ibaraki 305–8272, Japan.

Kitoh A, Hosaka M, Adachi Y, Kamiguchi K (2005) Future projections of precipitation characteristics in East Asia simulated by the MRI CGCM2. *Advances in Atmospheric Sciences* 22: 467–478.

Mearns LO, Rosenzweig C, Goldberg R (1992) Effects of changes in interannual variability on CERES-wheat yields: sensitivity and $2 \times CO_2$ General Circulation Model studies. *Agricultural and Forest Meteorology* 62:159–189.

Monteith JL, Szeicz G (1962) Radiative temperature in the heat balance of natural surfaces. *Quaterly Journal of the Royal Meteorological Society* 88:496–507.

Nakicenovic N, Swart R (2000) *Special Report on Emissions Scenarios: A Special Report of Working Group III of the Intergovernmental Panel on Climate Change.* New York: Cambridge University Press.

Olesen JE, Bindi M (2002) Consequences of climate change for European agricultural productivity, land use and policy. *European Journal of Agronomy* 16:239–262.

Raisanen J (2001) CO_2-induced climate change in CMIP2 experiments: Quantification of agreement and role of internal variability. *Journal of Climate* 14(9):2088–2104.

Randall DA, Wood RA, Bony S, Colman R, Fichefet T, Fyfe J, Kattsov V, Pitman A, Shukla J, Srinivasan J, Stouffer RJ, Sumi A, Taylor KE (2007) Climate Models and Their Evaluation. In Solomon S, Qin D, Manning M, Chen Z, Marquis M, Averyt KB, Tignor M, Miller HL (eds) *Climate Change: The Physical Science Basis. Contribution of Working Group I to the Fourth Assessment Report of the Intergovernmental Panel on Climate Change.* Cambridge and NewYork: Cambridge University Press.

Ritchie JT (1991) Wheat phasic development. p. 31–54. In Hanks J, Ritchie JT (eds) Modeling plant and soil systems. *Agronomy Monograph* 31, ASA, CSSSA, SSSA, Madison, WI.

Ritchie JT, Singh U, Godwin DC, Bowen WT (1998) Soil Water Balance and Plant Water Stress. In Tsuji GY, Hoogenboom G, Thornton PK (eds) *Understanding Options for Agricultural Production* pp 83–102. Dordrecht: Kluwer Academic Publishers.

Roeckner E, Arpe K, Bengtsson L, Christoph M, Claussen M, Dümenil L, Esch M, Gioretta M, Schlese U, Schulzweida U (1996) *The Atmospheric General Circulation Model ECHAM4: Model Description and Simulation of Present-Day Climate* (Report No. 218). Hamburg: Max-Planck Institute for Meteorology (MPI).

Rosenzweig C, Hillel D (1998) *Climate Change and the Global Harvest: Potential Impacts of the Greenhouse Effect on Agriculture.* Oxford: Oxford University Press.

Rosenzweig C, Jones JW, Hatfield JL, Ruane AC, Boote KJ, Thorburn P, Antle JM, Nelson GC, Porter C, Janssen S, Asseng S, Basso B, Ewert F, Wallach D, Baigorria G, Winter JM (2013) The Agricultural Model Intercomparison and Improvement Project (AgMIP): Protocols and pilot studies. *Agricultural and Forest Meteorology* 170:166–182.

Schulze R (2000) Transcending scales of space and time in impact studies of climate and climate change on agrohydrological responses. *Agriculture, Ecosystems and Environment* 82:185–212.

Southworth J, Randolph JC, Habeck M, Doering OC, Pfeifer RA, Rao DG, Johnston JJ (2000) Consequence of future climate change and changing climate variability on maize yields in the Midwestern United States. *Agriculture, Ecosystems and Environment* 82:139–158.

Supit I, Hooijer AA, van Diepen CA (1994) System Description of the WOFOST 6.0 Crop Simulation Model Implemented in CGMS. In Supit I, Hooijer AA, van Diepen CA (eds) *Volume 1: Theory and Algorithms.* Catno: CL-NA-15956-EN-C. EUR 15956, Office for Official Publications of the European Communities, Luxembourg.

Tubiello FN, Ewert F (2002) Simulating the effects of elevated CO_2 on crops: approaches and applications for climate change. *European Journal of Agronomy* 18:57–74.

Türkeş M (1996) Spatial and temporal analysis of annual rainfall variations in Turkey. *International Journal of Climatology* 16:1057–1076.

van Dam JC, Huygen J, Wesseling JG, Feddes RA, Kabat P, van Walsum PEV, Groendijk P, van Diepen CA (1997) *Theory of SWAP version 2.0. Simulation of Water Flow, Solute Transport and Plant Growth in the Soil-Water-Atmosphere-Plant Environment.* Technical Document 45, DLO Winand Staring Centre, Report 71, Dept. of Water Resources, Agricultural University: Wageningen.

Wolf J, van Oijen M, Kempenaar C (2002) Analysis of the experimental variability in wheat responses to elevated CO_2 and temperature. *Agriculture, Ecosystems and Environment* 93:227–247.

Yano T, Aydın M, Haraguchi T (2007) Impact of climate change on irrigation demand and crop growth in a Mediterranean environment of Turkey. *Sensors* 7(10):2,297–2,315.

Yano T, Koriyama M, Haraguchi T, Aydın M (2005) Prediction of future change of water demand following global warming in the Çukurova region of Turkey. *Proceedings of International Conference on Water, Land and Food Security in Arid and Semi-Arid Regions*, (in CD-ROM), Mediterranean Agronomic Institute Valenzano (Bari), CIHEAM-MAIB, Italy, September 6–11.

Yukimoto S, Noda A, Kitoh A, Sugi M, Kitamura Y, Hosaka M, Shibata K, Maeda S, Uchiyama T (2001) The new Meteorological Research Institute coupled GCM (MRI-CGCM2). *Papers in Meteorology and Geophysics* 51:47–88.

Zhao C, Liu B, Piao S, Wang X, Lobell DB, Huang Y, Huang M, Yao Y, Bassu S, Ciais P, Durand JP, Elliott J, Ewert F, Janssens IA, Li T, Lin E, Liu Q, Martre P, Müller C, Peng S, Peñuelas J, Ruane AC, Wallach D, Wang T, Wu D, Liu Z, Zhu Y, Zhu Z, and Asseng S (2017) Temperature increase reduces global yields of major crops in four independent estimates. *PNAS* 114(35):9326–9331. https://doi.org/10.1073/pnas.1701762114

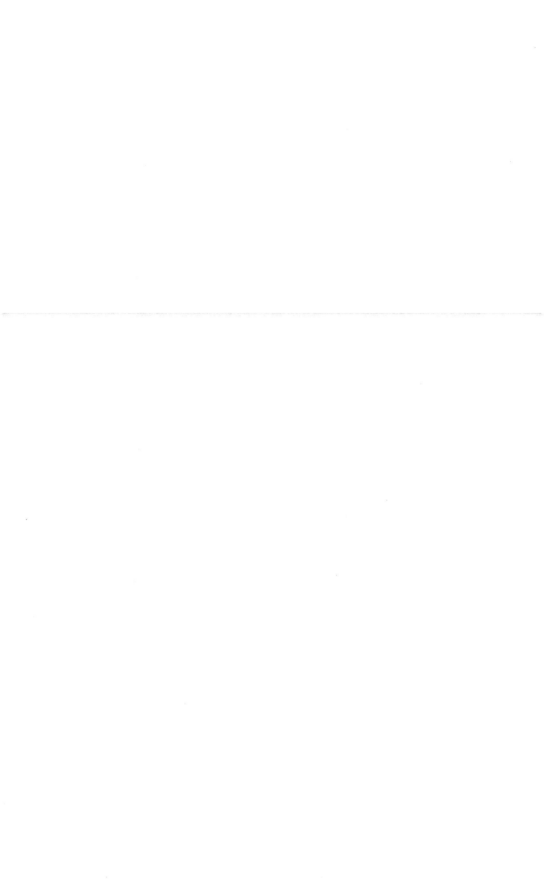

Chapter 14
Sensitivity of Reference Evapotranspiration and Soil Evaporation to Climate Change in the Eastern Mediterranean Region

Mehmet Aydın, Tsugihiro Watanabe and Selim Kapur

Abstract Climate data generated by a regional climate model (RCM) under the A2 scenario were used to quantify the evaporative demand of the atmosphere in the Mediterranean region of Turkey in a baseline period (1994–2003) and the future (2070–2079). The daily reference evapotranspiration and bare soil evaporation were computed using the FAO-56 Penman-Monteith and E-DiGOR models, respectively, for both periods. The sensitivity of Penman-Monteith type equations to the major climatic variables was determined. Based on decadal averages, solar radiation, air temperature, and wind-speed were projected to increase from 16.084 to 16.324 MJ m^{-2} day^{-1}, from 19.3 °C to 20.7 °C, and from 0.75 to 0.77 m s^{-1} respectively, by the period of 2070–2079 compared with the baseline period. By contrast, the relative humidity is expected to decrease from 68.1 to 67.5% (equivalent to a 0.9% reduction). The reference evapotranspiration (*Eto*) and potential soil evaporation (*Ep*) are projected to increase by 92.0 mm year^{-1} and 68.6 mm year^{-1} respectively by the 2070s. Conversely, the actual soil evaporation (*Ea*) is expected to decrease by 49.6 mm year^{-1} by the same period due to the decrease in rainfall and soil wetness.

The reference evapotranspiration was more sensitive to the net radiation in all seasons; followed by the air temperature in the summer months, and by the relative humidity in the winter months under both the present and future conditions. In terms of the sensitivity coefficients, the *Ep* responded better to the changes in climatic variables than the *Eto*. The sensitivity of *Ep* to the key climatic elements varied with the seasons: the net radiation was the most causative variable in the summer, whereas the air temperature and relative humidity were the most influential

M. Aydın, Retired Professor, Mustafa Kemal University, Department of Soil Science and Plant Nutrition, Antakya, Turkey; e-mail: maydin08@yahoo.com.

T. Watanabe, Professor, Kyoto University, Regional Planning Graduate School of Global Environmental Studies, Kyoto, Japan; e-mail: nabe@kais.kyoto-u.ac.jp.

S. Kapur, Retired Professor, Çukurova University, Department of Soil Science and Plant Nutrition, Adana, Turkey; e-mail: kapurs@cu.edu.tr.

© Springer Nature Switzerland AG 2019
T. Watanabe et al. (eds.), *Climate Change Impacts on Basin Agro-ecosystems*,
The Anthropocene: Politik—Economics—Society—Science 18,
https://doi.org/10.1007/978-3-030-01036-2_14

variables in the winter. The mean sensitivity coefficients for air temperature and wind-speed are projected to increase from 0.40 to 0.45 and from 0.15 to 0.19 respectively by the period of 2070–2079. A slight change in the sensitivity coefficient for relative humidity is projected. This could be explained by the expected air temperature rise and the increase in wind-speed, followed by a negligible decrease in humidity, which can increase the *Ep* rate. By contrast, the relative contribution of the net radiation to *Ep* would decrease in the future with a coefficient decreasing from 0.84 to 0.80. This outcome can be attributed to the proportionally higher increases in air temperature and wind-speed in the future, which would reduce the relative portion of the net radiation.

Keywords Climate change · Evapotranspiration · Sensitivity coefficients Soil evaporation

14.1   Introduction

One of the major issues of this century is global warming. Many impacts associated with climate change have become evident during recent decades; therefore, the issue of climate change has been the major interest of many interdisciplinary studies. Climate change means not only changes in globally averaged surface temperature, but also changes in atmospheric circulation, in the size and patterns of natural climate variations, and in local weather (US-NAS and Royal Society 2014). General circulation models (GCMs) are useful tools to project the possible future climate with increasing confidence, although many aspects of climate change are still uncertain. The projections of some GCMs are available on the IPCC website http://ipcc-ddc.cru.uea.ac.uk/; however, models vary in their projections for the magnitude of the expected warming. Nevertheless, all the models agree that the overall net effect of feedbacks is to amplify warming (US-NAS and Royal Society 2014). The adequate downscaling of GCM projections is necessary to study most impacts of climate change at local to regional scales (Yano et al. 2007). Therefore, finer resolution is essential to resolve climatic features and weather extremes and is also a necessity for any quantitative impact study. Thus, there is a need to make extensive use of regional climate models (RCMs) capable of resolving spatial scales as fine as possible http://prudence.dmi.dk.

Surface temperatures and precipitation in most regions vary greatly from the global average because of the geographical location, in particular latitude and continental position (US-NAS and Royal Society 2014). Based on the multi-model projections of the 21st century, precipitation is projected to decrease in the Mediterranean region. By contrast, the models projected a temperature rise by 2100 in the same basin (Fujihara et al. 2008; Giorgi/Lionello 2008; Sanchez-Gomez et al. 2009; IPCC 2013). In arid or semi-arid regions, minor variations in precipitation and temperature have easily induced significant changes in hydrological processes (Huo et al. 2013). In other words, climate change is likely to have a profound effect on the hydrological cycle, namely precipitation, evapotranspiration and soil moisture (Goyal 2004). An accurate estimation of evapotranspiration is essential to better understand land-atmosphere interactions, since it is an important variable in the hydrological cycle for many applications, such as hydrological modelling, water resources management, crop water requirement, and irrigation schedules. Usually, actual crop evapotranspiration can be derived from reference evapotranspiration (*Eto*) by means of appropriate crop and water stress coefficients. Recently, therefore, extensive research efforts have examined the potential impact of climate change on reference evapotranspiration. Briefly, *Eto* is a measure of the evaporative demand of the atmosphere independent of crop type, crop development, and management practices (Estevez et al. 2009; Kwon/Choi 2011; Zhang et al. 2011; Herrnegger et al. 2012; Yang et al. 2012; Huo et al. 2013; Li et al. 2013; Terink et al. 2013; Ebrahimpour et al. 2014).

Similarly, the evaporation from bare soils is the link between the atmosphere and soil surface in the water cycle; and it is a key issue in the hydrological processes, which are linked to water fluxes in the soil (Romano/Giudici 2009; Vanderborght et al. 2010). That is to say, quantification of the evaporation from bare soils is critical in the physics of land-surface processes, because soil evaporation is an important component of the water balance and the surface energy balance (Bittelli et al. 2008; Allen 2011; Xiao et al. 2011). Agam et al. (2004) concluded that during dry seasons, the latent heat flux played a major role in the dissipation of the net radiation. Potential evaporation from bare soil (hereafter, potential soil evaporation) is similar to evaporation from open-water surface, and it is independent of the soil water content. However, under natural conditions, the soil surface is usually not at, or near saturation. Therefore, the actual evaporation from the bare soil (in brief, actual soil evaporation) is largely dependent on the soil water content, in addition to meteorological conditions (Aydın et al. 2015). In this context, it is essential to know the impacts of climate change on the components of soil-water balance and to use the information for developing adaptations in water use and soil-water management. However, these impacts have not been conclusively determined.

On the other hand, sensitivity analysis is important in understanding the impacts of climatic variables on evaporation or evapotranspiration. Several leading studies have assessed the parameter of sensitivity for vegetation or open-water surfaces (McCuen 1974; Saxton 1975; Coleman/DeCoursey 1976; Beven 1979). Although many studies to determine the effects of climatic variables on evapotranspiration have been conducted for sensitivity analysis of the Penman-Monteith equation (Beven 1979; Goyal 2004; Gong et al. 2006; Estevez et al. 2009; Kwon/Choi 2011; Huo et al. 2013), studies on the sensitivity of the same model for potential soil evaporation are rare in the literature (Aydın/Keçecioglu 2010; Aydın et al. 2015). Although many aspects of climate change are still unpredictable, particularly at site scale and regional level (Zhang/Liu 2005), numerous studies have been carried out on the regional and seasonal behaviour of the sensitivity of reference evapotranspiration to climate change in recent years (Rana/Katerji 1998; Hupet/Vanclooster 2001; Goyal 2004; Huo et al. 2013; Xie/Zhu 2013; Tabari/Hosseinzadeh-Talaee 2014; Xing et al. 2014; Xu et al. 2014). However, to our knowledge, there is no (or very limited) data available on the sensitivity of evaporation and evapotranspiration to climate change in the Mediterranean region of Turkey. In the present chapter, therefore, an attempt has been made to study the sensitivity of reference evapotranspiration and soil evaporation to the major climatic variables such as radiation, air temperature, relative humidity, and wind-speed under the current and projected future conditions.

14.2 Materials and Methods

14.2.1 Study Area and Climate Data

The study station (36°59′N, 35°18′E) is located in the Adana province on the Eastern Mediterranean coast of Turkey. The soils in the region are predominantly fine-textured. A typical Mediterranean climate prevails in the study area, with mild rainy winters and hot dry summers. The long-term mean annual temperature at the site is 18.5 °C. The temperature extremes are −6.4 °C in February and 44.0 °C in July. The mean annual precipitation is about 700 mm, and the mean relative humidity is 66%. About 87% of precipitation occurs during the winter (November–May) (http://meteor.gov.tr/). The observed climate data for the study area were obtained from the Turkish State Meteorological Service (DMI). The mean monthly climatic data prevailing in the evaporative demand of the atmosphere for the period of 1994–2003 are given in Fig. 14.1. The mean monthly air temperature was calculated by adding up the daily mean temperatures for each day of a month and then dividing that sum by the number of days in the month. Solar radiation and the mean air temperature were higher, but precipitation was lower during the summer than the winter. A seasonal variation of relative humidity was observed, and the maximum and minimum wind-speed (2 m height) occurred in February and August respectively.

The climate data derived by a RCM have been used in the present chapter, where the RCM projections were based on a newly suggested approach (Pseudo Warming) using re-analysis data (Kimura/Kitoh 2007). The forcing data for the boundary condition of the RCM are given by two GCMs (MRI-CGCM2 and CCSR/NIES-CGCM) for the A2 scenario in the SRES (Yukimoto et al. 2001; Kitoh et al. 2005). Climate change scenarios were created by superimposing projected anomalies on observed climate data of the baseline period (1994–2003). Temperature changes have been added as absolute changes to the baseline series; the other climate parameters have been adapted according to their relative changes (Tao et al. 2003; Dhungana et al. 2006; Yano et al. 2007). The future (2070–2079)

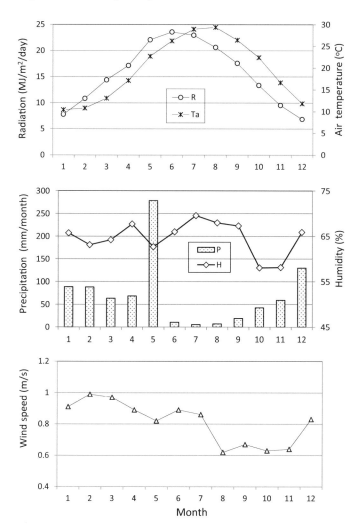

Fig. 14.1 Mean monthly-observed climatic data of the study area for the baseline period (1994–2003) [R: Radiation, Ta: Air temperature, P: Precipitation, H: Humidity]. *Source* The authors

climate changes have been applied to the available baseline climate series in a straightforward way by using the daily data for the baseline period.

14.2.2   Model Descriptions

Evaporation from bare soils can be estimated using the E-DiGOR model, which incorporates the quantification of runoff, drainage, actual soil evaporation, and soil water storage (Aydın 2008; Aydın et al. 2014). Potential soil evaporation is commonly computed using the classic Penman-Monteith equation, with a surface resistance of zero (Allen et al. 1994; Wallace et al. 1999; Aydın et al. 2005). The Penman-Monteith equation can be simplified to estimate the potential evaporation from a bare soil surface for a 24-hour time step by assuming a standardised height for wind-speed, temperature, and humidity measurements at 2.0 m (Aydın et al. 2015):

$$E_p = \frac{\Delta(R_n - G_s) + 0.1248\rho c_p u_2 e_s\left(1 - \frac{H}{100}\right)}{\lambda(\Delta + \gamma)} \tag{14.1}$$

where E_p is the potential soil evaporation (mm day^{-1}), Δ is the slope of saturated vapour pressure-temperature curve (kPa °C^{-1}), R_n is the net radiation (MJ m^{-2} day^{-1}), G_s is the soil heat flux (MJ m^{-2} day^{-1}), ρ is the air density (kg m^{-3}), c_p is the specific heat of air (kJ kg^{-1} °C^{-1} = 1.013), u_2 is the wind-speed (m s^{-1}) at 2.0 m height, e_s is the saturation vapour pressure (kPa), H is relative humidity (%), λ is the latent heat of vaporisation (MJ kg^{-1}), and γ is the psychrometric constant (kPa °C^{-1}).

Then, the actual soil evaporation can be calculated by the Aydın equation (Aydın et al. 2005; Bellot/Chirino 2013):

$$E_a = \frac{Log|\psi| - Log|\psi_{ad}|}{Log|\psi_{tp}| - Log|\psi_{ad}|} E_p$$
If $|\psi| \le |\psi_{tp}|$, then $E_a = E_p$ or $E_a/E_p = 1$ $\tag{14.2}$
For $|\psi| \ge |\psi_{ad}|$, $E_a = 0$
Note that $E_p \ge 0$

where E_a is the actual soil evaporation (mm day^{-1}), $|\psi_{tp}|$ is the absolute value of soil water potential (matric potential) at which actual evaporation starts to drop below potential one (cm of water), $|\psi_{ad}|$ is the absolute value of soil water potential at air-dryness (cm), which can be defined as the water potential of the soil dried to an air-dry state (Aydın et al. 2015), and $|\psi|$ is the absolute value of the soil water potential at the surface layer (cm).

The soil water potential at the top surface layer is estimated using the Aydın and Uygur equation (Aydın et al. 2008):

$$\psi = -\left[\frac{(1/\alpha)\left(10\sum E_p\right)^3}{2(\theta_{fc} - \theta_{ad})(D_{av}t/\pi)^{1/2}}\right] \tag{14.3}$$

where ψ is the soil water potential (cm of water) at the top surface layer, α is a soil specific parameter (cm) related to flow path tortuosity in the soil, $\sum E_p$ is the cumulative potential soil evaporation (cm), θ_{fc} and θ_{ad} are the volumetric water content (cm^3 cm^{-3}) at field capacity and air-dryness respectively, D_{av} is the average hydraulic diffusivity (cm^2 day^{-1}) determined experimentally, t is the time (day), and π is 3.1416.

The minimum water potential at the dry soil surface can be derived from the Kelvin equation (Brown/Oosterhuis 1992; Aydın et al. 2005; Aydın 2008):

$$\psi_{ad} = \frac{R_g T}{mg} \ln H_r \tag{14.4}$$

where ψ_{ad} is the water potential for air-dry conditions (cm of water), T is the absolute temperature (K), g is the acceleration due to gravity (981 cm s^{-2}), m is the molecular weight of water (0.01802 kg mol^{-1}), H_r is the relative humidity of the air (fraction), and R_g is the universal gas constant (8.3143 $\times$ 10^4 kg cm^2 s^{-2} mol^{-1} K^{-1}).

The Penman-Monteith method, considering aerodynamic resistance and surface resistance, has been successfully used to calculate evapotranspiration from different land covers (Gao et al. 2012). The FAO-56 Penman-Monteith (hereafter, FAO56-PM) equation (Allen et al. 1998) may be written as (Aydın et al. 2015):

$$Et_O = \frac{0.408\Delta(R_n - G_s) + \gamma \frac{900}{T_a + 273} u_2 e_s \left(1 - \frac{H}{100}\right)}{\Delta + \gamma(1 + 0.34u_2)} \tag{14.5}$$

where Et_O is the grass reference evapotranspiration (mm day^{-1}), and T_a is the mean daily air temperature (°C). Since the magnitude of daily soil heat flux beneath the grass reference surface is relatively small, G_s can be ignored (Allen et al. 1998).

14.2.3 Sensitivity Analysis

There are different ways of computing sensitivity coefficients for climate variables (Goyal 2004; Gong et al. 2006; Irmak et al. 2006; Estevez et al. 2009; Huo et al. 2013). In this study, the non-dimensional relative sensitivity coefficients were calculated following McCuen (1974) and Beven (1979):

$$S_i = \frac{\partial O}{\partial V_i} \cdot \frac{V_i}{O} \tag{14.6}$$

here S_i represents the fraction of change in variable V_i, transmitted to the change in output O.

Based on the quotient rule for derivates and Eq. (14.6), the derived formulas for the relative sensitivity coefficients (S) of the variables in Eq. (14.1) are as follows (Aydın et al. 2015):

$$S_{R_n} = \frac{\Delta R_n}{\Delta(R_n - G_s) + 0.1248\rho c_p u_2 e_s\left(1 - \frac{H}{100}\right)} \tag{14.7}$$

$$S_H = \frac{-H/100}{\frac{\Delta(R_n - G_s)}{0.1248\rho c_p u_2 e_s} - \frac{H}{100} + 1} \tag{14.8}$$

$$S_{u_2} = \frac{1}{\frac{\Delta(R_n - G_s)}{0.1248\rho c_p u_2 e_s\left(1 - \frac{H}{100}\right)} + 1} \tag{14.9}$$

The sensitivity function related to the air temperature (T_a) can be approximated by (see: McCuen 1974):

$$\frac{dE_p}{dT_a} = \Delta\frac{dE_p}{d\delta} \tag{14.10}$$

Here, δ is the vapour pressure deficit of the air (kPa). S_{T_a} can found as:

$$S_{T_a} = \frac{\Delta T_a}{\frac{\Delta(R_n - G_s)}{0.1248\rho c_p u_2} + e_s\left(1 - \frac{H}{100}\right)} \tag{14.11}$$

Aydın et al. (2015) reported that the contribution of the soil heat flux (G_s) to daily evaporation was negligible, and G_s could not be considered in the sensitivity analysis.

Similarly, the derivative approach can be used to calculate the relative sensitivity coefficients ($^*S$) of the variables in Eq. (14.5):

$$^*S_{R_n} = \frac{0.408\Delta R_n}{0.408\Delta R_n + \gamma\frac{900}{T_a + 273}u_2 e_s\left(1 - \frac{H}{100}\right)} \tag{14.12}$$

$$^*S_H = \frac{-H/100}{\frac{0.408\Delta R_n}{\gamma\frac{900}{T_a + 273}u_2 e_s} - \frac{H}{100} + 1} \tag{14.13}$$

$$^*S_{u_2} = \frac{1}{\frac{0.408\Delta R_n}{\gamma\frac{900}{T_a + 273}u_2 e_s\left(1 - \frac{H}{100}\right)} + 1} - \frac{0.34u_2\gamma}{\Delta + \gamma(1 + 0.34u_2)} \tag{14.14}$$

$$^*S_{T_a} = \frac{T_a\frac{df(T_a)}{dT_a}}{0.408\Delta R_n + \gamma\frac{900}{T_a + 273}u_2 e_s\left(1 - \frac{H}{100}\right)} - \frac{T_a\frac{dg(T_a)}{dT_a}}{\Delta + \gamma(1 + 0.34u_2)} \tag{14.15}$$

where $\frac{df(T_a)}{dT_a}$ and $\frac{dg(T_a)}{dT_a}$ are the derivatives of the numerator and denominator of Eq. (14.5) with respect to air temperature (T_a) respectively, and are given below in full (Aydın et al. 2015):

$$\frac{df(T_a)}{dT_a} = \frac{1021.29R_n\left(4098.17e^{\frac{17.27T_a}{T_a+237.3}} - 2(T_a+237.3)e^{\frac{17.27T_a}{T_a+237.3}}\right)}{(T_a+237.3)^4}$$
$$+ \frac{549.72\gamma(1-\frac{H}{100})u_2\left(\frac{4098.17(T_a+273)e^{\frac{17.27T_a}{T_a+237.3}}}{(T_a+237.3)^2} - e^{\frac{17.27T_a}{T_a+237.3}}\right)}{(T_a+273)^2}$$

(14.16)

$$\frac{dg(T_a)}{dT_a} = \frac{2503.16\left(4098.17e^{\frac{17.27T_a}{T_a+237.3}} - 2(T_a+237.3)e^{\frac{17.27T_a}{T_a+237.3}}\right)}{(T_a+237.3)^4}$$

(14.17)

The non-dimensional relative sensitivity coefficients, as defined by Eqs. (14.7)–(14.17), were calculated on a daily basis for net radiation, air temperature, relative humidity, and wind-speed, using climatic data of the baseline (1994–2003) and future (2070–2079) periods. Monthly average sensitivity coefficients were obtained, by averaging daily values. Changes in stomatal resistance to enhanced CO_2 concentration have not been considered in the sensitivity analysis.

In order to simplify the outline of main processes, particularly for readers, and to better understand the frame of our discussion in this chapter, a flow-chart is

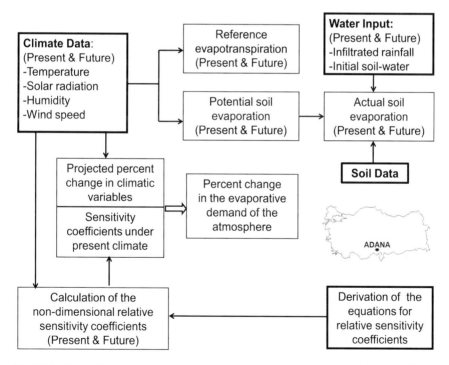

Fig. 14.2 The outline of the main processes and target elements under the Adana conditions. *Source* The authors

depicted in Fig. 14.2. It is a very well-known fact that the reference evapotranspiration (Et_O) is based on a hypothetical grass surface, where an actual land-surface and soil parameter is not used. Similarly, the potential soil evaporation (E_p) is based on a hypothetical uniform soil structure; and it is dependent on the climatic variables except albedo and soil heat flux, which are the only non-climatic parameters. This means that both Et_O and E_p can serve the purpose of comparison of the climatic conditions, as well as the sensitivity to the climatic variables. However, the actual soil evaporation (E_a) is a soil controlled process in addition to climatic conditions.

14.3 Results and Discussion

Daily climatic data necessary for computing the evaporative demand of the atmosphere were downscaled by the RCM. As mentioned in Sect. 14.2.1, the RCM projections were compared to the ten-year observed data of the baseline period (1994–2003). The created data for the baseline period and projected changes in their magnitudes in the future (2070–2079) are given in Fig. 14.3. Based on decadal averages, solar radiation, air temperature, and wind-speed are projected to increase from 16.084 to 16.324 MJ m^{-2} day^{-1}, from 19.3 to 20.7 °C, and from 0.75 to 0.77 m s^{-1} (equivalent to 1.5%, 7.3%, and 2.7%) respectively by the period 2070–2079. By contrast, the mean annual precipitation and relative humidity are expected to decrease from 658.9 to 356.2 mm and from 68.1 to 67.5% (equivalent to 45.9% and 0.9% reduction) respectively by the 2070s when compared with the baseline period. This reveals that the earlier version of the RCM with a grid distance of 25 km had projected a drastic decrease in precipitation which consequently may not be reliable.

The computations of the daily reference evapotranspiration (*Eto*), and potential (*Ep*) and actual (*Ea*) soil evaporations were made for each of the years during the decades of 1994–2003 and 2070–2079 using the data set of daily solar radiation, air temperature, relative humidity and wind-speed of each period. The graphical illustration of the daily data for the entire periods is not presented here since it would have required a lot of space. Instead, the magnitudes of the annual *Eto*, *Ep*, and *Ea* under the current and future climatic conditions are compared in Fig. 14.4. Both *Eto* and *Ep* are projected to increase in the future, in response to the changes in climatic variables such as solar radiation, air temperature, relative humidity, and wind-speed by the 2070s, as predicted by the RCM. Both potential rates (*Eto* and *Ep*) represent the evaporative demand of the atmosphere (Aydın et al. 2008). By contrast, the annual *Ea* is projected to decrease by the period of 2070–2079 compared with the period of 1994–2003 due to lower precipitation and consequently increased soil dryness in the future. The atmospheric evaporative demand and actual soil evaporation denoted noticeable inter-annual fluctuations.

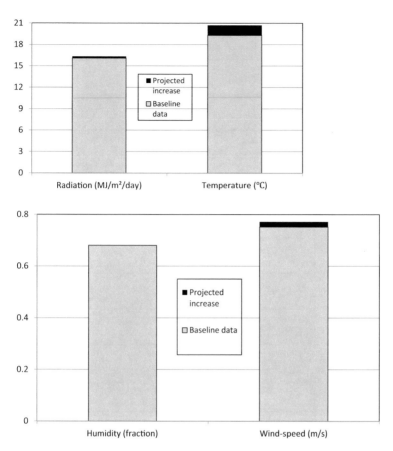

Fig. 14.3 The created climatic data for the baseline period (1994–2003) and projected changes in their magnitudes in the future (2070–2079), as the average over ten years. *Source* The authors

A comparison of the mean monthly *Eto*, *Ep*, and *Ea* for the periods of 1994–2003 and 2070–2079 is given in Fig. 14.5, along with mean annual values. The seasonal variations in each component revealed similar patterns in the baseline period and the future. The rates of *Eto* were greater than those of *Ep*. Although both *Eto* and *Ep* are used to determine the evaporative power of the atmosphere, the differences in their magnitudes can be attributed to the influences of the surface types. In other words, the dissimilarities in the aerodynamic resistance of the above surfaces, the surface resistance, radiation reflection (albedo), heat storage and transfer of the media, and leaf area index etc., may produce obvious differences in water-loss from the bare soils and from the soils beneath a canopy. Both *Eto* and *Ep* were higher during the summer months than the winter months under the present and future climate. By contrast, *Ea* rates were very low in the summer and high in the spring months, depending on soil wetness. That is to say, on the one hand, the size of rainfall events may play a key role in the magnitude of the actual soil

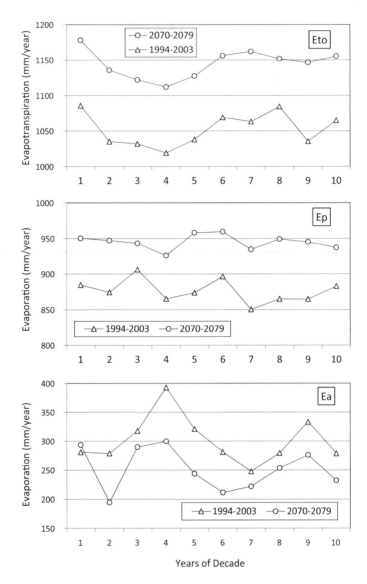

Fig. 14.4 Trends in annual reference evapotranspiration (*Eto*), and potential (*Ep*) and actual (*Ea*) soil evaporations under the climatic conditions of baseline and future decades. *Source* The authors

evaporation; but on the other hand, the variations in the actual soil evaporation may reflect the effects of the rainfall pattern. Unlike *Ep*, the *Ea* is largely influenced by soil wetness. In other words, the magnitude of *Ea* is strongly related to the temporal rainfall pattern, because the contribution of rainfall to the soil water content is considerably dependent on the size of the rainfall events (Aydın et al. 2015). The mean annual *Eto* and *Ep* are projected to increase by 92.0 mm and 68.6 mm

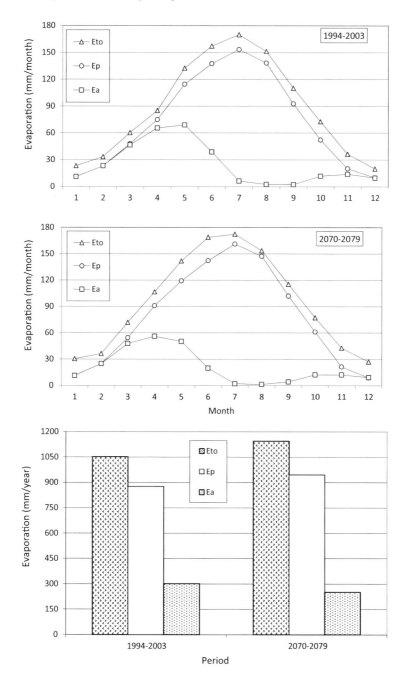

Fig. 14.5 Changes in mean monthly and ten-year average reference evapotranspiration (*Eto*), and potential (*Ep*) and actual (*Ea*) soil evaporations under the climatic conditions of baseline and future periods. *Source* Modified from Aydın et al. (2008: 128)

respectively in the future. By contrast, the mean annual *Ea* is projected to decrease by 49.6 mm, as also reported by Aydın et al. (2008) together with the high discrepancy observed between the *Ep* and *Ea* during the dry summer. Under wet conditions (i.e., winter months), however, the *Ea* rates were very close to the *Ep* rates.

The sensitivity coefficients of the major climatic variables are shown in Figs. 14.6 and 14.7. For example, a sensitivity coefficient of 0.4 for a climatic variable would suggest that a 20% increase of that variable, while the other variables are held constant, may increase a dependent variable (*Eto* or *Ep*) by 8%. Negative coefficients would indicate that a reduction in a dependent variable would result from an increase in that climatic variable. In evaporation studies, the sensitivity coefficients may vary day-by-day, being dependent on the current values of all the independent variables, and the value of the dependent variable (Beven 1979).

Reference evapotranspiration was more sensitive to the net radiation in all seasons, followed by the air temperature in the summer months, and by the relative humidity in the winter months under both the present and future conditions. The response of *Eto* to the same climatic variables revealed no considerable difference between the present and future, based on decadal averages (Fig. 14.6). The sensitivity of potential soil evaporation to the key climatic elements varied with seasons: the net radiation was the most causative variable in the summer, while the air temperature and relative humidity were the most influential variables in the winter (Fig. 14.7). The mean sensitivity coefficient related to air temperature is projected to increase from 0.40 to 0.45 by the period of 2070–2079. Similarly, the mean sensitivity coefficient for wind-speed is expected to increase from 0.15 to 0.19 by the 2070s. A slight change in the sensitivity coefficient for relative humidity is also noted. This could be explained by the expected air temperature rise and the increase in wind-speed, followed by a negligible decrease in humidity, which can increase the *Ep* rate. By contrast, the relative contribution of the net radiation to *Ep* would decrease in the future with a coefficient reducing from 0.84 to 0.80 (Fig. 14.7). This pattern can be attributed to the proportionally higher increases in the air temperature and wind-speed in the future, which would reduce the relative portion of the net radiation. The results for *Ep* are not directly comparable to those of the previous works, due to the differences in the analysed equations, whereas different approaches have been used in the literature to conduct the sensitivity analyses and to derive and define the coefficients (Estevez et al. 2009; Ambas/Baltas 2012). In spite of this fact, the order of sensitivity coefficients for major climatic variables transmitted to *Ep* in this study, is similar to that reported by Aydın et al. (2015) using climate data of seventeen meteorological stations of South Korea. Bearing in mind that our results are based on the Penman-Monteith type equations, in all decades the wind-speed was the least sensitive variable in both *Eto* and *Ep*. Gong et al. (2006), analysing the sensitivity of the FAO56-PM model, also found lower sensitivity coefficients for wind-speed. In the present study, based on the mean decadal coefficients, both *Eto* and *Ep* were more sensitive to net radiation, followed by air temperature, relative humidity and wind-speed. Contradictory results can be found in the literature, as is the case in the studies conducted by Huo et al. (2013) and Li

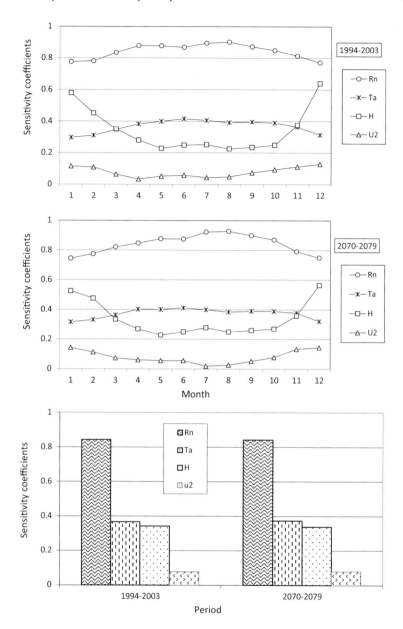

Fig. 14.6 The sensitivity coefficients for net radiation (Rn), air temperature (Ta), relative humidity (H), and wind-speed (u₂), transmitted to *Eto* in baseline and future periods (the coefficients for H were multiplied by −1, to facilitate visual comparison). *Source* The authors

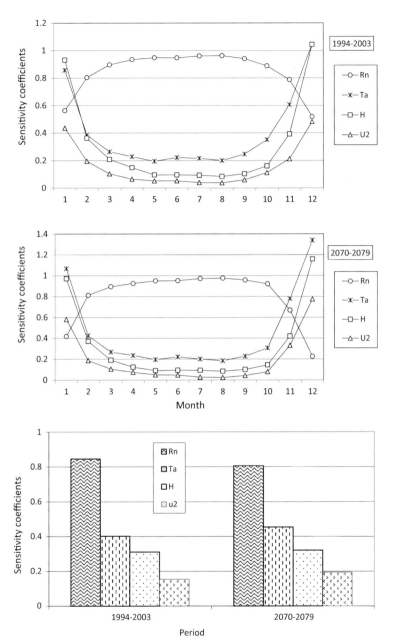

Fig. 14.7 The sensitivity coefficients for net radiation (Rn), air temperature (Ta), relative humidity (H), and wind-speed (u_2), transmitted to *Ep* in baseline and future periods (the coefficients for H were multiplied by -1, to facilitate visual comparison). *Source* The authors

et al. (2013), who report that in some regions of China the wind-speed was the most sensitive meteorological variable for *Eto*, followed by relative humidity, temperature, and radiation. In the same country, Xing et al. (2014) indicated that radiation was mainly responsible for the *Eto* change in the southern and eastern basin of the Haihe River, whereas relative humidity and wind-speed were the leading factors in the eastern coastal and northern area. Kwon/Choi (2011) found that in Korea, vapour pressure had the highest influence on *Eto*, followed by wind-speed and radiation. By contrast, Goyal (2004) noted that in an arid region of India *Eto* was sensitive to temperature and net radiation, followed by wind-speed and vapour pressure. Some other studies (McCuen 1974; Saxton 1975; Coleman/DeCoursey 1976; Beven 1979) showed that potential evaporation/evapotranspiration was much more sensitive to radiation, humidity, and temperature. According to the research findings of Tabari/Hosseinzadeh-Talaee (2014), the *Eto* was most sensitive to air temperature in an arid climate compared with semi-arid and humid climates; similarly, the sensitivity of *Eto* to wind-speed decreased from an arid to a humid climate. In contrast to the pattern of air temperature and wind-speed, the sensitivity of *Eto* to sunshine duration increased from an arid to a humid environment. In this regard, Estevez et al. (2009) indicated that the sensitivity of evapotranspiration to the same climatic variables could vary with location. Similarly, Aydın et al. (2015) reported that the influence of the variables to evapotranspiration (or evaporation) was not the same for each period; and the order of the variables changed with geographical positions. They emphasised that, in general, the *Ep* was more sensitive to net radiation, while the *Eto* was mostly affected by relative humidity.

In brief, the sensitivity coefficients change temporally and spatially, because they are, in themselves, sensitive to the relative magnitudes of all the independent (climatic) and dependent (*Ep* or *Eto*) variables. On the other hand, the sensitivity coefficients can be used to predict the change in the atmospheric demand for water vapour in response to the expected changes in climatic variables. As an example, the change in the evaporative demand of the atmosphere in the future was calculated by summing up the contribution of each climatic variable, based on their sensitivity coefficients and percentage change in their magnitudes. Significant effects can be produced by several combined variables acting in a concerted mode in sensitivity studies. Nonetheless, the combined effects can be larger than the sum of the individual effects of all the variables, because changes in more than one variable are likely to occur at the same time in nature (Gong et al. 2006). An increase of 44.3 mm year^{-1} appeared in the evaporative demand by the 2070s compared with the baseline period, regardless of changes in *Eto* and *Ep* with different rates. Unfortunately, the reliability of the sensitivity coefficients for the future period, depends on the accurate estimates of future climate data. Atmospheric temperature is the most widely used indicator of climatic change in both global and regional scales (Moratiel et al. 2010). However, according to the US-NAS and Royal Society (2014), "several major issues make it impossible to give precise estimates of how global or regional temperature trends will evolve decade by decade into the future." In addition, there has been a widespread expectation that evaporation would increase as air temperature increases. Paradoxically, despite the observed rises in average temperature, observations from the Northern Hemisphere show that the rate of pan evaporation has been steadily decreasing over the past fifty years (Roderick/Farquhar 2004; Pockley 2009). Therefore, the changes in all the major climatic variables must be taken into account in evaporation studies to make more robust predictions about future changes in the hydrological cycle (Roderick/Farquhar 2002).

14.4   Conclusions

Solar radiation, air temperature, and wind-speed are projected to increase, while the relative humidity is expected to slightly decrease by the period of 2070–2079 compared with the baseline period (1994–2003). The drastic decrease in future precipitation predicted by the RCM with a grid distance of 25 km may be questionable. Climate change may affect soil dryness doubly, since the evaporative demand of the atmosphere increases at the same time that precipitation decreases. Unlike the reference evapotranspiration (*Eto*) and potential soil evaporation (*Ep*) projected to increase due to the elevated evaporation requirements, the actual soil evaporation (*Ea*) is projected to decrease due to the increased soil dryness. Considerable discrepancy between the *Ep* and *Ea* was observed during the dry seasons. The sensitivity of the *Eto* and *Ep* showed significant differences between seasons in the same climatic variables. In general, both *Eto* and *Ep* were more sensitive to net radiation, followed by air temperature, relative humidity, and wind-speed; this was consistent for both periods, baseline and projected. The sensitivity coefficients could be used to obtain the accurate change in the atmospheric demand for water vapour. In terms of sensitivity coefficients, the *Ep* responded better to the changes in climatic variables than the *Eto*. Overall, the outcomes of this chapter can provide reference information to understand the implications of climate change for the future water balance in a Mediterranean environment and to improve the regional strategy for agricultural water management and irrigation practices.

References

Agam N, Berliner PR, Zangvil A, Ben-Dor E (2004) Soil water evaporation during the dry season in an arid zone. *Journal of Geophysical Research: Atmospheres* 109:16103.

Allen RG (2011) Skin layer evaporation to account for small precipitation events – An enhancement to the FAO-56 evaporation model. *Agricultural Water Management* 99(1):8–18.

Allen RG, Smith M, Perrier A, Pereira LS (1994) An update for the definition of reference evapotranspiration. *ICID Bulletin* 43:92.

Allen RG, Pereira LS, Raes D, Smith M (1998) *Crop evapotranspiration: Guidelines for computing crop water requirements*. Irrigation and Drainage paper No. 56. FAO, Rome.

Ambas VTh, Baltas E (2012) Sensitivity analysis of different evapotranspiration methods using a new sensitivity coefficient. *Global NEST Journal* 14(3):335–343.

Aydın M (2008) A model for evaporation and drainage investigations at ground of ordinary rainfed-areas. *Ecological Modelling* 217(1–2):148–156.

Aydın M, Keçecioglu SF (2010) Sensitivity analysis of evaporation module of E-DiGOR model. *Turkish Journal of Agriculture and Forestry* 34(6):497–507.

Aydın M, Yang SL, Kurt N, Yano T (2005) Test of a simple model for estimating evaporation from bare soils in different environments. *Ecological Modelling* 182:91–105.

Aydın M, Yano T, Evrendilek F, Uygur V (2008) Implications of climate change for evaporation from bare soils in a Mediterranean environment. *Environmental Monitoring and Assessment* 140:123–130.

Aydın M, Jung YS, Yang JE, Lee HI (2014) Long-term water balance of a bare soil with slope in Chuncheon, South Korea. *Turkish Journal of Agriculture and Forestry* 38:80–90.

Aydın M, Jung YS, Yang JE, Kim SJ, Kim KD (2015) Sensitivity of soil evaporation and reference evapotranspiration to climatic variables in South Korea. *Turkish Journal of Agriculture and Forestry* 39:652–662.

Bellot J, Chirino E (2013) Hydrobal: An eco-hydrological modelling approach for assessing water balances in different vegetation types in semi-arid areas. *Ecological Modelling* 266:30–41.

Beven K (1979) A sensitivity analysis of the Penman–Monteith actual evapotranspiration estimates. *Journal of Hydrology* 44:169–190.

Bittelli M, Ventura F, Campbell GS, Snyder RL, Gallegati F, Pisa PR (2008) Coupling of heat, water vapor, and liquid water fluxes to compute evaporation in bare soils. *Journal of Hydrology* 362(3–4):191–205.

Brown RW, Oosterhuis DM (1992) Measuring plant and soil water potentials with thermocouple psychrometers: Some concerns. *Agronomy Journal* 84:78–86.

Coleman G, DeCoursey DG (1976) Sensitivity and model variance analysis applied to some evaporation and evapotranspiration models. *Water Resources Research* 12(5):873–879.

Dhungana P, Eskridge KM, Weiss A, Baenziger PS (2006) Designing crop technology for a future climate: An example using response surface methodology and the CERES-Wheat model. *Agricultural Systems* 87:63–79.

Ebrahimpour M, Ghahreman N, Orang M (2014) Assessment of climate change impacts on reference evapotranspiration and simulation of daily weather data using SIMETAW. *Journal of Irrigation and Drainage Engineering* 140(2), Article No: 04013012.

Estévez J, Gavilán P, Berengena J (2009) Sensitivity analysis of a Penman–Monteith type equation to estimate reference evapotranspiration in southern Spain. *Hydrological Processing* 23:3,342–3,353.

Fujihara Y, Tanaka K, Watanabe T, Nagano T, Kojiri T (2008) Assessing the impacts of climate change on the water resources of the Seyhan River Basin in Turkey: Use of dynamically downscaled data for hydrologic simulations. *Journal of Hydrology* 353:33–48.

Gao G, Xu CY, Chen D, Singh VP (2012) Spatial and temporal characteristics of actual evapotranspiration over Haihe River basin in China estimated by the complementary relationship and the Thornthwaite water balance model. *Stochastic Environmental Research and Risk Assessment* 26(5):655–669.

Giorgi F, Lionello P (2008) Climate Change Projections for the Mediterranean Region. *Global Planet Change* 63:90–104.

Gong L, Xu C, Chen D, Halldin S, Chen YD (2006) Sensitivity of the Penman–Monteith reference evapotranspiration to key climatic variables in the Changjiang (Yangtze River) basin. *Journal of Hydrology* 329:620–629.

Goyal RK (2004) Sensitivity of evapotranspiration to global warming: A case study of arid zone of Rajasthan (India). *Agricultural Water Management* 69:1–11.

Herrnegger M, Nachtnebel HP, Haiden T (2012) Evapotranspiration in high alpine catchments – an important part of the water balance. *Hydrological Research* 43(4):460–475.

Huo Z, Dai X, Feng S, Kang S, Huang G (2013) Effect of climate change on reference evapotranspiration and aridity index in arid region of China. *Journal of Hydrology* 492:24–34.

Hupet F, Vanclooster M (2001) Effect of the sampling frequency of meteorological variables on the estimation of the reference evapotranspiration. *Journal of Hydrology* 243:192–204.

IPCC (2013) Summary for Policymakers. In Stocker TF, Qin D, Plattner GK, Tignor M, Allen SK, Boschung J, Nauels A, Xia Y, Bex V, Midgley PM (eds) Climate Change 2013: *The Physical Science Basis. Contribution of Working Group I to the Fifth Assessment Report of the Intergovernmental Panel on Climate Change.* Cambridge and New York: Cambridge University Press.

Irmak S, Payero JO, Martin DL, Irmak A, Howell TA (2006) Sensitivity analyses and sensitivity coefficients of standardized daily ASCE Penman-Monteith equation. *Journal of Irrigation and Drainage Engineering* 132(6):564–578.

Kimura F, Kitoh A (2007) *Downscaling by Pseudo Warming Method. Final Report of the ICCAP*, Research Institute for Humanity and Nature, Kyoto, Japan.

Kitoh A, Hosaka M, Adachi Y, Kamiguchi K (2005) Future projections of precipitation characteristics in East Asia simulated by the MRI CGCM2. *Advances in Atmospheric Sciences* 22:467–478.

Kwon, H, Choi M (2011) Error assessment of climate variables for FAO-56 reference evapotranspiration. *Meteorology and Atmospheric Physics* 112:81–90.

Li ZL, Li ZJ, Xu ZX, Zhou X (2013) Temporal variations of reference evapotranspiration in Heihe River basin of China. *Hydrology Research* 44(5):904–916.

McCuen RH (1974) A sensitivity and error analysis of procedures used for estimating evaporation. *Water Resources Bulletin* 10(3):486–498.

Moratiel R, Duran JM, Snyder RL (2010) Responses of reference evapotranspiration to changes in atmospheric humidity and air temperature in Spain. *Climate Research* 44:27–40.

Pockley P (2009) The evaporation paradox. *Australasian Science* Nov/Dec. 2009, 12–13.

Rana G, Katerji N (1998) A measurement based sensitivity analysis of the Penman-Monteith actual evapotranspiration model for crops of different height and in contrasting water status. *Theoretical and Applied Climatology* 60:141–149.

Roderick ML, Farquhar GD (2002) The cause of decreased pan evaporation over the past 50 years. *Science* 298:1410–1411.

Roderick ML, Farquhar GD (2004) Changes in Australian pan evaporation from 1970 to 2002. *International Journal of Climatology* 24:1077–1090.

Romano E, Giudici M (2009) On the use of meteorological data to assess the evaporation from a bare soil. *Journal of Hydrology* 372(1–4):30–40.

Sanchez-Gomez E, Somot S, Mariotti A (2009) Future changes in the Mediterranean water budget projected by an ensemble of regional climate models. *Geophysical Research Letters* 36: L21401.

Saxton KE (1975) Sensitivity analysis of the combination evapotranspiration equation. *Agricultural Meteorology* 15:343–353.

Tabari H, Hosseinzadeh-Talaee P (2014) Sensitivity of evapotranspiration to climatic change in different climates. *Global Planet Change* 115:16–23.

Tao F, Yokozawa M, Hayashi Y, Lin E (2003) Terrestrial water cycle and the impact of climate change. *Agriculture, Ecosystems and Environment* 95:203–215.

Terink W, Immerzeel WW, Droogers P (2013) Climate change projections of precipitation and reference evapotranspiration for the Middle East and Northern Africa until 2050. *International Journal of Climatology* 33(14):3055–3072.

US-NAS and Royal Society (2014) *Climate Change: Evidence and Causes* (An overview from the Royal Society and the US National Academy of Sciences. The primary writing team: Eric Wolff FRS, Inez Fung, Brian Hoskins FRS, John Mitchell FRS, Tim Palmer FRS, Benjamin Santer, John Shepherd FRS, Keith Shine FRS, Susan Solomon, Kevin Trenberth, John Walsh, Don Wuebbles).

Vanderborght J, Graf A, Steenpass C, Scharnagl B, Prolingheuer N, Herbst M, Franssen HJH, Vereecken H (2010) Within-Field Variability of Bare Soil Evaporation Derived from Eddy Covariance Measurements. *Vadose Zone Journal* 9(4):943–954.

Wallace JS, Jackson NA, Ong CK (1999) Modelling soil evaporation in an agroforestry system in Kenya. *Agricultural and Forest Meteorology* 94:189–202.

Xiao X, Horton R, Sauer TJ, Heitman JL, Ren T (2011) Cumulative soil water evaporation as a function of depth and time. *Vadose Zone Journal* 10(3):1016–1022.

Xie H, Zhu X (2013) Reference evapotranspiration trends and their sensitivity to climatic change on the Tibetan Plateau (1970–2009). *Hydrological Processes* 27(25):3685–3693.

Xing W, Wang W, Shao Q, Peng S, Yu Z, Yong B, Taylor J (2014) Changes of reference evapotranspiration in the Haihe River Basin: Present observations and future projection from climatic variables through multi-model ensemble. *Global Planet Change* 115:1–15.

Xu YP, Pal SL, Fu GT, Tian Y, Zhang XJ (2014) Future potential evapotranspiration changes and contribution analysis in Zhejiang Province, East China. *Journal of Geophysical Research: Atmospheres* 119(5):2,174–2,192.

Yang XL, Ren LL, Singh VP, Liu XF, Yuan F, Jiang SH, Yong B (2012) Impacts of land use and land cover changes on evapotranspiration and runoff at Shalamulun River watershed, China. *Hydrological Research* 43(1–2):23–37.

Yano T, Aydın M, Haraguchi T (2007) Impact of climate change on irrigation demand and crop growth in a Mediterranean environment of Turkey. *Sensors* 7(10):2297–2315.

Yukimoto S, Noda A, Kitoh A, Sugi M, Kitamura Y, Hosaka M, Shibata K, Maeda S, Uchiyama T (2001) A new Meteorological Research Institute coupled GCM (MRI-CGCM2) – its climate and variability. *Papers in Meteorology and Geophysics* 51:47–88.

Zhang XC, Liu WZ (2005) Simulating potential response of hydrology, soil erosion, and crop productivity to climate change in Changwu tableland region on the Loess Plateau of China. *Agricultural and Forest Meteorology* 131:127–142.

Zhang Q, Xu CY, Chen XH (2011) Reference evapotranspiration changes in China: Natural processes or human influences? *Theoretical and Applied Climatology* 103(3–4):479–488.

Part V
Climate Change Impacts on Regional Agricultural Production

Chapter 15
The Role of Efficient Management of Water Users' Associations for Adapting to Future Water Scarcity Under Climate Change

Chieko Umetsu, Sevgi Donma, Takanori Nagano and Ziya Coşkun

Abstract Under climate change, the role of water management organisations is becoming critical for mitigating future water scarcity in Arid Regions. During the last couple of decades, many government-managed water allocation schemes were transferred to private organisations such as water users' associations (WUAs). The transfer of the water management authority from the government to WUAs had significant impacts on improving operation and maintenance of irrigation facilities as well as increasing the water fee collection rates. However, recently some WUAs are experiencing difficulties in management because of their small-scale operation sizes. This chapter attempts to address the relative efficiency of WUA management by suggesting an alternative composite efficiency index. We observed the case studies earlier conducted on the WUAs in the Lower Seyhan Irrigation Project in Adana, Turkey. And we applied DEA (data envelopment analysis) to compare the efficiency levels with management-, engineering- and welfare-focused models. The analysis revealed that some WUAs are suffering from unfavourable management practices and there is a scope for major reorganisation. Concerning the future climate change and water scarcity in the region, the role of WUAs for efficient management of water resources seems important.

Keywords Composite index · DEA · Irrigation water management
Middle East · Turkey · Water users association

C. Umetsu, Professor, Kyoto University, Graduate School of Agriculture, Kyoto, Japan, Research Institute for Humanity and Nature (RIHN), Kyoto, Japan; e-mail: umetsu.chieko.5e@kyoto-u.ac. jp. She would like to deeply acknowledge water users' associations in the Lower Seyhan Irrigation Project area for providing valuable information on water management.

S. Donma, Senior Officer, 6th Regional Directorate of State Hydraulic Works, Adana, Turkey; e-mail: sevgidonma@hotmail.com.

T. Nagano, Associate Professor, Kobe University, Graduate School of Agricultural Science, Japan, Research Institute for Humanity and Nature (RIHN), Kyoto, Japan; e-mail: naganot@ruby.kobe-u.ac.jp.

Z. Coşkun, Agricultural Economist, 6th Regional Directorate of State Hydraulic Works, Adana, Turkey; e-mail: ziyacoskun2008@gmail.com.

© Springer Nature Switzerland AG 2019
T. Watanabe et al. (eds.), *Climate Change Impacts on Basin Agro-ecosystems*,
The Anthropocene: Politik—Economics—Society—Science 18,
https://doi.org/10.1007/978-3-030-01036-2_15

15.1 Introduction

Under climate change and expected water scarcity in the near future, local water management organisations play a critical role in mitigating water scarcity. The efficient management of water resources has been debated among stakeholders, including policy-makers and end users, ever since the large-scale irrigation schemes started to deteriorate due to the decay of the infrastructure. Because of the ageing of the infrastructure in addition to the Government's budgetary problems which have led to insufficient investment in irrigation maintenance and the decline of system-wide efficiency due to water leakages, most irrigation schemes have been facing the challenge of improving water efficiency for a long time. Evaluation of the investment in irrigation critically depends on how efficiently water is used on farms (Molden et al. 1998; Chakravorty/Umetsu 2003; Ward 2010). The role of social organisation for efficient use and equitable distribution of water resources and the evolution and sustainability of such an organisation over time have been described in detail in various studies (Ostrom 1990; Lam 1998; Berkes 2002; Dayton-Johnson 2003). Transferring management authority of irrigation water from government to end users such as a water users' association was considered particularly promising for improving the overall water efficiency in the command area.

Turkey is considered one of the countries that has achieved successful transfer of government water management systems to water users' associations (hereafter WUAs). Since 1994, the government accelerated the transfer programme, and water management of nearly one million hectares of publicly irrigated land was rapidly transferred to local WUAs within three years. By 2002, the transfer of the management authority reached roughly two million hectares of irrigated land. If this trend were to continue, the present and future role of the WUAs in irrigation water management would most likely have a greater impact.

The Lower Seyhan Irrigation Project (hereafter LSIP) in the Adana Province (Figs. 15.1 and 15.2) was initiated by the Turkish government as one of the important irrigation projects located in southern Turkey. The construction of the Seyhan Dam was completed in 1956 for the purposes of irrigation, power generation and flood protection. The reservoir can store 1.2 billion cubic metres that supply water to LSIP. The construction plan of irrigation and drainage networks of the Seyhan Plain has four stages. So far, only the areas up to Phase III have been completed and the area for Phase IV downstream was left aside without a canal infrastructure. The completion of Phase IV is facing the problems of a high water table, salinity and insufficient drainage in addition to the government budgetary shortage (Tekinel 2001; Mert 2003; Donma et al. 2004). WUAs were established in LSIP from 1994 and currently there are eighteen WUAs managing operations and maintenance of canal networks in the command area of 120,000 ha. However, recently some WUAs in LSIP are having difficulties in management because of relatively small-scale operation sizes of less than 3,000 ha, although the size of WUAs in LSIP is much larger than most of the WUAs in other countries such as those in Asia. It has been suggested by the General Directorate of State Hydraulic

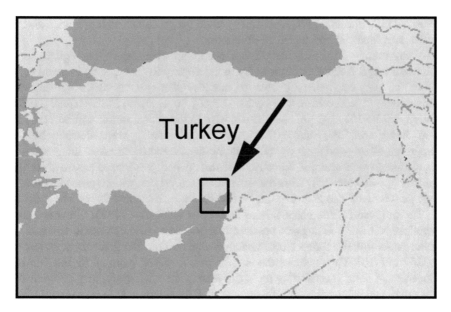

Fig. 15.1 Location of the Lower Seyhan Irrigation Project, Adana, Turkey. *Source* The authors

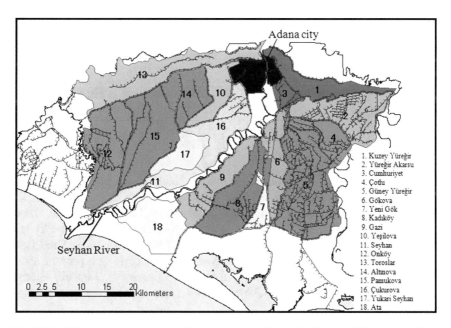

Fig. 15.2 WUAs in the Lower Seyhan Irrigation Project. *Source* 6th Regional Directorate of State Hydraulic Works

Works (DSİ[1]) that some WUAs in LSIP should merge into a larger operation size to solve their financial and logistical problems.

Water scarcity in the distant future is becoming a concern in this region due to climate change. According to the study in the Seyhan River Basin by Fujihara et al. (2008) utilising two general circulation models (GCMs) under SRES A2 scenario, the increase in temperature will be 2.0–2.7 °C, the decrease in the annual precipitation will be 157–182 mm and the decrease in the annual runoff will be 118–139 mm during the 1990s and 2070s. Also the increase in water demand by the expansion of irrigated areas as well as domestic and industrial use in urban areas is accelerating water scarcity. Already more than twenty years have passed since the transfer of water authority from the Government to WUAs, which started in 1994. It may be worthwhile to assess the status and the future scenarios for WUAs in LSIP.

The purpose of this chapter is to assess the efficiency of WUA management practices in LSIP and suggest possible improvements in organisation to adapt to water scarcity under future climate change. The chapter first gives an overview of WUAs in LSIP. This includes the environmental review required by the Turkish Government before transferring the water management authority, in addition to the objectives and responsibilities of water users' associations in irrigation projects. We also compared the impacts of transferring authority to WUAs and the current problems faced by the WUAs. The second section explains the method of analysis and the data sets. The third section shows the results of efficiency analysis for the current eighteen WUAs as well as for the artificially merged six new WUAs. The fourth section describes the results of the analysis. The last section concludes with some policy implications.

15.2 Overview of WUAs in the Lower Seyhan Irrigation Project

15.2.1 Establishment of WUAs in Turkey

During the early 1950s, the Turkish government slowly started transferring the role of irrigation water management to water users. Three laws became the basis for transferring authority of water management to water users' associations. These were the 1953 DSİ Establishment law[2] (Law number 6200), the 1954 Municipality law (Law number 1580), and the 1960 Cooperative law (Law number 1163). Until 1993 small-scale irrigation systems were transferred to water users at a pace of about 2,000 ha per year. The DSİ encouraged farmers to organise Irrigation Groups (IGs) or Water User Groups (WUGs) with limited responsibility for operation and

---

[1]DSİ stands for Devlet Su İşleri Genel Müdürlüğü in Turkish, the General Directorate of State Hydraulic Works, Ministry of Forestry and Water Affairs, Government of Turkey.

[2]Enacted 18 December 1953; effective 28 February 1954.

maintenance. After 1994, the government started to transfer large-scale irrigation systems including the Lower Seyhan Irrigation Project (LSIP) to WUAs (Tekinel 2001).

The main reasons that the Turkish Government accelerated the transfer of water management authority are as follows. First, the government budgetary problems made it difficult to pay overtime salaries after 5 pm, which became a statewide problem. Thus, the cost of operation and maintenance became a huge burden to DSİ. Second, since the 1980s small government budgets were preferred, in turn cutting budgets and freezing new employment to achieve this goal (Svendsen/Nott 2000). Third, because of the budget cuts, DSİ was not able to provide sufficient services to beneficiary farmers. Not only were the operation and management of public irrigation systems costly, but also the water fee collection rate by DSİ was quite low (42%), making operations consequently unsustainable. Therefore, the establishment of WUAs and the transfer of management authority to WUAs was the policy tool to decentralise water management authority and perform more economically efficient operation and maintenance services.

Between the 1960s and the 1980s, its was mainly small irrigation projects that were transferred to WUAs. During the 1990s, because DSİ failed to provide sufficient services, farmers themselves were willing to take responsibility for water management. One DSİ official mentioned that farmers were more ready and eager to take on responsibilities while DSİ was not yet ready to transfer them officially. The acceleration of transferring water management authority after 1994 proceeded rapidly beyond DSİ's expectations. During the initial phase, 10,000 ha were transferred to WUAs compared to the annual average of 2,000 ha before the acceleration programme. By 1995, DSİ had already transferred 800,000 ha to WUAs nationwide, the level that was expected to be reached in 2000. DSİ had already achieved the goal of 2000 five years earlier.

Svendsen/Nott (2000) point out the specific characteristics of transfer programme in Turkey. First, the transfer programme utilised the existing local government organisations and leaders rather than local farmers' grass-roots organisations. Local organisations are village and municipality governments and their heads. Second, the scale of transferred units and the number of beneficiary farmers involved is quite large and the average size of WUAs is 6,500 ha. This average unit size to be transferred was much larger than those in South-East and South Asia. The staff of the regional DSİ operations and maintenance division played a major role in implementing the transfer programme at local level.

15.2.2   The Role of WUAs in Water Management

Currently in Turkey about 91% of transferring organisations are WUAs. The remaining 9% include municipalities, cooperatives, water user groups (WUGs) or irrigation groups (IGs). Before 1994, WUGs or IGs, led by a village head, took responsibility for the operation and maintenance of tertiary distribution canals and were thus considered appropriate intermediate organisations for WUAs. The following are the types of various transferring organisations based on local government in the irrigation scheme (Tekinel 2001):

   (i)   An irrigation scheme can be transferred to *WUAs* where there is more than one local administrative unit (village, legal entities, and municipality) within one irrigation scheme.
  (ii)   An irrigation scheme can be transferred to *Municipality* where the irrigation scheme serves only a single village. The mayor is the natural chairman of this organisation.
 (iii)   An irrigation scheme can be transferred to a *Village organisation* where the scheme serves only a single village. The *Muhtar* (village head) is the natural chairman of this organisation.
  (iv)   An irrigation scheme can be transferred to *Cooperatives* where a legal cooperative can be formed on the request of a minimum of fifteen farmers before a scheme is undertaken.

When WUAs were established, the irrigation facilities were turned over based on the turnover contract and protocols made between DSİ and WUAs. While DSİ owned the irrigation facilities and was responsible for carrying water through the main canals, i.e. secondary canals, operation and maintenance was transferred to WUAs, and they were responsible for the main and the tertiary canals leading to the crop fields (Mert 2003). Water rights, on the other hand, were not transferred to WUAs (Scheumann 1997), where the government still possessed the rights over water resources in the irrigation project. On the other hand, the operation and maintenance of conveyance canals were the responsibility of DSİ. Figure 15.3 shows the irrigation facilities in the Lower Seyhan Irrigation Project.

The objectives of the WUAs are as follows (Svendsen/Nott 2000): (a) providing adequate and timely irrigation water supplies to all farmers in the unit; (b) providing irrigation services in a reliable and sustainable manner; (c) contracting operation and maintenance costs; (d) collecting service fees from all benefiting farmers; (e) acquiring mechanical equipment for maintenance and repair.

Responsibilities of WUAs include: (a) scheduling and delivering water within the WUA unit; (b) monitoring deliveries to farms; (c) collecting operational monitoring data; (d) resolving disputes; (e) paying irrigation pumping costs.

15.2.3   The Impacts of Transferring Authority to WUAs

The Lower Seyhan Irrigation Project is located in the south of Adana city, stretching to the Mediterranean coast (Fig. 15.2). A Mediterranean climate prevails in the region, with hot and dry summers and mild and rainy winters. The average annual rainfall is approximately 650 mm and most precipitation occurs in May and December (Donma et al. 2004). The average temperature is 18 °C with max 45.6 °C

Fig. 15.3 Irrigation facilities in the Lower Seyhan Irrigation Project (Top left: Conveyance canal; Top right: Diversion weir; Bottom left: Secondary canal; Bottom right: Tertiary canal). *Source* Photos by Chieko Umetsu

and min −8.1 °C (Mert 2003). The main crops in LSIP are maize (52%), citrus (14%), cotton (7%), vegetables (6%) in terms of area planted, and citrus (39%), corn (33%), melon (10%), vegetables (6%) in terms of production value (DSİ 2003c). The dominant irrigation technology is gravity irrigation.

In the Lower Seyhan Irrigation Project (LSIP) area, eighteen WUAs were established during 1994–1996 (Fig. 15.2).

Although the available data is limited, the impacts of transferring authority from DSİ to WUAs can be mainly summarised in four points, namely as (i) a reduction in operation and maintenance costs, (ii) a reduction of water fees, (iii) increased fee collection rate by WUAs and (iv) more equitable distribution of water among head and tail farmers compared to the DSİ regime.

The assessment of the irrigation scheme in the Yüreğir Plain, the left bank of the Seyhan River, during 1994–5 indicated that the total operation and maintenance costs by WUA was only 41% of the cost paid by DSİ (Scheumann 1997). Svendsen/Nott (2000) reported that the water fee initially doubled when WUAs were established during the early 1990s. In the case of the LSIP, the water fee became less than the fee previously assessed by DSİ, as indicated in Table 15.1. For example, the water fee for corn reduced from 10 TL/da (100 TL/ha) to 5.5 TL/da (55 TL/ha). Since the water fee is generally not only the cost of water but also a

Table 15.1 Water fee assessed by DSİ and WUA in Region VI in 1994 and 2003. *Source* Mert (2003: 26)

Crop	Fee by DSİ before 1994 (TL/da)	Fee by WUA 2003 (TL/da)
Maize	10	5.5
Soybean	8	4.5
Cotton	15.5	5.5
Melons	8	5.5

da (decare) = 0.1 ha, cf. New Turkish Lira was introduced after devaluation in 2005 by deleting 6 zeros

service fee of WUAs to farmers, there is regional variation depending on the endowment of the WUAs so the above statement of Svendsen/Nott (2000) should be examined carefully. Also the water fee for each crop was determined by the WUAs before the next irrigation season. A wide range of water fees depending on the WUA may raise a question of equity among farmers in the irrigation project.

The fee collection rate, on the other hand, increased drastically. From 1989 to 1994, the average fee collection rate by DSİ was 37.6% (Yazar 2002), while the average collection rate of the assessed fee was 65% in 2002 (Table 15.2). Also, farmers believe that water allocation became more equitable among head and tail farmers compared to the times of the DSİ regime.[3]

15.2.4 The WUAs in the Lower Seyhan Irrigation Project: Current Issues

Table 15.2 shows the general information of eighteen WUAs during the 2002 irrigation season. The Seyhan right bank, Tarsus Plain, has eight WUAs and the Seyhan left bank, Yüreğir Plain, has ten WUAs (Umetsu et al. 2006). At the time of transferring authority, the basic concept of making WUA boundary was to assign management responsibility of one main canal to one WUA and also to show concern for ethnic groups in the command area. However, in reality many right bank WUAs share the same main canal TS-3, and this situation is causing frequent water sharing problems which are still ongoing. Having learned from the lesson of the right bank, WUAs in the left bank basically do not share the main canals. Their total service area includes irrigated areas with concrete canal infrastructure and irrigated areas with earthen canals. The total irrigated area ranges from 1,651 ha for Cumhuriyet WUA to 16,528 ha for Güney Yüreğir WUA. The service area includes the area of uncompleted LSIP project Phase IV that does not have canal infrastructure. For example, Ata WUA's service area is totally without any concrete canal infrastructure. Some portion of the land is irrigated by groundwater for citrus and

---

[3]From authors' interview survey in summer 2003.

vegetable cultivation. Groundwater use is high in Çukurova WUA for citrus cultivation. The number of irrigators ranges from 283 for Ata WUA to 4,731 for Toroslar WUA. The number of parcels is also highest in Toroslar. The average water fee ranges from the lowest in Yeni Gök, TL3.20/da[4] (US$19.5/ha) to the highest in Cumhuriyet, TL9.10/da (US$55.5/ha), and this is due to the pumping cost for irrigation. The fee collection rate in Table 15.2 shows that six WUAs could not collect even 60% of the total water fee expected in 2002. Water fee is based on crop type and each WUA charges a different water fee for each crop. For example, the water fee per decare for corn in LSIP ranged from TL3.5 to TL6.7 during the 2002 irrigation season.[5] The highest water fee is charged for citrus, fruit trees and vegetable cultivation.

Table 15.3 shows the financial information of the eighteen WUAs in LSIP. The major revenue of WUAs comes from the water fee collected from irrigators (DSİ 2003a, b). The total expected WUA fee is estimated by WUAs based on the actual cropping pattern of their irrigated land. However, the WUA revenue is short of the expected amount because for the following reasons. First, the collection rate is on average 65% and there is a substantial amount of delayed payment. Six WUAs had more than a 40% share of delayed payment in expected fee revenue. Therefore, the actual amount of fee collected is the sum of the collected fee for this year and the collected fee from previous years as shown in column (d) of Table 15.3. The collection rate of many WUAs exceeds 80% only when the fees from the past years are included, as shown in column (f). Staff salary indicates that a share of personnel cost is overwhelmingly exceeding 60% of its actual budget in the case of Cumhuriyet and Kadıkoy. Compared to staff salary, operation and maintenance costs provided by WUAs ranged from 9% (Altınova, Yeni Gök) to 86% (Güney Yüreğir). According to the regulations, WUAs are supposed to allocate 30% of their annual budget to staff salary and 40% to repair and maintenance. However, small scales of operation and high staff salaries are causing financial difficulties in some WUAs, thus making it difficult to conform to the regulations.

Table 15.4 shows the water demand and irrigation efficiency of eighteen WUAs. Claimed demand of water indicates that this amount of water is requested from DSİ in March by each WUA before the irrigation season. Before the cropping season starts, WUAs collect information on cropping patterns for the coming season from farmers in the service area and estimate the total water demand for the coming irrigation season. DSİ plans the annual water allocation based on this request from WUAs. Net demand of water is estimated based on the actual cropping pattern during the irrigation season. DSİ uses annual irrigation water requirement tables for each crop, estimated by precipitation, residual soil moisture and evapotranspiration of crops. Gross demand of water is estimated by DSİ by multiplying around the factor of 1.8 to take into account conveyance and other physical losses of the irrigation system. This multiplying factor reflects the system-wide conveyance

---

[4]da (decare) = 0.1 ha. 1 US$ = 1.64 TL (October 2002).

[5]In normal years, the irrigation season starts in April and ends in October.

Table 15.2 General information of eighteen WUAs in the Lower Seyhan Irrigation Project, 2002. *Source* Umetsu et al. (2006)

Map no	WUA name	Established year	Main canal	Irrigated area with infrastructure (ha)	Irrigated area without infrastructure (ha)	Total irrigated area (ha)	Non-irrigated area (ha)	Groundwater irrigated area (ha)	Number of irrigators	Number of irrigated parcels	Average water fee (TL/da)	Fee collection rate %
13	Toroslar	1995	TS1,2	13,719	0	13,719	1,891	190	4,731	9,795	7.20	50
10	Yeşilova	1994	TS3	2,577	0	2,577	853	310	413	881	6.20	57
14	Altınova	1995	TS5	5,478	0	5,478	252	420	694	1,465	5.40	73
16	Çukurova	1995	TS3	6,133	0	6,133	920	799	1,757	3,123	6.10	78
17	Yukarı Seyhan	1996	TS3	4,001	0	4,001	960	74	734	1,578	5.80	94
11	Seyhan	1994	TS3	3,060	0	3,060	540	120	651	1,288	6.50	75
12	Onköy	1994	TS8,9,10	8,697	0	8,697	3,096	190	1,589	4,044	6.90	75
15	Pamukova	1995	TS6,7	10,688	0	10,688	1,194	155	2,070	4,956	6.10	64
2	Yüreğir Akarsu	1995	YS2	7,523	0	7,523	1,812	0	918	1,666	5.50	65
3	Cumhuriyet	1994	YSO	1,651	0	1,651	934	160	602	1,161	9.10	51
1	Kuzey Yüreğir	1994	YSI	3,066	539	3,606	1,971	0	1,141	1,133	5.00	47
4	Çötlu	1994	YS4	1,850	790	2,640	607	0	310	971	5.00	59
6	Gökova	1994	YS6	4,139	0	4,139	974	0	435	734	5.60	72
5	Güney Yüreğir	1994	YS5,3	14,354	2,175	16,528	4,965	0	1,620	3,419	4.40	66
8	Kadıköy	1994	YS8	8,839	1,570	10,409	898	322	1,275	1,640	3.90	73
7	Yeni Gök	1994	(YS8) YS9	1,822	2,867	4,688	42	44	519	735	3.20	60
9	Gazi	1994	YS7	5,516	1,470	6,985	1,328	278	569	1,270	3.50	84
18	Ata	1996	(YS7)	0	4,360	4,360	5,700	300	283	996	4.50	64

Source DSI (2003a, b) Transferred Irrigation Association Year 2002 Observation and Evaluation Report and interview survey in 2003
Main canal in bracket indicates that this canal is shared with other WUA for water distribution. Irrigated area without infrastructure means area irrigated by earthen canals without concrete canal infrastructure. Total irrigated area does not include the area irrigated by groundwater. Fee collection rate is the collected 2002 fee out of expected fee in 2002

Table 15.3 The financial information of the eighteen WUAs in the Lower Seyhan Irrigation Project, 2002. *Source* DSI (2003a, b) Transferred Irrigation Association Year 2002 Observation and Evaluation Report

Map no	WUA name	(a) WUA fee revenue expected	(b) WUA fee revenue from past years	(c) WUA fee revenue collected	(d) = (b) + (c) WUA fee revenue actual	(e) delayed payment 2002	(f) = (d)/(a) actual/ expected %	(g) = (e)/(a) delayed/ expected %	(h) staff salary (TL)	(i) = (h)/(d) % of WUA staff salary in (d)	(j) O&M cost (TL)	(k) = (h)/(d) % of WUA O&M expenditure
13	Toroslar	1,082,591	341,878	537,943	879,821	544,648	81.27	50.31	336,389	38	231,800	26
10	Yeşilova	209,133	20,163	119,003	139,166	90,130	66.54	43.10	91,459	66	20,560	15
14	Altınova	330,820	14,686	240,946	255,632	89,874	77.27	27.17	101,017	40	22,456	9
16	Çukurova	368,111	35,792	288,794	324,586	79,317	88.18	21.55	166,247	51	60,512	19
17	Yukarı Seyhan	115,636	9840	108,202	118,042	7434	102.08	6.43	74,902	63	73,363	62
11	Seyhan	188,581	16,597	140,735	157,332	47,846	83.43	25.37	79,399	50	51,712	33
12	Onköy	638,100	67,899	476,832	544,731	161,268	85.37	25.27	336,205	62	279,908	51
15	Pamukova	708,000	40,805	456,050	496,855	251,950	70.18	35.59	372,141	75	385,260	78
2	Yüreğir Akarsu	366,538	47,514	237,499	285,013	129,039	77.76	35.20	148,284	52	69,178	24
3	Cumhuriyet	149,127	20,255	75,361	95,616	73,766	64.12	49.47	58,394	61	29,812	31
1	Kuzey Yüreğir	178,469	31,855	83,620	115,475	94,849	64.70	53.15	47,703	41	60,883	53
4	Çötlu	128,281	54,792	76,198	130,990	52,083	102.11	40.60	50,730	39	36,113	28
6	Gökova	213,412	32,258	153,543	185,801	59,869	87.06	28.05	97,560	53	41,055	22
5	Güney Yüreğir	662,199	95,375	440,224	535,599	221,975	80.88	33.52	161,766	30	460,234	86
8	Kadıköy	382,408	38,697	277,651	316,348	104,757	82.73	27.39	203,348	64	84,870	27
7	Yeni Gök	143,765	198,182	85,748	283,930	58,017	197.50	40.36	61,106	22	26,398	9
9	Gazi	230,056	27,481	194,245	221,726	35,811	96.38	15.57	125,567	57	106,240	48
18	Ata	182,536	23,067	117,244	140,311	65,292	76.87	35.77	37,531	27	41,587	30

Source DSI (2003a, b) Transferred Irrigation Association Year 2002 Observation and Evaluation Report
TL: Turkish Lira; (i) and (k) does not sum to one because of other incomes and shortfalls

Table 15.4 The water demand and irrigation efficiency of eighteen WUAs in the Lower Seyhan Irrigation Project. *Source* DSI (2003a, b) Transferred Irrigation Association Year 2002 Observation and Evaluation Report

Map no	WUA name	(a) claimed demand (MIL m³)	(b) net demand (MIL m³)	(c) gross demand (MIL m³)	(d) actual water (MIL m³)	(d)/(c) water use efficiency	(b)/(d) * 100 irrigation efficiency %	(a)/(c) claimed demand/ gross	Net (b)/ irrigated area (m³/ha)	Actual (d)/ irrigated area (m³/ha)	Gross revenue (TL/m³)
13	Toroslar	129.51	71.16	128.79	156.39	1.21	45.50	1.01	5187	11,400	0.452
10	Yeşilova	31.40	13.86	25.08	35.89	1.43	38.62	1.25	5378	13,927	0.263
14	Altınova	62.57	30.20	54.65	70.46	1.29	42.86	1.14	5513	12,862	0.257
16	Çukurova	60.29	32.71	59.20	66.29	1.12	49.34	1.02	5333	10,809	0.356
17	Yukarı Seyhan	40.22	21.70	39.28	46.73	1.19	46.44	1.02	5424	11,680	0.333
11	Seyhan	33.25	16.27	29.44	38.05	1.29	42.76	1.13	5317	12,435	0.233
12	Onköy	105.47	44.95	81.35	114.21	1.40	39.36	1.30	5168	13,132	0.213
15	Pamukova	100.92	57.61	104.27	114.54	1.10	50.30	0.97	5390	10,717	0.284
2	Yüreğir Akarsu	89.61	41.99	76.00	80.04	1.05	52.46	1.18	5582	10,640	0.435
3	Cumhuriyet	22.52	8.77	15.87	25.44	1.60	34.48	1.42	5311	15,406	0.273
1	Kuzey Yüreğir	43.15	17.58	33.05	55.96	1.69	31.41	1.31	4874	15,518	0.187
4	Çötlu	23.51	13.20	23.89	21.01	0.88	62.84	0.98	5001	7958	0.284
6	Gökova	44.00	21.94	39.70	49.38	1.24	44.42	1.11	5300	11,930	0.397
5	Güney Yüreğir	159.39	87.40	158.20	212.60	1.34	41.11	1.01	5288	12,863	0.265
8	Kadıköy	107.69	54.28	98.24	117.63	1.20	46.14	1.10	5214	11,301	0.335
7	Yeni Gök	81.77	25.01	45.26	77.41	1.71	32.31	1.81	5334	16,512	0.103
9	Gazi	79.10	37.83	68.47	87.99	1.29	42.99	1.16	5416	12,597	0.314
18	Ata	46.76	20.09	36.35	54.50	1.50	36.85	1.29	4607	12,500	0.235

Source DSI (2003a, b) Transferred Irrigation Association Year 2002 Observation and Evaluation Report
MIL m³: Million cubic meters; TL: Turkish Lira

efficiency and may increase as the canal infrastructure deteriorates over time. Theoretically this gross demand satisfies the demand for water at the farm level; however, the actual water release by DSİ is shown in the last column, which is larger than the amount of claimed demand. Irrigation efficiency, i.e., percentage of net demand out of actual release of water, indicates that Çotlu has the highest irrigation efficiency (63%) while Kuzey Yüreğir has the lowest (31%) irrigation efficiency (DSİ 2003b). It is customary for WUAs to overestimate the cultivation area at the time of aggregating farmers' water requests and report larger amounts of claimed water to the DSİ office than the actual gross demand for water.

The current issues of WUAs in LSIP can be summarised as follows: (a) a large number of delayed fee payments; (b) low fee collection rate; (c) high staff salary; (d) low operation and maintenance expenditure; (e) artificially high water demand being claimed by WUAs; and (f) small operational scale. In engineering and other courses on resource management there is a considerable focus on the concept of engineering efficiency, labour analysis, and welfare analysis. There also is a presumption that involving farmers actively in the day-to-day management of irrigation systems may increase overall efficiency. In the following section, we perform an efficiency analysis of eighteen WUAs and compare them to find out how their overall efficiency can be improved further.

15.3 Method and Data

15.3.1 Method of Analysis

The input-orientated CCR (Charnes et al. 1978) efficiency is the radial measure of technical efficiency in which the efficiency is obtained by radially reducing the level of inputs relative to the frontier technology holding the level of output constant. The input-orientated model implicitly assumes cost-minimising behaviour, and the output-orientated DEA (data envelopment analysis) model, on the other hand, assumes revenue-maximising behaviour of organisations. It is more reasonable to assume that organisations have a budget constraint and thus minimise costs. In general, DEA efficiency measure requires input and output quantity information and is independent of input prices as well as behavioural assumptions about producers. Also CCR efficiency measure assumes constant returns to scale. Figure 15.4 illustrates the input-orientated CCR measure and distance function for a two-input case. Vertical and horizontal axes indicate the level of inputs, x_1 and x_2. The frontier technology is given by the piecewise linear isoquant, ABCDE, i.e., the unit output y_1 from the combination of x_1 and x_2. Efficient production activity occurs at the extreme point of the convex hull of this frontier (BCD). Line segments extending from B and D, AB and DE indicate strong disposability of inputs i.e., disposal of surplus inputs is free. Production activity F is inside of the input requirement set and thus inefficient. In terms of distance, the CCR technical

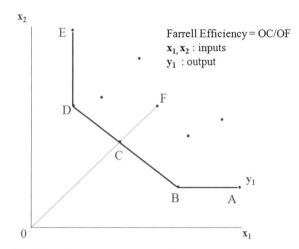

Fig. 15.4 Production frontier and Farrell efficiency. *Source* Revised from Charnes et al. (1978)

efficiency at period t is given by OC/OF. The CCR efficiency measure varies between zero and one and equals one when the observation is efficient, i.e., the observed Decision Making Unit (DMU) is on the frontier technology (C).

The production possibility set P is defined by the set of feasible activities as follows:

$$P = \{(x, y)|x \geq X\lambda, y \geq Y\lambda, \lambda \geq 0\}, \tag{15.1}$$

CCR model is estimated as a linear programming model as follows (Cooper et al. 2000).

$$\begin{aligned} max \quad & uy_0 \\ \text{subject to} \quad & vx_0 = 1 \\ & -vX + uY \leq 0 \\ v \geq 0, u \geq 0 \end{aligned} \tag{15.2}$$

where row vector v is input multipliers and row vector u is output multipliers. The dual problem of the above equations (15.2) can be expressed as the following linear programming problem where θ is a real variable, and $\lambda = (\lambda_1, \ldots, \lambda_2)$ is a non-negative vector of variables.

$$\begin{aligned} min \quad & \theta \\ \text{subject to} \quad & \theta x_0 - X\lambda \geq 0 \\ & Y\lambda \geq y_0 \\ & \lambda \geq 0 \end{aligned} \tag{15.3}$$

The dual LP problem has a feasible solution $\theta = 1$, $\lambda_0 = 1$, $\lambda_j = 1(j \neq 1)$. Hence the optimal θ^* is not greater than 1. Also the non-negative constraint for the data forces λ to be non-zero. Thus θ must be greater than zero. Thus $0 < \theta^* \leq 1$.

The constraints of dual LP (15.3) requires the activity $(\theta\, x_0, y_0)$ to remain in the production possibility set P, while the LP minimises θ that contracts the input vector x_0 radially to $\theta\, x_0$. Dual LP problem seeks the activity in P that maintains at least the output level y_0 while reducing the input level x_0 radially at a minimum level.

Umetsu et al. (2003) applied an input-orientated DEA (CCR) model for constructing Malmquist indices to analyse agricultural productivity changes and compare differences at a regional level. In this chapter, we use a similar method for analysing management efficiencies by considering WUA as DMU and comparing their efficiency levels.

15.3.2   Data Sets

For performing the efficiency analysis of WUAs, we consider three models for different focus, i.e., management efficiency, engineering efficiency and welfare focus.

The first model focuses on *management efficiency*. The management efficiency model has two outputs – WUA fee, total irrigated area served – and five inputs – actual water supply (gross water), operation and maintenance costs, staff salary, a number of technical staff and delayed water fee payment. Thus we consider WUA's management efficiency and the second model focuses on *engineering efficiency*[6] that tried to capture water efficiency of water distribution in the service area. Engineering efficiency has two outputs – total irrigated area and net water demand – and three inputs – actual water demand, maintenance and repair costs and a number of technical staff. The third model considers *farmer welfare* by including the value of agricultural production. Thus, the welfare-orientated model has three outputs, namely the WUA fee, total irrigated area served, gross revenue from production, and five inputs: actual water supply, operation and maintenance costs, staff salary, a number of technical staff and the delayed water fee payment. Detailed data description follows. Various costs and operational information of WUAs for 2002 irrigation season are taken from the *Transferred Irrigation Association Year 2002 Observation and Evaluation Report* supplemented by the information collected from the authors' interview survey.

The *total irrigated area* (ha) is the sum of the areas with canal infrastructure and without canal infrastructure that were irrigated by WUAs and subject to water charges. This total irrigated area of WUAs changes every year because there are some farmers who decide not to irrigate in a particular year. The *WUA fee* (Turkish Lira: TL) is the annual total water charge actually collected in 2002 by each WUA.

---

[6]Criteria are often found for performance evaluation focused on engineering. For example, Kanber et al. (2005) analysed the irrigation system performance of various water basins by Turkey using the following criteria: (1) hydraulic performance indicators, (2) economic performance indicators, (3) agricultural performance indicators.

This amount includes the fee collected from past years and the fee collected for 2002. The fee that was not collected for the year 2002 is the *delayed payment* that causes WUAs difficulty in planning their annual budget. *Operation and mainte-nance costs* (TL) include electricity charges, machinery costs, and other operation expenses, such as communication, office rental fees and utility charges, and maintenance and repair expenses. *Maintenance costs* (TL) done in that year includes concrete repair works for canals, canal cleaning, *kanalet*[7] repairs, painting, maintenance of the underground structure, service roads, building, and others. *Number of technical staff* is the sum of irrigation engineer, operation and mainte-nance technician, water distribution technician, pump operator, electric technician and machine operator. *Staff salary* (TL) includes staff expenses, president's salary and travel expenses and money paid to committee members for meetings.

The Gross revenue from production (1,000 Turkish Lira: TL) for each WUA is calculated by area cultivated in 2002 reported by WUAs (DSİ 2003b) and the average gross revenue/da in 2002 for each crop in the Lower Seyhan region (DSİ 2003a). WUAs keep a record of irrigated area by crops for charging purposes; however, non-irrigated crops, such as wheat and other agricultural revenue such as livestock, are not considered in our analysis. Information reported in the *Briefing of WUA and Year 2002 Management Activity Report* (DSİ 2003b) were used for actual water supply, net water demand, and claimed water demand (million cubic metres) for each WUA.

15.4   Estimation Results

15.4.1   Efficiency Scores of Eighteen WUAs

We performed the efficiency analysis by estimating CCR efficiency scores for three models – management efficiency, engineering efficiency and welfare-focused – as mentioned in the previous section. The efficiency score shows the efficiency level of each WUA relative to the efficient frontier.

Table 15.5 indicates the results of the efficiency scores for these three models with different focuses. For the management efficiency (ME), ten WUAs scored one and they are on the efficient frontier. The least efficient DMUs in this category include Cumhuriyet (0.71) and Kuzey Yüreğir (0.76). Cumhuriyet is one of the WUAs that has financial difficulties because of its small operation size. On average, the right bank management efficiency (0.97) is slightly better than the left bank (0.93).

The second column shows the engineering efficiency (EE) scores. Eight WUAs scored 1 and are on the frontier, and Cumhuriyet (0.70) and Kuzey Y. (0.74) again showed low performance in engineering efficiency because of the large number of

---

[7]Kanalet is a small open concrete-tube canal system used in Turkey that conveys water to the farm.

Table 15.5 Efficiency scores of Eighteen WUAs in the Lower Seyhan Irrigation Project. *Source* The authors

Map no	DMU	ME score	EE score	W score	Composite index
13	Toroslar (R)	1	0.97	1	0.99
10	Yeşilova (R)	0.93	0.79	0.93	0.88
14	Altınova (R)	1	1	1	1
16	Çukurova (R)	1	1	1	1
17	Yukarı Seyhan (R)	1	1	1	1
11	Seyhan (R)	0.88	0.87	0.88	0.87
12	Onköy (R)	0.95	0.75	0.95	0.88
15	Pamukova (R)	1	1	1	1
2	Y. Akarsu (L)	0.98	0.86	1	0.94
3	Cumhuriyet (L)	0.71	0.70	0.72	0.71
1	Kuzey Y. (L)	0.76	0.74	0.77	0.76
4	Çotlu (L)	1	1	1	1
6	Gökova (L)	0.92	0.89	1	0.94
5	Güney Y. (L)	1	0.97	1	0.99
8	Kadıköy (L)	1	1	1	1
7	Yeni Gök (L)	1	1	1	1
9	Gazi (L)	0.98	0.94	1	0.97
18	Ata (L)	1	1	1	1
	Right Bank average	0.97	0.92	0.97	0.95
	Left Bank average	0.93	0.90	0.94	0.92
	18 WUAs average	0.95	0.91	0.95	0.94

Key ME: management efficiency; EE: engineering efficiency; W: welfare
(R): right bank; (L): left bank

technical staff employed by the WUAs. Önköy's low performance in engineering efficiency (0.75) is largely because they employ the largest number of technical staff for water distribution among all WUAs. On average, the right bank engineering efficiency (0.92) is slightly better than the left bank (0.90) in spite of the old canal infrastructure.

The third column shows the welfare-focused efficiency scores that take into account agricultural revenue from the command area. Thirteen WUAs formed a frontier and Cumhuriyet (0.72) and Kuzey Y. (0.77) are again low performers. Cumhuriyet WUA has a command area in proximity to the city of Adana and the average parcel size is only 1.3 ha and the smallest among all WUAs after Toroslar (1.2 ha). Again, on average, the right bank welfare score (0.97) is higher than the left bank (0.94). This may be due to the fact that the right bank includes Toroslar, which specialises in high-value crops, such as vegetables and citrus.

The last column shows the composite index, which is estimated by taking the geometric mean of the three efficiency scores. The results indicate that eight WUAs scored a composite index of 1, namely Altınova, Çukurova, Yukarı Seyhan,

Table 15.6 Projected input levels to reach efficient frontier for Cumhuriyet and Kuzey Y. WUAs. *Source* The authors

DMU input/output	Score data	Projection	Difference	% change (%)
Cumhuriyet (L)	0.719			
Gross water/WUA (M m³)	25.44	18.28	−7.16	−28.13
Operation and maintenance costs (TL)	29,812	21425.72	−8386.28	−28.13
Staff salaries (TL)	58,394	41967.44	−16426.56	−28.13
Number of technical staff	5	2.36	−2.64	−52.74
Delayed payments (TL)	73,766	39762.81	−34003.19	−46.10
Gross revenue from production (1000 TL)	6941.30	6941.30	0	0.00
WUA fee revenue (TL)	95,616	95,616	0	0.00
Total irrigated area (ha)	1651	1675.26	2426	1.47
Kuzey Y. (L)	0.768			
Gross water/WUA (M m³)	55.959	42.98	−12.98	−23.19
Operation and maintenance costs (TL)	60,883	36090.13	−24792.87	−40.72
Staff salaries (TL)	47,703	36639.74	−11063.26	−23.19
Number of technical staff	6	2.03	−3.97	−66.19
Delayed payments (TL)	94,849	56216.06	−38632.94	−40.73
Gross revenue from production (1000 TL)	10479.05	10479.05	0	0.00
WUA fee revenue (TL)	11,5475	123933.31	8458.31	7.32
Total irrigated area (ha)	3606	3606	0	0.00

Key M m³: Million cubic meters; TL: Turkish Lira; (L): left bank

Pamukova, Çotlu, Kadıkoy, Yeni Gök and Ata. It is surprising to see that Ata, which does not have a concrete canal infrastructure in its entire area, is on the efficiency frontier, indicating that its water users are utilising their limited resources most efficiently compared to other inefficient WUAs.

Table 15.6 shows the projected input levels to reach the efficient frontiers of the welfare model for Cumhuriyet and Kuzey Y., which are the WUAs that demonstrated the lowest performance in all categories. The projection shows the level of input that can be reduced to reach the same level of output by comparing other efficient DMUs. For example, the delayed payments of Cumhuriyet can be reduced by 46% or by 34,003 TL, thus the efficient level of delayed payments are 39,763 TL. Similarly, the actual water supply, operation and maintenance costs, staff salaries and the number of technical staff can be reduced by 28%, 28%, 28%, and 53% respectively. In the case of Kuzey Y., the major reduction of input should come from operation and maintenance costs (41%), technical staff (66%) and delayed payments (41%). Thus DEA analysis provides the target input levels for a major reorganisation.

Following the management difficulties reported by some WUAs, it has been suggested that the eighteen WUAs in the LSIP should reorganise into smaller

Table 15.7 Information about merged WUAs in the Lower Seyhan Irrigation Project based on data from 2002. *Source* Authors' calculation from DSI (2003a, b) Transferred Irrigation Association Year 2002 Observation and Evaluation Report

Merged WUA	Main canal	Total services area (ha)	Irrigated area with infrastructure	Irrigated area without infrastructure	Total irrigated area (ha)	Non-irrigated area (ha)	Groundwater irrigated area (ha)	Number of irrigators	Number of irrigated parcels	Average water fee (TL/da)	Fee collection rate (%)
R-1	TS1,2	13,700	13,719	0	13,719	1,891	190	4,731	9,795	7.20	50
R-2	TS3,5	23,516	21,249	0	21,249	3,525	1,723	4,249	8,335	5.93	74
R-3	TS6,7,8,9,10	20,869	19,385	0	19,385	4,290	345	3,659	9,000	6.46	69
L-1	YS0,1,2	12,783	12,241	539	12,780	4,717	160	2,661	3,960	5.82	57
L-2	YS3,4,5,6	23,059	20,343	2,965	23,307	6,546	0	2,365	5,124	4.68	67
L-3	YS7,8,9	27,756	16,177	10,267	26,442	7,968	954	2,646	4,641	3.77	72

Source DSI (2003a, b) Transferred Irrigation Association Year 2002 Observation and Evaluation Report

Total irrigated area does not include the area irrigated by groundwater. Fee collection rate is the fee collected 2002 fee out of expected fee in 2002. TL: Turkish Lira

number of WUAs with larger command areas. We tentatively merged the current eight WUAs on the right bank into three and the ten WUAs on the left bank into three – six WUAs in total – for further analysis. Before the transfer in 1994, the right bank and left bank of the Seyhan River had four and three DSİ field offices respectively. This suggested that the aggregation level of the six newly merged WUAs is similar to the previous operation scale under DSİ administration before the transfer in the early 1990s. Table 15.7 shows the characteristics of the newly merged WUAs, named tentatively R-1 (1 WUA), R-2 (5 WUAs), R-3 (2 WUAs) for the right bank and L-1 (3 WUAs), L-2 (3 WUAs), L-3 (4 WUAs) for the left bank. The main advantage of this merging is that none of the new WUAs shares the same main canal within its command area, which avoids conflicts over water allocation. Thus in the following section, we try to consider efficiency analysis for newly merged WUAs.

Table 15.8 Efficiency scores of merged WUAs. *Source* The authors

Map no	DMU	W Score	Rank
13	Toroslar (R-1)	1	1
10	Yeşilova (R-2)	0.93	19
14	Altınova (R-2)	1	1
16	Çukurova (R-2)	1	1
17	Yukarı Seyhan (R-2)	1	1
11	Seyhan (R-2)	0.88	21
12	Onköy (R-3)	0.95	17
15	Pamukova (R-3)	1	1
2	Y. Akarsu (L-1)	1	1
3	Cumhuriyet (L-1)	0.72	24
1	Kuzey Y. (L-1)	0.77	23
4	Çotlu (L-2)	1	1
6	Gökova (L-2)	1	1
5	Güney Y. (L-2)	1	1
8	Kadıköy (L-3)	1	1
7	Yeni Gök (L-3)	1	1
9	Gazi (L-3)	1	1
18	Ata (L-3)	1	1
19	R-1	1	1
20	R-2	0.92	20
21	R-3	0.94	18
22	L-1	0.87	22
23	L-2	1	16
24	L-3	1	1

Key W: welfare; R: right bank; L: left bank

Table 15.9 Projected input levels to reach efficient frontier for L-1 WUA. *Source* The authors

DMU input/output	Score data	Projection	Difference	% change (%)
L-1	0.87			
Gross water/WUA (M m^3)	161.44	140.00	−21.44	−13.28
Operation and maintenance costs (TL)	159,873	128172.85	−31700.15	−19.83
Staff salary (TL)	254,381	220601.85	−33779.15	−13.28
Number of technical staff	21	13.89	−7.11	−33.85
Delayed payments (TL)	297,654	231434.39	−66219.61	−22.25
Gross revenue from production (1000 TL)	52205.07	52205.07	0	0.00
WUA fee revenue (TL)	496,104	496,104	0	0.00
Total irrigated area (ha)	12,780	12,780	0	0.00

Key M m^3: million cubic meters; TL: Turkish Lira

15.4.2   Efficiency Scores of Merged WUAs

In the second stage, we performed efficiency analysis of a welfare model for artificially merged WUAs for R-1, R-2, R-3, L-1, L-2 and L-3. First data sets of all eighteen WUAs were merged into six WUAs. The newly created six WUAs (DMUs) were included in estimating the efficiency scores together with the current eighteen WUAs. Thus we have twenty-four DMUs altogether and could estimate the efficiency scores of new DMUs in reference to the existing DMUs. Table 15.8 shows the efficiency scores of merged WUAs with current WUAs. R-1, L-2 and L-3 scored 1 because they consisted of originally efficient WUAs as shown above. On the other hand, L-1 showed lowest scores (0.87) among the new WUAs, because it consists of the originally inefficient Cumhuriyet and Kuzey Y. WUAs. It is obvious that simply merging inefficient WUAs will result in inefficient WUAs.

Table 15.9 shows the projected input levels to reach the frontier for L-1. The reduction level required is more moderate compared to the reduction revel in Table 15.6. However, L-1 needs to reduce its technical staff and delayed payments by 34% and 22% respectively. By merging WUAs, the average efficiency score improved slightly from 0.95 to 0.97. However, simply merging to create a lesser number of WUAs does not improve the efficiency level significantly. In order for new WUAs to reach the frontier, significant reorganisation, i.e., reduction of some inputs, is required.

15.5 Conclusion

The WUAs that were established rapidly after 1994 became the major actors of water management in Turkey. The benefits of reducing the operation and maintenance costs and alleviating inequality of water distribution are considered large. However, recently some WUAs are having difficulties in management because of their relatively small-scale operation size. This chapter attempted to address ways to improve the relative efficiency of WUA management to adapt for future water scarcity under climate change in the region. We suggested the alternative composite efficiency index. We observed the case study of WUAs in the Lower Seyhan Irrigation Project in Adana, Turkey. We applied data envelopment analysis to compare efficiency levels with three models, i.e., management-, engineering- and welfare-focused models. The composite index was estimated by taking the geometric mean of the three efficiency indices. The analysis revealed that some WUAs are suffering from unfavourable management practices and there is a scope for major reorganisation to improve water efficiency. In particular, the reorganisation should come from the reduction of technical staff and delayed payments of water fees. The current eighteen WUAs are grouped into six artificially created WUAs to find the effects of the merger. Merging results shows that the average efficiency score improved slightly from 0.954 to 0.966. However, simply reducing the number of WUAs does not automatically improve the efficiency of DMU significantly. In order for newly created WUAs to reach the efficiency frontier, significant reorganisation, i.e., reduction of some inputs, such as staff salary, is required.

For further analysis, comprehensive assessment of WUAs management and productivity in the Seyhan River Basin with reference to other regions of Turkey may be necessary to understand and predict future scenarios for WUAs. In-depth analysis of both efficient and inefficient WUAs may be useful for identifying other factors that would affect water management practices and long-term efficiency that would need to be taken into account. Also, due to data availability, environmental factors, such as soil quality, gradient and salinity conditions in each WUA, were not considered. It may be worth separating the external environment that may be affecting management practices when the data set is available. WUAs' contribution to improving water efficiency and their basin-wide impact on water use and allocation are still in need of further investigation. Although the transfer of WUAs played a significant role in reducing cost and increasing fee collection rates, the efficiency analysis revealed that some WUAs are suffering from their small-scale management. It is suggested that a major reorganisation is necessary to further improve the management of WUAs for the efficient use of water resources and farmer welfare.

When the WUAs were established by the Municipality Law in 1954 under the Ministry of Interior, the DSİ could not directly control WUAs. In the summer of 2011, the government passed new legislation for the WUAs under the new Ministry of Forestry and Water Affairs that administered DSİ, so that DSİ can directly make recommendation to WUAs about their technical and financial work. The government institutional support is also important for improving the efficiency of WUAs for basin-wide water management.

In the face of future climate change and water scarcity in the region, the role of WUAs for efficient management of water resources seems quite important. This chapter suggested that not only improving engineering efficiency but also managerial efficiency is important. There is still scope for major reorganisation and efficiency improvement in managerial practices so as to prepare for future water scarcity under climate change. The results of this paper are also applicable to other regions of Turkey as well as regions in other countries where the efficient use of water resources is critical.

Acknowledgements   This is a partial contribution of Impact of Climate Change on Agricultural Systems in Arid Areas (ICCAP) Project, administered by the Research Institute for Humanity and Nature (RIHN) and the Scientific and Technical Research Council of Turkey (TÜBİTAK). Please address all correspondence to Chieko Umetsu, Graduate School of Agriculture, Kyoto University, Kitashirakawa Oiwake-cho, Sakyo-ku, Kyoto, 606-8502 Japan. Authors appreciate DSİ Region VI Adana for providing valuable information, and the staff of WUAs in LSIP for their kind assistance during our interview survey. Also we appreciate Rıza Kanber, Bülent Özekici and Tsugihiro Watanabe for facilitating the research and Thamana Lekprichakul for technical suggestions. Insightful and detailed comments from Elinor Ostrom are gratefully acknowledged. This paper does not reflect the view of DSİ, and the usual disclaimer applies.

References

Berkes F (2002) Cross-scale institutional linkages: perspectives from the bottom up. In Ostrom E, Dietz T, Dolšak N, Stern PC, Stonich S, Weber EU (eds) *The drama of the commons. Committee on the human dimensions of global change*, National Research Council. Washington, DC: National Academy Press.

Chakravorty U, Umetsu C (2003) Basinwide water management: a spatial model. *Journal of Environmental Economics and Management* 45(1):1–23.

Charnes A, Cooper WW, Rhodes E (1978) Measuring the efficiency of decision-making units. *European Journal of Operational Research* 2:429–444.

Cooper WW, Seiford LM, Tone K (2000) *Data Envelopment Analysis: a comprehensive text with models, applications, references and DEA-Solver Software*. Boston: Kluwer Academic Publishers.

Dayton-Johnson J (2003) Small-holders and water resources: a review essay on the economics of locally-managed irrigation. *Oxford Development Studies* 31(3):335–339.

Donma S, Pekel M, Kapur S, Akça E (2004) Integrated rural development in river basin management: the Seyhan river basin example. Paper presented at Pilot River Basin Management Conference in 22–24 September, Brindisi, Italy.

DSİ (2003a) Briefing of WUA and year 2002 management activity report, DSİ VI Region, Adana.

DSİ (2003b) Transferred irrigation association year 2002 observation and evaluation report. DSİ VI Region, Lower Seyhan Irrigation Project, Operation and Maintenance Department.

DSİ (2003c) Year 2002 yield census results for areas constructed, operated and reclaimed by DSİ. DSİ Operation and Maintenance Department, Ankara.

Fujihara Y, Tanaka K, Watanabe T, Nagano T, Kojiri T (2008) Assessing the impacts of climate change on the water resources of the Seyhan river basin in Turkey: Use of dynamically downscaled data for hydrologic simulations. *Journal of Hydrology* 353:33–48.

Kanber R, Ünlü M, Çakmak EH, Tüzün M (2005) Country Report: Turkey Irrigation Systems Performance. *Options méditerranéennes, Series B* 52:205–226.

Lam WF (1998) *Governing irrigation systems in Nepal: institutions, infrastructure, and collective action.* Oakland, California: ICS Press.

Mert H (2003) *Introduction of General Directorate of State Hydraulic Works and 6th Regional Directorate.* Ministry of Energy and Natural Resources, DSİ, Adana, November 2003.

Molden D, Sakthivadivel R, Perry CJ, de Fraiture C, Kloezen WH (1998) *Indicators for comparing performance of irrigated agricultural systems*, IWMI Research Report 20, Colombo, Sri Lanka: International Water Management Institute.

Ostrom E (1990) *Governing the commons: the evolution of institutions for collective action.* Cambridge: Cambridge University Press.

Scheumann W (1997) *Managing salinization: institutional analysis of public irrigation systems.* Berlin: Springer.

Svendsen M, Nott G (2000) Irrigation management transfer in turkey: process and outcomes. In Groenfeldt D, Svendsen M (eds) *Case studies in participatory irrigation management.* Washington, DC: The World Bank.

Tekinel O (2001) Participatory approach in planning and management of irrigation schemes. Advanced Short Course – Appropriate Modernization and Management of Irrigation Systems. Kahramanmaraş, Turkey.

Umetsu C, Lekprichakul T, Chakravorty U (2003) Efficiency and technical change in the Philippine rice sector: A Malmquist total factor productivity analysis. *American Journal of Agricultural Econics* 85(4):943–963.

Umetsu C, Donma S, Nagano T, Coşkun Z (2006) The efficiency of WUA management in the lower seyhan irrigation project. *Journal of Rural Economics*: Special Issue 2005. 440–444.

Ward FA (2010) Financing irrigation water management and infrastructure: a review. *Water Resources Development* 26(3):321–349.

Yazar A (2002) Participatory Irrigation Management (PIM) in Turkey: a case study in the Lower Seyhan Irrigation Project. In Hamdy A, Lacirignda C, Lamaddalena N (eds) *Water valuation and cost recovery mechanisms in the developing countries of the Mediterranean region.* Bari: CIHAM, pp 191–210.

Chapter 16
An Econometric and Agro-meteorological Study on Rain-fed Wheat and Barley in Turkey Under Climate Change

Hiroshi Tsujii and Ufuk Gültekin

Abstract The objective of our study is to project the impacts of climate change on rain-fed wheat and barley production in Turkey for 2070–2079 and identify their policy implications. We first estimate the wheat and barley yield and area sown functions for the Konya and Adana provinces, which have been representative wheat and barley production areas in Turkey. Most of the wheat and barley in Turkey have been produced in rain-fed conditions. Rain-fed arable land in Turkey is either planted to wheat or barley, leading Turkish farmers to make planting decisions according to the relative price of these crops. The relative price reflects the previous year's Turkish demand for and supply of both crops, and of animal products produced by using barley as feed. As expected from the rain-fed, traditional and low input wheat and barley production in Turkey, most of the estimated yield and area sown functions for these crops have statistically significant correlation coefficients to spring heat-damage variables and drought-damage variables, as well as to the cumulative monthly rainfall variables. Iterative estimation methods were used for selecting the best correlation coefficients of these variables. These coefficients not only appropriately reflect the severe fragility in the rain-fed wheat and barley production on the rain-fed arable land in Konya and Adana, but also provide a good basis for estimating the 2070s' wheat and barley production using the monthly temperature and rainfall projected by a regional circulation model (RCM) in our study for that period. We think that the impact of climate change on crop yield and area sown in the real world is determined not only by the crop responses, but also by farmers' adaptations and agricultural experiment stations' research adaptations to climate change. This is affected by the changes in demand for and supply of wheat, bread, barley and animal products, reflected in the relative price between wheat and barley in the previous year, as well as by policy and institutional changes. Our model incorporates most of these aspects explicitly and

H. Tsujii, Ph.D. University of Illinois, USA, Professor Emeritus, Kyoto University, Agricultural Economics, 104-1 Higashianshincho, Okamedani, Fukakusa, Fushimi, Kyoto, Japan 612-0832; email: tsujii1809press@yahoo.co.jp.

U. Gültekin, Associate Professor, Çukurova University, Department of Agricultural Economics, Adana, Turkey; e-mail: ugultekin@gmail.com.

© Springer Nature Switzerland AG 2019
T. Watanabe et al. (eds.), *Climate Change Impacts on Basin Agro-ecosystems*,
The Anthropocene: Politik—Economics—Society—Science 18,
https://doi.org/10.1007/978-3-030-01036-2_16

implicitly. Thus, we can conclude that the process to assess the impacts of climate change on wheat and barley production in Turkey using our model better emulates the real world process than the physiological plant growth model that focuses the impacts of climatic change on mainly the growth of wheat and barley. Consequently, we insert the 2070–2079 monthly rainfall and temperature data projected by the RCM into the estimated yield and area sown functions in order to project wheat and barley production for that period. Then, adding the FACE 2 * CO_2 fertilisation effect of 13% to the projected yields, the final change rates in the wheat and barley production projected for the 2070–2079 period are −14% for wheat and −28% for barley in Konya, and −0.46% for barley and +3.5% for wheat in Adana. The projected impacts of climate change on wheat and barley production in Turkey can be calculated as weighted averages of these impacts with production weights for Central Anatolia and its peripheral region, where the regions of Konya and Adana are the typical representative areas. The impact of climate change on wheat production in Turkey is −12.06% of current production. For barley this impact is −14.39%. Given the self-sufficient wheat and barley market of Turkey, these impacts may cause a food crisis in the case of wheat, and severe shortage of feed and livestock products in the case of barley in Turkey. We suggest the development and use of new wheat and barley varieties that are resistant to spring heat damage and drought and preparation of the economic and political capabilities to import the needed wheat and barley that have been staple foods for Turkish people for thousands of years.

Keywords Wheat and barley · Rain-fed production · Climate change impact Econometric approach

16.1   Introduction

Under the A2 scenario of the SRES, AR4 (4th assessment report) of IPCC (the Intergovernmental Panel on Climate Change) predicts that the temperature of the Earth will increase by 2.8 °C by 2100 and rainfall will decrease in most subtropical regions including Turkey and drought-stricken areas. In turn, heat damage to crops, frequency of heavy rainfall and flood events will increase. AR4 also points out that an increase in these weather anomalies will significantly reduce production of crops and livestock products. These large decreases in the production of crops and livestock products of the low latitude areas covering Turkey could lead to a severe collapse of food security in the relevant areas.

The objectives of our study are to identify the impacts of climate change on wheat and barley production in Turkey and to assess the policy implications of the impacts. Our projection method of 2070–2079 wheat and barley production in Turkey consists of three steps. We first econometrically estimate wheat and barley yield and area sown functions for Konya and Adana provinces using annual production and monthly weather data for the period from 1959 to 2002. This period was chosen because of the data availability when our study started. We attempt to estimate the effects of spring heat-damage, drought, monthly cumulative rainfall and technological and institutional change to wheat and barley yields. For the wheat and barley area sown functions, we estimate the effects of cumulative monthly rainfall, the relative prices of wheat and barley, and technological and institutional change variables. This relative price reflects the previous year's Turkish demand for and supply of wheat and bread, and animal products produced by using barley as feed.

Second, we remove the model bias from the projected monthly rainfall and temperature data calculated by a regional circulation model (RCM) developed by Fujio Kimura (Sato et al. 2006) for the period from 2070 to 2079. Then, we insert the bias-removed projected monthly weather values to spring heat-damage dummies, a drought dummy, and cumulative monthly rainfall variables, sample averages of the price variables, technological change variables, and institutional change variables of the estimated wheat and barley yield and area sown functions. This is done in order to obtain the predicted values of 2070–2079 yield and area sown in Konya and Adana. Since the yield projection for 2070–2079 does not reflect the FACE 2 * CO_2 fertilisation effect,[1] (Long et al. 2006; Taub 2010; Lobell/Gourdji 2012), we add a 13% yield increase as the 2 * CO_2 fertilisation effect to the yield obtained in the second step of our method. As we assume that Konya represents the Central and South East Anatolian Regions, whereas Adana represents the remaining region of Turkey, we estimate the change from now to 2070–2079 in wheat and barley production in the country as the weighted average of production changes in Konya and Adana, with productions in these two regions as weights. We finally seek to evaluate their economic and policy implications based on the projected wheat and barley production changes obtained for Turkey.

16.2   Econometric Method versus Crop Growth Model for Estimating the Climate Change Impacts on Crop Production of a Country

The econometric study of the effects of climate and weather on agricultural production was initiated during the 1970s and 1980s by Hiroshi Tsujii, M. M. Yoshino, and others. Tsujii (1977) conducted an econometric study of the impact of rainfall on rice production in Thailand. A statistically significant effect of rainfall on rice

---

[1]Increase in atmospheric CO_2 concentration directly affects crop yield through photosynthesis and stomata conductance. FACE stands for Free-Air Carbon Dioxide Enrichment. It is a technique to identify the effect of elevated concentration of CO_2 to the crop yield by allowing natural or agricultural ecosystems to be fumigated by pipes with elevated concentrations of CO_2 in the field. The FACE 2 * CO_2 experiments from 1995 to 2010 consistently found a larger yield effect than the yield effects of the old greenhouse or chamber experiments. The FACE yield effects were 5–7% on rice and 8–20% on wheat. The simple average of these effects was 13%, which we assumed as the yield effect of 2 * CO_2 fertilisation to wheat and barley.

production was estimated using the rice production function for the Mae-Nam Chaopraya Delta of Thailand. The abundance and scarcity of water in the vast delta was measured by the positive and negative distances of water level from the normal level in November each year, determined by the historical experiences on the tall water measure pole established by a king at Ayuthaya, the centre of the delta. A significant positive correlation between this distance and rice production was found in the delta. This distance is the result of the rainfall in the delta and its upstream during a year before November each year. Tsujii (1987) and Tsujii/ Yoshino (1988) conducted econometric studies on the effects of temperature and climatic variation on Japanese rice and the agricultural market. An economic analysis of the effect of the Japanese rice insurance policy on rice production was conducted in 1986. The insurance policy paid indemnity to the farmers who suffered considerable weather and other natural damages in return for a premium which was heavily subsidised by the Japanese government. It was found that the rice insurance policy had a considerable positive effect on rice production in Japan. In these econometric studies concerning the impacts of climatic conditions and insurance policies on crop production, the supply function or production function is estimated with their usual explanatory and climate change variables, along with policy, technological change and institutional variables. In these econometric studies, in order to find out the impacts of weather and climate change on crop production or supply, time series data are used. This means that the estimated impacts of climate change on crop production or supply are long-run impacts covering not only the short-run but also the long-run impacts caused by farmers and experiment station adaptations to technological improvements, policy and structural changes. Moreover, the consumers' demand response and farmers' supply response to changes in the relative prices of wheat and barley are incorporated in the area-planted functions of our model.

Agronomic crop growth simulation is often used in the impact study of climate change on crop production. It is a rigorous and gradual increment crop growth model responding to changing moisture, temperature and atmospheric CO_2 content in controlled conditions, such as greenhouses and phytotrons. But this rigorous physiological method cannot handle the certain aspects of the total climate change impacts on wheat and barley production. These aspects are the adaptations by farmers and experimental stations to climate change, technological, policy and structural changes, and the consumers' demand response and farmers' supply response that are caused by the changes in the prices of wheat and barley. As we used the relative price of these crops as an explanatory variable in the area sown functions, we can conclude that the econometric model developed in this study is more appropriate to project the impacts of climate change on crop production in a specific country or region in the real world.

As an ideal approach, the integrated model of the crop level biological and physical models, the farm level economic model (Tsujii 1986), and the local, national, and global level economic and policy models was proposed by Parry (1990). However, it was impossible for Parry (1990) to conduct the integrated impact study due to the lack of detailed and realistic climate scenarios in the 1990s.

So climate change aspects, their impact on agricultural production, and adaptation to these impacts were studied separately at that time. Further, Parry (1990) stated that at crop level, the agronomic and econometric method should accurately reflect the knowledge of crop physiology, and the biological and physical model should describe dynamically the crop growth at various stages (Uchijima/Seino 1987).

We think the total impacts of climate change on crop yield and area sown are determined by the crop responses and the adaptations by farmers and agricultural experiment stations and by consumers to climate change in the long-run. Agricultural production is done in a region of a country, as many farmers, agricultural experiment stations and consumers react to changes in weather and climate, in the prices of outputs and inputs, in agricultural policy and technology, and in the availability of natural resources. Farmers choose crops, the variety of the crops, and farm management elements in response to the aspects of changes listed just above. Agricultural experiment stations determine their research strategies and develop new crop varieties and agricultural technologies in response to the changes stated just above.

We think yield and area sown functions estimated in our study reflect not only the crop response to climate change but also farmers' and experiment stations' adaptations to climate change and to other relevant economic, policy, technology and structural change variables because they are long-run response functions estimated using time series inputs, outputs, economic data, weather data, and institutional data (Tsujii 2005). As we will describe thoroughly later in this paper, estimating the yield function, we will find statistically significant coefficients for weather anomalies, such as spring heat damage, drought and monthly cumulative rainfall, and for the technological change variables. For the area sown function we estimate significant coefficients for relative prices between wheat and barley, monthly cumulative rainfall, and structural change variables using annual data for production and crop prices and monthly weather data from 1959 to 2002. We believe the estimated coefficients of our model based on these long-run and multi-sector data can appropriately reflect not only the crop response to climate change but also farmers' and experiment stations' adaptation to climate change and to other relevant economic, policy, technology and structural change variables. Our econometric and agro-meteorological method cannot rigorously estimate the crop response and the adaptive responses of farmers and experiment stations to climate change separately. But its heuristic nature of simulating real world impacts of climate change on crop production and the adaptations to climate change by farmers, experimental stations and consumers is a good basis for realistically projecting the impact of climate change on 2070–2079 wheat and barley production in Turkey.

16.3   Wheat and Barley in Turkey and the General Framework for the Projection of 2070–2079 Wheat and Barley Production

Wheat and barley in Turkey have mostly been produced under rain-fed conditions for human consumption and animal feed from 1961 to the present (FAO 2013). About 70–80% of wheat in Turkey has been produced for making bread, and the rest for animal feed, whereas most of the barley produced is consumed as animal feed (FAO 2013). The main production area of wheat and barley in Turkey is Central Anatolia, where the Konya province is located. Central Anatolia has been producing, on average, 36% of the total production of wheat in Turkey from 2010 to 2012 (Foreign Agricultural Service, USDA 2012). Central Anatolia is located on the vast Anatolian Plateau, where the average height is about 1,200 m above sea level and rainfall is about 300 mm per year, with dry summers and cold and wet winters. The other large wheat and barley production area is the South East of Turkey, which is located to the east of Central Anatolia. The area is mountainous with not much rainfall, and is dry in summer and cold in winter. In these two regions, wheat and barley and animal grazing are the major agricultural activities. On average, 51% of the total wheat production in Turkey was harvested in these regions from 2010 to 2012, when Konya produced about 8% of the total wheat production of Turkey. Since these regions produce more than half of Turkey's wheat production and have similar climatic and agricultural conditions, we group these areas as the core wheat and barley production sites in Turkey. We consider Konya to be the representative wheat and barley production province in the core area. The other target province of our research is the Adana province, which is located in the Çukurova region characterised by a Mediterranean climate. The Mediterranean region produces about 14% of the total wheat production in Turkey, and Adana alone produces about 7%. Such regions as the Aegean, Aydın and Hatay also belong to this climatic area, where it rains considerably in winter and is dry and hot in summer. The remaining regions, such as Thrace, the Black Sea and the North East, are considerably rainy and mild in winter and dry and hot in summer. Since these and the Mediterranean region have similar climatic conditions, we group them as the peripheral area for wheat and barley production in Turkey. We assume that the Adana province is a representative province from the perimeter area. The locations of the Konya and Adana provinces in Turkey are shown in Fig. 16.1.

Annual production of wheat and barley in Turkey varies greatly, as they are produced in rain-fed conditions. From 1961 to 2012 the annual rate of change in wheat production has ranged from −18 to +39%. For barley the range was from −30 to +35%. Belaid/Morris (1992) assert that chronic water deficiency and high temporal rainfall variability, too low temperature in winter and too high temperature in spring decrease wheat and barley yield in marginal West Asian areas, including Turkey.

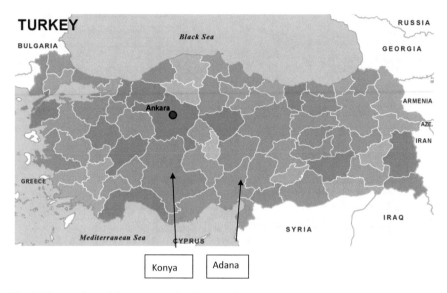

Fig. 16.1 Location of the Konya and Adana provinces. *Source* The authors

As described above, under the A2 scenario, AR4 of IPCC predicts that the surface temperature of the Earth will increase by 2.8 °C by 2100. AR4 expects that global warming will decrease rainfall in most subtropical regions. AR4 also predicts that global warming will increase the drought-stricken area, heat damage to crops, and the frequency of heavy rainfall and flooding. AR4 also says that a rise in these weather anomalies will significantly reduce production of crops and livestock products, especially the grain production of subsistence peasant in low latitudes, and may have major negative effects on food security.

The general framework for projection of 2070–2079 wheat and barley production in Konya and Adana is the three-stage process described in Fig. 16.2. We first estimate yield and area sown functions of wheat and barley for these provinces using time series data of annual yield and area sown of wheat and barley and monthly weather data from 1959 to 2002, focusing on the impacts of spring heat damage, drought, and monthly cumulative rainfall and temperature to yield and area sown. Second, we eliminate the model bias in the 2070–2079 RCM (Regional Circulation Model) monthly projection, then insert the bias-removed monthly weather projection in the estimated yield and area sown functions in order to project the 2070–2079 yield and area sown for Konya and Adana. Third, we add the 2 * CO_2 fertilisation effect to the projected wheat and barley yield, and multiply the adjusted yield and area sown to obtain the projected annual production of wheat and barley in Konya and Adana for 2070–2079.

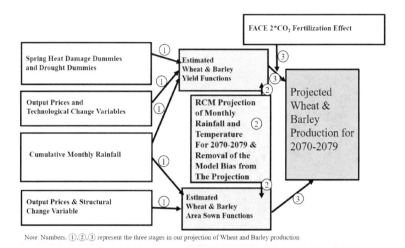

Fig. 16.2 The three stage projection of 2070–2079 wheat and barley production in Konya and Adana. *Source* The authors

16.4 Theoretical Description of Yield and Area Sown Functions of Wheat and Barley

The formulations of yield and area sown functions are based on the past studies of Hirohsi Tsujii on the relationship between rice, weather and climate in Thailand and Japan (Tsujii 1977, 1987; Tsujii/Yoshino 1988). The yield and area sown functions are assumed to be linear functions.

We first explain formulation of the yield function. The yield function has monthly spring heat damage dummies, monthly spring drought dummies, cumulative monthly rainfall variables that affect soil moisture, wheat and barley prices, and a technological change variable as its explanatory variables. All the estimated coefficients of drought dummies and heat damage dummies are assumed to be negative. The coefficients of a cumulative rainfall variable and prices are assumed to be positive.

Many Turkish agronomists mentioned that the spring heat damage to wheat and barley yield has been a very serious problem in Turkey. The heat damage occurs when temperature rises more than a threshold level at the flowering period of wheat or barley, since pollens lose fertility. Each of the heat damage dummies for April, May and June in our yield function is given the value of one when the average temperature in each of these spring months rises more than a threshold level within the month, and zero otherwise. We estimate many yield functions for each crop and each province, with all the three heat damage dummies changing the threshold temperature for each dummy by 1 °C in each estimation within the agricultural and meteorological probable range. We select a yield function with significant monthly heat damage dummies for each crop and each province. An example heat damage dummy is represented by DHAK(t)20. This means that the heat damage dummy takes the value of one when the April average temperature in Konya in the present year (t) rises higher than a threshold temperature of 20 °C and zero otherwise. In this study many heat damage dummies are tested by changing the monthly threshold temperature 1 °C in each estimation to find significant monthly threshold temperatures.

A drought dummy is represented, for example, by DDAK(t)20. This means that the drought dummy takes the value of one when the April average rainfall in the year of study (t) is less than 20% of the sample April average rainfall in Konya, and zero otherwise. We estimated many yield functions in order to find out the most significant drought dummy in April, May or December by changing the threshold percentage by one per cent each time in the agricultural and meteorological probable range. Some significant drought dummies are found in our estimated yield functions.

We assume that the cumulative monthly rainfall affects soil moisture and, in turn, the yield and area sown of wheat and barley in the country. We estimated many yield functions in order to reveal the significant duration of cumulative rainfall by changing the monthly duration of cumulative rainfall within the agricultural and meteorological probable range. This variable is shown as CRKO(t − 1) My(t), which covers the cumulative rainfall in Konya from October in the previous year to May in the present year. We found one cumulative rainfall variable was significant in each yield function of wheat and barley in Konya.

As an economic explanatory variable in the yield function, we used the annual relative price of wheat and barley, the annual growth ratio in the nominal price of the crop, and the annual real price of the crop. We assume the estimated coefficient of each price variable to be positive, since we think farmers respond positively to an increase in the price of the crop they grow.

We found a standard pattern in the change of wheat and barley yield over time in Konya and Adana. The pattern reflected a fast increase in the yield during the 1960s and 1970s, then stagnation after 1980 until the recent year. We assumed that this fast increase in the wheat and barley yield during the 1960s and 1970s reflected the impacts of the green revolution in Turkey. In order to capture the impact of the green revolution in the country, we introduced a technological change variable in each of our yield functions that takes the values of the years of the 1960s and 1970s until the starting year of yield stagnation, then takes the value of the year when stagnation started.

Next we explain the formulation of the area sown function of wheat and barley in our model. We assume the area sown function is explained by the cumulative monthly rainfall, price variables such as the relative price of wheat and barley, the growth rate of the nominal price of a crop, the real price of a crop and the structural change variable. We assume that the cumulative monthly rainfall before a crop is planted influences soil moisture, and thus affects the area sown with wheat or barley. We estimated many areas sown functions for each crop and each province with many cumulative monthly rainfall variables by changing the duration of months before the crop is planted in the agricultural and meteorological probable range. We tried to find the most significant cumulative rainfall variable, and found

Table 16.1 Variable definitions of the wheat yield function for Konya. *Source* The authors

Dependent variable	
YWK(t)	Average wheat yield in Konya in year (t)
Explanatory variables	
CRKO(t − 1) My(t)	Cumulative monthly rainfall from October in year (t − 1) to May in year (t)
DDAK(t)20	Drought dummy in April in year (t) (1 if April rain $\leq$ 20% of long-run average April rainfall, 0, otherwise)
DHAK(t)128	Heat damage dummy for April in year (t) (1 if April temperature $\geq$ 12.8 °C, 0, otherwise)
DHMyK(t)163	Heat damage dummy for May in year (t) (1 if May temperature $\geq$ 16.3 °C, 0, otherwise)
DHJeK(t)207	Heat damage dummy for June in year (t) (1 if June temperature $\geq$ 20.7 °C, 0, otherwise)
TCWYKR79	Technological improvement variable (1979 if year > 1978, year number, otherwise)

one in the area sown function for the Konya barley (Table 16.1). The expression of the cumulative rainfall variable is the same as the cumulative rainfall variable explained for the yield function just above.

In the rain-fed area in Turkey farmers can plant either wheat or barley. Thus we assume that the farmers decide to grow wheat or barley depending on the relative price of the two crops in the previous year. Wheat and barley are sown in the later parts of each calendar year in Turkey, where the areas sown are listed in the following calendar year of Turkish government statistics. Thus we assume the relative price between wheat and barley in the previous year affects the wheat area sown in the present year. An example expression of the relative price in the wheat area sown function is Pw/Pb(t − 1), meaning farm-level relative price between wheat and barley in the previous year (t − 1). For the barley area sown function the relative price Pb/Pw(t − 1) is obtained by reversing the numerator and the denominator of the relative price from the relative price of wheat area sown function. We assume the estimated coefficients of the relative prices in the area sown functions for wheat and barley are positive. We also assume that the coefficients of the growth rate of the nominal price and the real price of wheat and barley in the area sown function are positive.

We found one or two distinctive kinks in the long-run trends of the area sown of wheat and barley in Konya and Adana. There is no standard pattern in the kinks in the long-run trends of the area sown, but we think there must have been some structural changes behind the kinks, such as the policy changes. So we introduced a structural change variable in order to handle these kinks in each area sown function.

16.5 Estimated Yield and Area Sown Functions of Wheat and Barley

16.5.1 Wheat

We estimated the yield and area sown functions of wheat and barley in Konya and Adana provinces using the annual production and economic data and monthly weather data for the period from 1959 to 2002 by the OLS (the Ordinary Least Squares method) assuming the linear functional form. The results for wheat are presented in the current subsection.

Table 16.1 just above lists the definitions of the dependent and explanatory variables of the wheat yield function in Konya. Table 16.2 shows the best OLS estimation results of the function. The AR^2 (adjusted coefficient of determination) shows how well the estimated function explains the variation in the dependent variable, which ranges from 0 to 1 (the higher the value the better the explanation). This value is 0.676 and this means the estimated yield function explains 68% of the total variation in the wheat yield and this is acceptable in this study. The Durbin-Watson (DW) test informs us whether there is autocorrelation in the disturbance term of the Konya wheat yield function, or not – in other words, whether or not the current formulation of the function misses some important explanatory variables. The DW statistic tells us there is no autocorrelation at a 5% significance level. Thus our OLS coefficients are BLUE (the best linear unbiased estimates) and reliable.

We presumed in the theoretical discussion of the yield function in Sect. 16.4 above that the past cumulative monthly rainfall increases soil moisture and contributes to yield increase. The estimated coefficient of the cumulative rainfall during October in the previous year and May in the current year, that is CRO $(t − 1)$My(t), in our Konya wheat yield function is positive and highly significant. This coefficient explains that a 1 mm increase in the cumulative rainfall significantly contributes to a

Table 16.2 Estimated best wheat yield function for Konya. *Source* The authors

Variables	MODEL		
	$R^2 = 0.721$	$AR^2 = 0.676$	DW = 1.650*
	Coefficient	t-value	Significant at
CONSTANT	−89782.36	−6.76	0.00
CRKO(t − 1)My(t)	1.49***	2.41	0.02
DDAK(t)20	−78.63	−0.52	0.61
DHAK(t)128	−239.69***	−2.05	0.05
DHMyK(t)163	−122.50*	−1.37	0.18
DHJeK(t)207	−164.77*	−1.66	0.11
TCWYKR79	46.17****	6.86	0.00

Notes Significance levels of t test; **** 1%, *** 5%, ** 10%, * 20%
DW test * no autocorrelation at 1%, +* positive autocorrelation at 1%, −* negative autocorrelation at 1%, Ind. Indeterminate

1.5 kg/ha increase in wheat yield in Konya. The estimated coefficient of the drought dummy for April (that is DDAK(t)20) is negative as assumed but is not significant. All of the estimated coefficients of the three spring heat damage dummies for April, May and June in this yield function are significant and have negative values. This indicates that spring heat damage to wheat yield in Konya is very severe. The severest heat damage is estimated to occur in April in Konya, and it amounts to a 240 kg/ha decrease in the wheat yield when the average April temperature exceeds a threshold temperature of 12.8 °C. The technological change variable, TCWYR79, is used to capture the fast wheat yield increase in Konya up to 1979, which was assumed to reflect the green revolution in wheat in Turkey in Sect. 16.4 above. Its estimated coefficient is positive and highly significant. Thus our assumption of the wheat green revolution is confirmed by our econometric estimation of Konya's wheat yield function.

Table 16.3 lists variable definitions of the wheat area sown function for Konya. Table 16.4 shows the best estimated results of the function. The adjusted R^2 is 0.600 and acceptable. The Durbin-Watson test tells us that positive autocorrelation exists at a 5% significance level. Thus our OLS coefficients are not BLUE, or the formulation of the function may not be good. But all of the estimated coefficients are statistically significant. The sign of the estimated coefficient of relative price (Pw/Pb(t − 1)) is positive and this shows that farmers in Konya increase their wheat area sown as wheat price relative to barley price increases, as assumed in the theoretical explanation of the area sown function in Sect. 16.4 above. The coefficient means that a 0.1 increase in the relative price may make farmers increase the wheat area sown by 9,299 ha. The wheat area sown in Konya was stable up to 1979,

Table 16.3 Variable definitions of the wheat area sown function for Konya. *Source* The authors

Dependent variable	
ASWK(t)	Wheat area sown in Konya (sown in October in year (t − 1), and listed in year (t))
Explanatory variables	
Pw/Pb(t − 1)	Relative farm gate price between wheat and barley in year (t − 1)
SCASWK79F	Structural change variable to show stable wheat area sown before 1979 to declining wheat area sown after 1979

Table 16.4 The estimated best wheat area sown function for Konya. *Source* The authors

Variables	MODEL 2		
	$R^2 = 0.619$	$AR^2 = 0.600$	DW = 0.580⁺***
	Coefficient	t-value	Significant
CONSTANT	15793614.49****	7.76	0.00
Pw/Pb(t − 1)	92994.68*	1.33	0.19
SCASWK79F	−7516.76****	−7.42	0.00

Notes Significance levels of t test; **** 1%, *** 5%, ** 10%, * 20%
DW test *** no autocorrelation at 5%, +*** positive autocorrelation at 5%, −*** negative autocorrelation at 5%, Ind. Indeterminate

Table 16.5 Variable definitions of the wheat yield function for Adana. *Source* The authors

Dependent variable	
YWA(t)	Average wheat yield in Adana in year (t)
Explanatory variables	
DDMyA(t) 10	Dummy for drought in May at Adana year (t) (1 if May rainfall $\leq$ 10% of the long-run average May rainfall, 0, otherwise)
DHAA(t)162	Dummy for heat damage in April at Adana in year (t) (1 if April temperature $\geq$ 16.2 °C, 0, otherwise)
DHMyA(t) 235	Dummy for heat damage in May at Adana in year (t) (1 if March temperature $\geq$ 23.5 °C, 0, otherwise)
TCRWYA83	Technological change variable to show its improvement before 1983, then stagnation after 1983

and declined afterwards. There must have been some structural change such as a policy change to cause such a sharp change in the area sown trend. So we introduced a structural change variable, SCASWK79F, in order to handle this subject matter. The coefficient is assumed to be negative and this assumption is confirmed by the estimation result. We could not find any significant coefficient to the cumulative monthly rainfall variables, despite the numerous estimations applied to the variables in order to find their effect on the area sown through the soil moisture increase within this area sown function.

Table 16.5 shows the variable definitions of the wheat yield function for Adana and Table 16.6 shows the best estimated results of the function. The adjusted R^2 is high (0.727) and this means that the estimated function explains 73% of the variation in the Adana wheat yield. The Durbin-Watson test reliably revealed that an autocorrelation does not exist at a 5% significance level, and our OLS coefficients are BLUE and reliable. The estimated coefficients of one drought dummy and two spring heat damage dummies in the wheat yield function in Adana are not significant at a normal significance level of 5%, but are negative, as assumed in Sect. 16.4

Table 16.6 The estimated best wheat yield function for Adana. *Source* The authors

Variables	MODEL 1		
	$R^2 = 0.752$	$AR^2 = 0.727$	DW = 1.757***
	Coefficient	t-value	Significant
CONSTANT	−165180.62	−10.39	0.00
DDMyA(t)10	−75.92	−0.30	0.77
DHAA(t)162	−109.40	−0.54	0.60
DHMyA(t)235	−155.62	−0.36	0.72
TCWYAR83	85.00****	10.58	0.00

Notes Significance levels of t test; **** 1%, *** 5%, ** 10%, * 20%
DW test *** no autocorrelation at 5%, +*** positive autocorrelation at 5%, −*** negative autocorrelation at 5%, Ind. Indeterminate

Table 16.7 Variable definitions of the wheat area sown function for Adana. *Source* The authors

Dependent variable	
ASWA(t)	Area sown in Adana (sown in November, year (t − 1), and listed in year (t))
Explanatory variables	
NPCW(t − 1)	Change rate in nominal farm gate price of wheat from year (t − 1) to year (t − 2)
CRAJy(t − 1)S(t − 1)	Cumulative monthly rainfall from July in year (t − 1) to September in year (t − 1)
SCWASAR80	Structural change in the wheat area sown in Adana (1979, if year ≥ 1979, year number, otherwise)

above. We found above that all of the estimated coefficients of the three spring heat damage dummies for April, May and June in the Konya wheat yield function were negative and significant. Thus the estimated coefficients of spring heat damage dummies in Adana wheat yield function show that the spring heat damage to wheat yield in Adana is less severe than in Konya. The Adana wheat yield data increased until 1983, and remained stable afterwards. This is assumed to reflect the green revolution of the Adana wheat. TCWYAR83 was introduced in order to handle this green revolution effect. The estimated coefficient is highly significant and positive, and this confirms the green revolution effect.

Table 16.7 shows the variable definitions of the wheat area sown function for Adana, and Table 16.8 shows the best estimation results of the function. The adjusted R^2 is high (0.807), and this means that the estimated function explains 81% of the total variation in the wheat area sown for Adana. The Durbin-Watson test reflects an indeterminate autocorrelation at 1% significance level, and this means that the estimated coefficients may not be reliable. Anyway, the wheat area sown in Adana responds positively and significantly to the nominal wheat price change rate from the t − 2 year to the t − 1 year. In the other estimations of wheat area sown functions for Adana, the estimated coefficient of the relative price of wheat and barley was not significant. This may reflect the fact that the Adana farmers respond more to nominal wheat price changes than the relative price

Table 16.8 The estimated best wheat area sown function for Adana. *Source* The authors

Variables	MODEL		
	$R^2 = 0.8200$	$AR^2 = 0.8065$	DW = 1.3287[ind.]
	Coefficient	t-value	Significant
CONSTANT	−20108458****	−9.008	0.00
NPCW(t − 1)	362.79**	1.6730	0.10
CRJy(t − 1)S(t − 1)	154.0565	0.6485	0.52
SCWASAR80	10319.75****	9.1024	0.00

Notes Significance levels of t test; **** 1%, *** 5%, ** 10%, * 20%
DW test *** no autocorrelation at 1%, +*** positive autocorrelation at 1%, −*** negative autocorrelation at 1%, Ind. Indeterminate at 1%

because of the high inflation, ranging from 20 to 120% from 1972 till 2002 and even today. The estimated coefficient of the cumulative rainfall from July to September in the previous year is not significant, but has a positive sign, as assumed in Sect. 16.4 above. The structural change variable, SCWASAR79, is introduced to the wheat area sown function in order to capture a rising trend up to 1997 and the stagnation afterwards, where its estimated coefficient is highly significant.

16.5.2   Barley

Table 16.9 shows the variable definitions of the barley yield function for Konya, and Table 16.10 shows the best estimated result of the function. The adjusted R^2 is 0.670 and is acceptable. The Durbin-Watson test indicates that the autocorrelation does not exist at 1% significance level. Consequently, our OLS coefficients are BLUE and reliable. Table 16.10 first shows that the estimated coefficient of the nominal barley price change rate from year $t - 1$ to year t has a positive sign, as assumed in Sect. 16.4 above, but it is insignificant. The cumulative rainfall in Konya from October in the previous year to June in the current year, CRKO($t - 1$) Je(t), has a very significant coefficient at 1% significance level. This means, that the 100 mm increase in the cumulative rainfall (CRKO($t - 1$)Je(t)) adds to the barley yield in Konya by 300 kg/ha. The estimated coefficient of the May drought dummy is negative, as assumed in Sect. 16.4, above, but it is not significant. The estimated coefficients of the two spring heat damage dummies are negative, as assumed in Sect. 16.4 above, and significant. Among them, the estimated coefficient of DHAK (t)137 is highly significant at 5% level and its size is 468 and very large. This means that when the average April temperature rises higher than a threshold temperature of 13.7 °C, the barley yield in Konya decreases by a large amount (468 kg/ha). These

Table 16.9 Variable definition of the barley yield function for Konya. *Source* The authors

Dependent variable	
YBK(t)	Average barley yield in Konya in year (t)
Explanatory variables	
NPCB(t)	Change rate of nominal farm gate price of barley from year $(t - 1)$ to year (t)
CRKO($t - 1$) Je(t)	Cumulative monthly rainfall in Konya from October in year $(t - 1)$ to June in year (t)
DDMyK(t)17	Dummy for drought in May in year (t) in Konya (1 if rain $\leq$ 17% of long-run average May rainfall, 0, otherwise
DHAK(t)137	Dummy for heat damage in April in Konya in year (t) (1 if April temperature $\geq$ 13.7 °C, 0, otherwise)
DHMyK(t) 163	Dummy for heat damage in May in Konya in year (t) (1 if May temperature $\geq$ 16.3 °C, 0, otherwise)
TCBKR76	Technological change variable in barley yield (year number, if year < 1976, 1976, otherwise)

Table 16.10 Best estimation results of barley yield function for Konya. *Source* The authors

Variables	MODEL 1		
	$R^2 = 0.716$	$AR^2 = 0.670$	DW = 1.772***
	Coefficient	t-value	Significant
CONSTANT	−107552.86	−4.84	0.00
NPCB(t − 1)	1.38	1.03	0.31
CRKO(t − 1)Je(t)	3.00****	4.35	0.00
DDMyK(t)17	−154.26	−0.44	0.66
DHAK(t)137	−468.11***	−2.01	0.05
DHMyK(t)163	−176.56*	−1.63	0.11
TCBKR76	55.06****	4.88	0.00

Notes Significance levels of t test; **** 1%, *** 5%, ** 10%, * 20%
DW test *** no autocorrelation at 1%, +*** positive autocorrelation at 1%, −*** negative autocorrelation at 1%, Ind. Indeterminate at 1%

estimated significant coefficients of cumulative rainfall variables and spring heat damage variables mean that rainfall shortage and spring heat damage are very severe to barley in Konya, as was the case for wheat yield in Konya (Table 16.2). The technological change variable, TCBKR76, is assumed to reflect the wheat green revolution in Konya. Its coefficient is highly significant at 1% significance level.

Table 16.11 shows the variable definitions of the barley area sown function for Konya. Table 16.12 shows the best estimation results of the function. The adjusted R^2 is 0.922 and this is very high. The Durbin-Watson test tells us that the auto-correlation test is indeterminate at 1% significance level. Thus our OLS coefficients may not be reliable. The barley area sown in Konya responds highly significantly to the relative price of barley and wheat (Pb/Pw(t − 1)). The estimated coefficient of the relative price is highly significant at 1% level, and this indicates that a 0.1 increase in the relative price of the previous year causes Konya farmers to increase area sown to barley by about 10,000 ha in the current year. A similar size response by Konya farmers in their wheat area sown decision against a 0.1 increase in the relative price of wheat and barley was estimated in Table 16.4. We think this

Table 16.11 Variable definition of the barley area sown function for Konya. *Source* The authors

Dependent variable	
ARSWADA(t)	Area sown in Konya (sown in September, year (t − 1), and listed in year (t))
Explanatory variables	
Pb/Pw(t − 1)	Relative farm gate price between barley and wheat in year (t − 1)
CRKO(t − 2)S(t − 1)	Cumulative monthly rainfall from June in year (t − 1) to September in year (t − 1)
SCASBKR7582	Structural changes variable in barley area sown (1975, if year < 1976, year number, if 1975 < year < 1982, and 1982, otherwise)

Table 16.12 Best estimation results of the barley area sown function for Konya. *Source* The authors

Variables	MODEL		
	$R^2 = 0.928$	$AR^2 = 0.922$	$DW = 1.253$ Ind.
	Coefficient	t-value	Significant
CONSTANT	−61717564.18	−21.15	0.00
Pb/Pw(t − 1)	99897.08****	2.55	0.01
CRKO(t − 2)S(t − 1)	116.61**	1.94	0.06
SCASBKR7582	31371.95****	21.22	0.00

Notes Significance levels of t test; **** 1%, *** 5%, ** 10%, * 20%
DW test *** no autocorrelation at 1%, +*** positive autocorrelation at 1%, −*** negative autocorrelation at 1%, Ind. Indeterminate at 1%

similarity confirms the reliability of area sown statistics of wheat and barley and their relative price data. The estimated coefficient of the cumulative rainfall variable from October in year t − 2 to September in year t − 1 is significant. This tells us that a 1 mm increase in this cumulative rainfall increases soil moisture in Konya, and contributes to an increase in barley area sown of 117 ha. The estimated coefficient of our structural change variable, SCASBKR7582, is highly significant. It emulates the fast increase in the barley area sown during 1975 and 1982, and stagnation in other periods.

Table 16.13 shows variable definitions of the barley yield function for Adana. Table 16.14 shows the best OLS estimation results of the function. The adjusted R^2 is 0.568 and this is low but acceptable. The Durbin-Watson test tells us that the autocorrelation does not exist at 5% significance level. Thus our OLS coefficients are BLUE and reliable. The estimated coefficient of the relative price of barley to wheat in the year t − 1 (Pb/Pw(t − 1)) has a positive sign, as assumed in Sect. 16.4 above, but it is not significant. The estimated coefficients of two spring heat damage dummies in the barley yield function in Adana are negative and significant. But the estimated coefficients of two spring heat damage dummies in the wheat yield function in Adana were not significant, as shown in Table 16.6. We think this

Table 16.13 Variable definitions of the barley yield function for Adana. *Source* The authors

Dependent variable	
YBA(t)	Average barley yield in Adana in year (t)
Explanatory variables	
Pb/Pw(t − 1)	Relative farm gate price of barley to wheat at year (t − 1)
DDDA (t − 1)23	Dummy for drought in December in year (t − 1) (1 if December rainfall ≤ 23% of the long run average December rainfall, 0, otherwise)
DHAA(t)189	Dummy for heat damage in April in year (t) (1 if April temperature ≥ 18.9 °C, 0, otherwise)
DHMyA(t)234	Dummy for heat damage in May in year (t) (1 if May temperature ≥ 23.4 °C, 0, otherwise)
TCBAR79	Yield trend (Year number, if year < 1979, 1979, otherwise)

Table 16.14 Best estimation results of the barley yield function for Adana. *Source* The authors

Variables	$R^2 = 0.618$	$AR^2 = 0.568$	$DW = 2.349$***
	Coefficient	t-value	Significant
CONSTANT	−63940.1****	−6.09	0.00
Pb/Pw(t − 1)	107.44	0.39	0.70
DDDA(t − 1)23	−267.992*	−1.64	0.11
DHAA(t)189	−186.512*	−1.60	0.12
DHMyA(t)234	−287.568*	−1.50	0.14
TCBAR79	33.458****	6.23	0.00

Notes Significance levels of t test; **** 1%, *** 5%, ** 10%, * 20%
DW test *** no autocorrelation at 1%, +*** positive autocorrelation at 1%, −*** negative autocorrelation at 1%, Ind. Indeterminate at 1%

difference reflects the fact that barley is more suited to colder climates than wheat, and the Adana spring climate is too hot for barley but roughly appropriate for wheat. The estimated coefficient of the December drought dummy is −268. This means that if rainfall in December in year t − 1 decreases below the threshold of 23% of the long-run average of the December rainfall, the Adana barley yield decreases by 268 kg/ha. The larger heat damage of the two heat damage coefficients is a decrease in barley yield in Adana by 289 kg/ha when the May temperature rises more than the threshold temperature of 23.4 °C. The technological change variable, TCBAR79, is assumed to capture the wheat green revolution in Adana, and has an estimated highly significant positive coefficient.

Table 16.15 shows the variable definitions of the barley area sown function for Adana. Table 16.16 shows the OLS estimation results of the function. The adjusted R^2 is 0.925, which means appropriate conformity to the data. The Durbin-Watson test tells us there is a positive autocorrelation in the disturbance terms at 5% significance level. Thus our OLS coefficients are not BLUE and may not be reliable. The barley area sown in Adana does not respond significantly to any type of price variables, such as the real price of barley, the relative price of barley and wheat, and the nominal barley price change in year t − 1. This implies irrational behaviour on

Table 16.15 Variable definitions of the barley area sown function for Adana. *Source* The authors

Dependent variable	
ASBA(t)	Barley area sown in Adana (sown in October, year (t − 1) and listed in year (t))
Explanatory variables	
CTAO(t − 1)N(t − 1)	Cumulative monthly temperature in Adana from October in year (t − 1) to November year (t − 1)
CRAO(t − 1)N(t − 1)	Cumulative monthly rainfall in Adana from January in year (t − 1) to October in year (t − 1)
SCASBAF6880	Structural change variable for falling area sown during 1968 and 1980, and stagnant area sown in other times

Table 16.16 Best estimation results of the barley area sown function for Adana. *Source* The authors

Variables	$R^2 = 0.9252$	$AR^2 = 0.9195$	$DW = 0.5571^{+}$***
	Coefficient	t-value	Significant
CONSTANT	5327167****	21.234	0.00
CTAO(t − 1)N(t − 1)	226.1948	0.6859	0.4967
CRAO(t − 1)N(t − 1)	−4.96788	−0.6184	−0.5398
SCASBAF6880	−2686.85****	−20.34	0.00

Notes Significance levels of t test; **** 1%, *** 5%, ** 10%, * 20%
DW test *** no autocorrelation at 5%, +*** positive autocorrelation at 5%, −*** negative autocorrelation at 5%, Ind. Indeterminate at 5%

the part of the Adana farmers, indicating that they do not increase the barley area sown when the barley price increases. This implication may be related to the result of the Durbin-Watson test stated just above. The estimated coefficient of the cumulative temperature from October in year t − 1 to November in year t − 1 is negative but not significant. The estimated coefficient of the cumulative rainfall variable from October in year t − 1 to November in year t − 1 is negative but not significant. We intended to estimate the retardation in the expansion of barley area sown if rainfall in these months in the previous year increases, but was not successful statistically. The estimated coefficient of our structural change variable, SCASBAF6880, is highly significant, and the variable reflects well the structural change such as the sharp decline in barley area sown in Adana from 1968 to 1980.

If we summarise the results of our estimation of the yield and area sown functions of wheat and barley in Konya and Adana, most of the estimated wheat and barley yield functions are very good, having significant coefficients to spring heat damage dummies and drought dummies, except the Adana wheat yield function. The cumulative monthly rainfall variable has a significant positive coefficient in wheat and barley yield functions in Konya. Most of the estimated functions have relatively high adjusted R^2 values, and their disturbance terms are not auto correlated. Thus it is important to emphasise that our econometric estimation of yield and area sown functions for wheat and barley in Konya and Adana found that severe weather events such as spring heat damage and drought, which are predicted by IPCC to increase as global warming continues, have highly depressed wheat and barley production in Konya and Adana during the past five decades. As global temperature increases and precipitation decreases until 2070–2079, with climate change predicted by IPCC, it is expected that wheat and barley production in Turkey will be highly reduced, mainly through spring heat damage and drought. How much Turkish wheat and barley production will be reduced by climate change is the next topic addressed in this paper.

16.6 Removal of Model Bias in the 2070–2079 RCM Projection of Monthly Rainfall and Temperature and the Final Projection of Wheat and Barley Production in Turkey

In the previous section, we completed the first stage of our projection, i.e. estimation of yield and area sown functions of wheat and barley in Konya and Adana. In the current section, we conduct the following. We describe the second stage of our projection where first the model bias in our RCM projection of monthly rainfall and temperature for 2070–2079 in Konya and Adana is removed. Secondly, the model bias removed projection of monthly rainfall and temperature data for 2070–2079 are inserted into spring heat damage dummies, drought dummies, cumulative monthly rainfall variables of the estimated yield and area sown functions for wheat and barley in order to obtain the 2070–2079 projection of the yield and area sown. In the third stage of our projection, we increase the projected yields by 13% as the FACE 2 * CO_2 fertilisation effect. Then we modify yield and projected area sown in order to obtain the final projection of 2070–2079 production of wheat and barley in Konya and Adana (Long et al. 2006).

Figures 16.3, 16.4, 16.5 and 16.6 show how the model biases in the projected RCM 2070–2079 average monthly temperature and rainfall are removed in the second stage of our projection. Taking Fig. 16.3, that describes the procedure of model bias removal from the RCM projected 2070–2079 monthly temperatures in

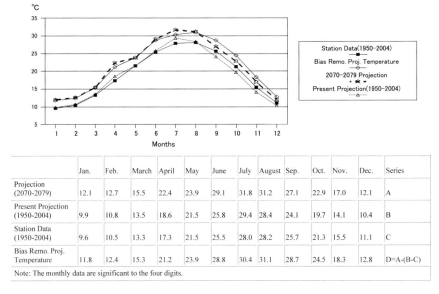

	Jan.	Feb.	March	April	May	June	July	August	Sep.	Oct.	Nov.	Dec.	Series
Projection (2070-2079)	12.1	12.7	15.5	22.4	23.9	29.1	31.8	31.2	27.1	22.9	17.0	12.1	A
Present Projection (1950-2004)	9.9	10.8	13.5	18.6	21.5	25.8	29.4	28.4	24.1	19.7	14.1	10.4	B
Station Data (1950-2004)	9.6	10.5	13.3	17.3	21.5	25.5	28.0	28.2	25.7	21.3	15.5	11.1	C
Bias Remo. Proj. Temperature	11.8	12.4	15.3	21.2	23.9	28.8	30.4	31.1	28.7	24.5	18.3	12.8	D=A-(B-C)

Note: The monthly data are significant to the four digits.

Fig. 16.3 Removal of the model bias from the projected Adana average monthly temperature for 2070–2079. *Source* The authors

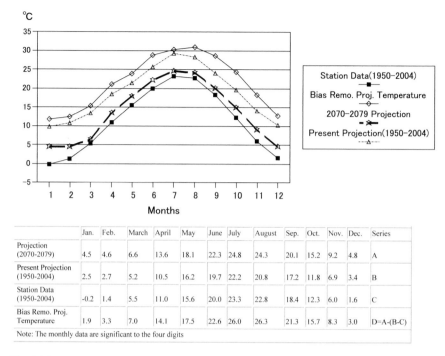

	Jan.	Feb.	March	April	May	June	July	August	Sep.	Oct.	Nov.	Dec.	Series
Projection (2070-2079)	4.5	4.6	6.6	13.6	18.1	22.3	24.8	24.3	20.1	15.2	9.2	4.8	A
Present Projection (1950-2004)	2.5	2.7	5.2	10.5	16.2	19.7	22.2	20.8	17.2	11.8	6.9	3.4	B
Station Data (1950-2004)	-0.2	1.4	5.5	11.0	15.6	20.0	23.3	22.8	18.4	12.3	6.0	1.6	C
Bias Remo. Proj. Temperature	1.9	3.3	7.0	14.1	17.5	22.6	26.0	26.3	21.3	15.7	8.3	3.0	D=A-(B-C)

Note: The monthly data are significant to the four digits

Fig. 16.4 Removal of the RCM Model bias from the predicted Konya average monthly temperature for 2070–2079. *Source* The authors

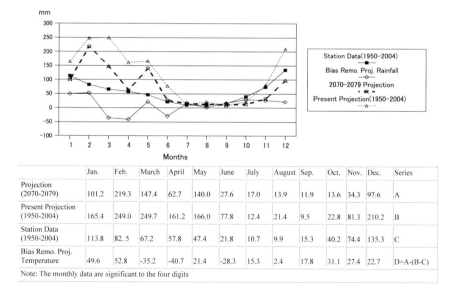

	Jan.	Feb.	March	April	May	June	July	August	Sep.	Oct.	Nov.	Dec.	Series
Projection (2070-2079)	101.2	219.3	147.4	62.7	140.0	27.6	17.0	13.9	11.9	13.6	34.3	97.6	A
Present Projection (1950-2004)	165.4	249.0	249.7	161.2	166.0	77.8	12.4	21.4	9.5	22.8	81.3	210.2	B
Station Data (1950-2004)	113.8	82. 5	67.2	57.8	47.4	21.8	10.7	9.9	15.3	40.2	74.4	135.3	C
Bias Remo. Proj. Temperature	49.6	52.8	-35.2	-40.7	21.4	-28.3	15.3	2.4	17.8	31.1	27.4	22.7	D=A-(B-C)

Note: The monthly data are significant to the four digits

Fig.16.5 Removal of the RCM Model bias from the predicted Adana average monthly rainfall for 2070–2079. *Source* The authors

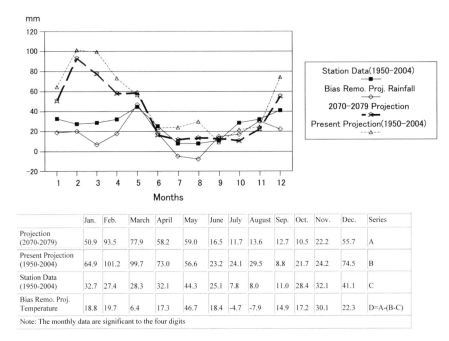

	Jan.	Feb.	March	April	May	June	July	August	Sep.	Oct.	Nov.	Dec.	Series
Projection (2070-2079)	50.9	93.5	77.9	58.2	59.0	16.5	11.7	13.6	12.7	10.5	22.2	55.7	A
Present Projection (1950-2004)	64.9	101.2	99.7	73.0	56.6	23.2	24.1	29.5	8.8	21.7	24.2	74.5	B
Station Data (1950-2004)	32.7	27.4	28.3	32.1	44.3	25.1	7.8	8.0	11.0	28.4	32.1	41.1	C
Bias Remo. Proj. Temperature	18.8	19.7	6.4	17.3	46.7	18.4	-4.7	-7.9	14.9	17.2	30.1	22.3	D=A-(B-C)

Note: The monthly data are significant to the four digits

Fig. 16.6 Removal of the RCM Model bias from the predicted Konya average monthly rainfall for 2070–2079. *Source* The authors

Adana as an example, we explain how the model bias is removed. The RCM of Fujio Kimura (Sato et al. 2006), based on the GCM of MRI that projected 2070–2079 monthly rainfall and temperature, has model biases. We assume the model bias is the difference between the present time (1950–2004) estimation of the average RCM monthly temperature (B) and the 1950–2004 station monthly average temperature (C) in Fig. 16.3. Thus we remove the bias by subtracting (B − C) from the RCM future (2070–2079) monthly estimation to obtain the bias removed projection of monthly temperature in Adana, i.e. A − (B − C). Figures 16.4, 16.5 and 16.6 show how the model biases are removed from the RCM projected 2070–2079 Konya monthly temperature, Adana monthly rainfall, and Konya monthly rainfall respectively.

Now we have the model-bias-removed projection of monthly rainfall and temperature data for 2070–2079. As described just above, we insert these bias-removed projected weather variables to spring heat damage dummies, drought dummies, and cumulative monthly rainfall variables of the estimated yield and area sown functions for wheat and barley in Konya and Adana in order to obtain the 2070–2079 projection of the yield and area sown. This is the next step in the second stage of our projection. In the third stage of our projection, we increase projected yields by 13% as the FACE 2 * CO_2 fertilisation Effect (Long et al. 2006). We then multiply

Table 16.17 The 2070–2079 projection of Konya wheat production. *Source* The authors

Wheat yield in Konya

Exp. variables	Coefficients	Var. value	2070s yield	59-02 yield			
CONSTANT	−89782.36	1	−89782.36				
TCWYKR79	46.17	1974.227	91150.07				
CRKO (t − 1)My(t)	1.49	218.65	325.79				
DDKA(t)20	−78.63	0	0.00				
DHKA(t)128	−239.69	1	−239.69				
DHKMy(t)163	−122.50	1	−122.50				
DHKJe(t)207	−164.77	1	−164.77				
					a (%)	**c (%)**	
			1.167	**1.619**	**−27.9**	**13**	

Wheat area sown in Konya

Exp. variables	Coefficients	Var. value	2070s Ar. sown	59-02 Ar. sown			
CONSTANT	15793614.00	1	15793614.00				
Pw/Pb(t − 1)	92994.68	1.3	120893.08				
SCASWK79F	−7516.76	1984.82	−14919401.92				
					b (%)		
			995.105	**988.510**	**0.7**		**a + b + c**
The total impact to Konya wheat production							**−14.3**

the modified yield and projected area sown in order to obtain the final projection of 2070–2079 production of wheat and barley in Konya and Adana.

All three stages of our projection of 2070–2079 wheat and barley production in Konya and Adana due to climate change can be summarised in the following Tables 16.17, 16.18, 16.19 and 16.20.

Assuming Y = yield, AS = area sown, Op = projected production, a = percentage change in Y caused by drought, heat damage, rainfall decrease, and temperature increase due to climate change, b = percentage change in AS caused by rainfall decrease and temperature increase due to climate change, and c = percentage change in Y caused by $2 * CO_2$ fertilisation, the effect of climate change to future crop production can be shown algebraically as follows:

$$Op = (1+a) * (1+c) * Y * (1+b) * AS \fallingdotseq (1+a+b+c) * Y * AS$$

As would be perceived by the readers of this paper, the values of a and b are different for each crop and each province, and the value of c is common to both

Table 16.18 The 2070–2079 projection of Konya barley production. *Source* The authors

The projection of barley yield in Konya

Exp. variables	Coefficients	Var. value	2070s yield	59-02 yield			
CONSTANT	−107552.86	1	−107552.86				
NPCB(t − 1)	1.38	32.8	45.26				
CRKO(t − 1)Je(t)	3.00	236.52	709.56				
DDKMy(t)17	−154.26	0	0.00				
DHKA(t)137	−468.11	1	−468.11				
DHKMy(t)163	−176.56	1	−176.56				
TCBKR76	55.06	1972.5227	108607.10		**a (%)**	**c (%)**	
			1.164	**1.896**	**−38.6**	**13**	

The projection of barley area sown in Konya

Exp. variables	Coefficients	Var. value	2070s Ar. sown	59-02 Ar. sown			
CONSTANT	−61717564.00	1	−61717564.00				
Pb/Pw(t − 1)	99897.08	0.79	78918.69				
CRKO(t − 2)S(t − 1)	105.62	259.86	27446.41				
SCASBKR7582	31371.95	1978.8182	62079385.06		**b (%)**		
			468.186	**481.044**	**−2.7**		**a + b + c**
The total impact of climate change to Konya barley production							**−28.3**

wheat and barley and Konya and Adana. The approximation sign, ≒, follows the fact that the values of a, b. c range from − 0.39 to + 0.13 and the elements in the formula of Op including the products of two or three of a, b and c are very small. The projected values of a, b and c for wheat and barley and for Konya and Adana are shown in the specified cells in Tables 16.17, 16.18, 16.19, and 16.20.

Tables 16.17 and 16.18 project the 2070–2079 wheat and barley productions in Konya respectively, whereas Tables 16.19 and 16.20 project the 2070–2079 wheat and barley productions in Adana respectively.

Using Table 16.17 as a model, we describe how the 2070–2079 wheat production in Konya is projected through the three projection stages mentioned above. The estimated coefficients of wheat yield function and of wheat area sown function in Konya found in estimating these functions in the first stage of our projection are inserted in the right hand cell of each of the corresponding explanatory variables of these functions. In the second stage of our projection, the model-bias-removed 2070–2079 monthly temperature and rainfall projections are inserted into spring heat damage dummies, drought dummy, and the cumulative monthly rainfall

Table 16.19 The 2070–2079 projection of Adana wheat production. *Source* The authors

Wheat yield in Adana							
Exp. variables	Coefficients	Var. value	2070s yield	59-02 yield			
CONSTANT	−165180.62	1	−165180.62				
TCWYAR83	85	1975.05	167879.25				
DDMay(t)10	−75.92	0	0				
DHDApril(t) 162	−109.4	1	−109.4				
DHDMay(t)235	−155.62	1	−155.62				
					a (%)	c (%)	
			2.434	2.684	−9.3	13	
Wheat area sown in Adana							
Exp. variables	Coefficients	Var. value	2070s Ar. sown	59-02 Ar. sown			
CONSTANT	−20108458.00	1	−20108458.00				
NPCW-1	362.79	31.2	11319.06				
CRAJy(t − 1) S(t − 1)	154.06	73.5	11323.15				
SCWASAR79	10319.75	1974.23	20373560.04		b (%)		
			287.744	288.207	−0.2		a + b + c
The total impact to Adana wheat production							3.5

variable of the estimated yield and area sown functions. This was done in order to get the values of these dummies and cumulative rainfall, and insert these values into the second cell to the right of each corresponding weather variable. For example, the spring heat damage dummy, DHKA(t)128 takes the value of 1 because the projected 2070–2079 temperature in April is higher than 12.8 °C. The values of the cumulative rainfall, the drought dummy, and the spring heat damage dummies thus determined are 218.65, 0, 1, 1, and 1, as shown in the values of variables column in the Konya wheat yield part of Table 16.17. We insert sample average values here for the technological change variable, relative price, and structural change variable in the Konya wheat yield and area sown functions and the value of 1 to the constant term. Then we get the sum of the products between the estimated coefficients and the projected values of the corresponding explanatory variables for the yield function to be 1,167, and for the area sown function to be 995,105 as the projected 2070–2079 yield and area sown for Konya wheat. The 1959–2002 sample average yield and area sown are 1,619 and 988,510 respectively. Thus the change rates from sample averages to 2070–2079 projected values are −27.9% for the wheat yield, i.e. a%, and +0.7% for wheat are sown, i.e. b%. In the third stage of our projection, we increase wheat yield by 13% as the FACE 2 * CO_2 fertilisation effect, i.e. c%. Consequently the total projected change in the Konya wheat production from the

Table 16.20 The 2070–2079 projection of Adana barley production. *Source* The authors

Barley yield in Adana

Exp. variables	Coefficients	Var. value	2070s Yield	59-02 yield			
CONSTANT	−63940.1	1	−63940.10				
Pb/Pw(t − 1)	107.44	0.79	84.88				
DDDA(t)23	−267.99	0	0.00				
DHAA(t)189	−186.51	1	−186.51				
DHAMy(t)234	−287.57	1	−287.57				
TCABR79	33.458	1974.227273	66053.70				
			1.724	**2.152**	**a (%)**	**c (%)**	
					−19.9	**13**	

Barley area sown in Adana

Exp. variables	Coefficients	Var. value	2070s Ar. sown	59-02 Ar. sown			
CONSTANT	5327167	1	5327167.00				
CTAO(t − 1)N (t − 1)	226.1949	42.79	9678.88				
CRO(t − 1)N(t − 1)	−4.96788	55.49	−275.67				
SCASBAF6880	−2686.85	1975.772727	−5308604.95		**b (%)**		
			27.965	**26.277**	**6.4**		**a + b + c**
The total impact to Adana barley production							**−0.5**

sample period to 2070–2079 is a + b + c = −27.9 + 0.7 + 13 = −14.3%. Similar processes were taken in order to project the 2070–2079 production under the climate change for barley in Konya, wheat in Adana, and barley in Adana, as shown in Table 16.18, Table 16.19, and Table 16.20 respectively.

Here, we would like to summarise the results of our econometric estimation of wheat and barley yield and area sown functions for Konya and Adana using the time series data from 1959 to 2002, and of our projection of the impacts of climate change to wheat and barley production in Konya and Adana during 2070–2079 based on the estimated functions and the projected monthly temperature and rainfall data for 2070–2079 by our RCM. Our econometric and agro-meteorological estimation of yield and area sown functions for wheat and barley revealed that the wheat and barley yield in Konya and the barley yield in Adana had been significantly reduced by spring heat damage. The same estimation also revealed that the cumulative monthly rainfall had significantly increased the barley area sown in Konya. Our model-bias-removed RCM projection of 2070–2079 monthly temperature and rainfall caused by the future climate change informs us that the monthly temperature in Konya and Adana for 2070–2079 will increase and the monthly rainfall for the same provinces will decrease considerably. Combining these monthly rainfall and temperature projections and our econometric estimations of wheat and barley yield and area sown functions, we expect that wheat and barley production in Konya and Adana for 2070–2079 will decrease considerably under climate change. This expectation was rigorously calculated for production of Konya wheat, Konya barley, Adana wheat, and Adana barley in Tables 16.17, 16.18, 16.19 and 16.20 respectively. Large projected decreases in the wheat and barley production of Konya are determined respectively by −14% and −28% in 2070–2079, mainly because of spring heat damage, as shown in Tables 16.17 and 16.18. But wheat and barley production in Adana were projected to change slightly by +3.5% for wheat and −0.5% for barley respectively, as shown in Tables 16.19 and 16.20. We think this difference is partly due to the much larger decreases (a%) in wheat and barley yields caused by the climate change in Konya by −28% for wheat and −39% for barley, as shown in Tables 16.17 and 16.18 and the smaller yield decreases due to climate change in Adana −9% for wheat and −20% for barley, as shown in Tables 16.19 and 16.20. This large difference in the yield decreases of wheat and barley between Konya and Adana probably reflects the fact that Adana is much warmer than Konya in the spring so that wheat and barley in Konya are more susceptible to spring heat damage caused by the higher spring temperature due to climate change than wheat and barley in Adana. Figures 16.7, 16.8, 16.9 and 16.10 graphically show projected percentage changes in yield, a%, in area sown, b%, in yield due to the FACE 2 * CO_2 fertilisation effect, c%, and in production, (a + b + c) %, of wheat and barley in Konya and Adana from the present time to 2070–2079 caused by climate change. These four Figures show the following points.

1. The projected impacts of climate change on wheat and barley yield, i.e. a%, are large decreases ranging from −10% for Adana wheat to −40% for Konya barley. Among them, the larger yield decreases occur in Konya barley (−40%, Fig. 16.8) and Konya wheat (−30%, Fig. 16.9). Much smaller decreases occur

Fig. 16.7 Effects of climate change on Adana wheat. *Source* The authors

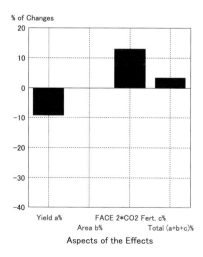

Fig. 16.8 Effects of climate
change on Konya barley.
Source The authors

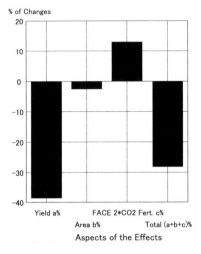

Fig. 16.9 Effects of climate
change on Konya wheat.
Source The authors

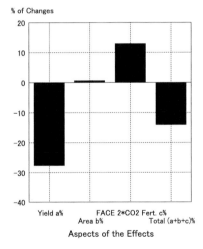

in Adana barley (−20%, Fig. 16.10) and Adana wheat (−10%, Fig. 16.7). This is because the projected 2070–2079 annual average temperature in Konya will be 21% (2.4 °C) higher than the station average, while in Adana it will be 15% (2.8 °C) higher than the station average, and this larger percentage increase in temperature in Konya than in Adana projected by our RCM contributes to greater spring heat damage and a larger decrease in wheat and barley yields in Konya than Adana.

2. The projected impacts of climate change on area sown with wheat and barley in Konya and Adana, i.e. b% are small, ranging from −2.7% for Konya barley to 6.4% for Adana barley (Figs. 16.7, 16.8, 16.9 and 16.10).

Fig. 16.10 Effects of climate change on Adana barley. *Source* The authors

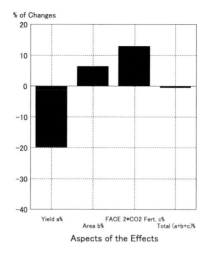

3. We assume the FACE 2 * CO_2 fertilisation effect on the yield of wheat and barley to be 13%, based on our literature survey (Figs. 16.7, 16.8, 16.9 and 16.10).
4. The total impact of climate change on wheat and barley production in Konya and Adana, i.e. (a + b + c)% are large decreases, such as −28.3% for Konya barley and −14.3% for Konya wheat, and smaller changes such as −0.5% for Adana barley and +3.5% for Adana wheat (Figs. 16.7, 16.8, 16.9 and 16.10).
5. In order to calculate the impacts of climate change on wheat production in Turkey, we computed the weighted average of the total impacts on the wheat of Konya and Adana, using wheat production of the core area, consisting of Central Anatolia and the South East region and wheat production of the peripheral region. We explained in Sect. 16.3 that we assume Konya represents the wheat and barley production in the core region and Adana represents the wheat and barley productions in the peripheral region of Turkey. Thus the impact of climate change on wheat production in Turkey is calculated as:

(The weight of wheat production in the core region) * (a + b + c for Konya wheat) + (The weight of wheat production in the peripheral region) * (a + b + c for Adana wheat) = 0.51 * (−14.28) + 0.49 * (+3.5) = −12.06%.

We calculated a similar formula for the impact of climate change on barley production for Turkey, and find the value of the impact to be: 0.51 * (28.3) + 0.49 * (−0.5) = −14.5%.

16.7   Conclusions

The objectives of our study are to project the impacts of climate change on rain-fed wheat and barley production in Turkey for 2070–2079 and identify their policy implications. Our projection consists of three steps. In the first step, we econometrically estimate the wheat and barley yield and area sown functions for Konya and Adana, which represent the core region and the peripheral region respectively for wheat and barley in Turkey. Most of the arable land in Turkey is rain-fed, and it is planted with either wheat or barley, and thus Turkish farmers decide which crop to plant depending on the relative price of the two crops. Thus this relative price is used as an explanatory variable of area sown functions of wheat and barley, and it reflects the previous year's Turkish demand for and supply of these crops and animal products produced by barley as feed. Rainfall variables and structural variables are also used as explanatory variables in the area sown function. The estimated wheat and barley yield functions are formulated as the functions of spring heat damage dummies, drought dummies, cumulative monthly rainfall variables and technological change variables.

Most of the estimated coefficients of yield and area sown functions in the first step are statistically significant and acceptable. These estimated functions have high or acceptable adjusted R^2s, and many of these functions show no autocorrelation in their disturbance terms. The estimated wheat and barley yield functions are significantly and dominantly explained by spring heat damage dummies, drought dummies, and cumulative monthly rainfall variables. This shows wheat and barley yields in Konya and Adana are mostly determined by temperature and rainfall. In other words, nature determines the yields of wheat and barley in traditional, low-input and rain-fed areas like Konya and Adana. Two of the four area sown functions estimated in our study have significant coefficients to the relative price. Adjusted R^2 of all the area sown functions are high but the autocorrelation tests are not perfect.

Second, we removed the model bias from the 2070s' monthly rainfall and temperature for Konya and Adana, projected using the RCM developed by F. Kimura (Sato et al. 2006). We then inserted the bias-removed monthly rainfall and temperature into heat damage dummies, drought dummies and other rainfall and temperature variables, and sample averages of the other variables of the estimated yield and area sown functions in order to project the bias-removed 2070s' yield and area sown of wheat and barley in Konya and Adana. These estimated yield values did not reflect the $2 * CO_2$ fertilisation effect in climate change. Thus, as the third step, a 13% yield increase as the $2 * CO_2$ fertilisation effect obtained from our literature study (Long et al. 2006) was added to the yield projections obtained in the second step of our method.

The agronomic crop growth simulation has often been used in the impact study of climate change on crop production. It is a rigorous and gradual increment crop growth model responding to changing moisture, temperature and atmospheric CO_2 content in controlled conditions such as a greenhouse and a phytotron. But this rigorous physiological method cannot handle certain important aspects of the total climate change impacts on wheat and barley production in the real world. These aspects are the adaptations by farmers and experimental stations to climate change, technological change and policy changes, and the consumers' demand response and farmers' supply response that can be captured by prices of wheat and barley, so we used the relative price of these crops as an explanatory variable in the area sown functions in our model.

The yield and area sown functions of our econometric model were estimated by time series data. So the estimated functions were long-run functions. The estimated coefficients of spring heat damage dummies in yield functions, for example, were average long-run effects of the damage during the sample period of 1959–2002 after the farmers and the experiment stations in Konya and Adana had adapted to the damage in the real world. We think this is the real world coefficient we need when we project the impact of global warming to crop yield. The agronomic crop growth simulation in a greenhouse cannot achieve this. Similar statements can be stated for other explanatory variables in the yield functions and area sown function as well. Consequently, we concluded that in order to project future production of crops in the real world, it would be better to use the econometric method similar to our model rather than the agronomic crop growth simulation.

The yield and area sown of wheat and barley in Konya and Adana projected under the assumption of global warming of A2 scenario of IPCC for the 2070s by the three steps explained above were then multiplied, and the 2070s' production of the two crops for Konya and Adana were projected. The final projected rates of changes in the production of wheat and barley from the sample period of 1952–2002 to the 2070s were −14% for wheat and −28% for barley in Konya, and +3.50% for wheat and −0.46% for barley in Adana. Our estimated impact of the A2 scenario global warming on wheat and barley production in Turkey as a whole can be calculated as weighted averages of these impacts with production weights for the core region and the peripheral region represented by Konya and Adana respectively. Our estimated impact of climate change on wheat production in Turkey is −12.06% compared with the 1952–2002 production. For barley this impact is −14.39%. Regarding the long-standing self-sufficient wheat and barley markets in Turkey these impacts may cause a food crisis in the case of wheat, and severe shortage of feed and livestock products in the case of barley. Finally, since wheat, barley and animal products have been the staple foods of Turkish people for thousands of years, we suggest the development and cultivation of new wheat and barley varieties resistant to spring heat damage and drought. Moreover, economic and political capabilities to import wheat and barley should be established in order to cope with the large shortages forecast by our model under climate change.

Acknowledgements This study is a part of the economic research sub-project of the ICCAP (Impact of Climate Change on Agricultural Production System in Arid Area). It is a collaborative research between Japanese and Turkish researchers in many disciplines. This project was supported by the RIHN (Research Institute for Humanity and Nature) in Japan and TÜBİTAK (The Scientific and Technical Research Council of Turkey) in Turkey.

References

Belaid A, Morris ML (1992) *Wheat and Barley Production in Rainfed Marginal Environments of West Asia and North Africa: Problems and Prospects*, CIMMYT Economics Working Paper 91/02, pp. 23–24.

FAO (2017), FAOSTAT.

Lobell DB, Gourdji SM (2012) The Influence of Climatic Change on Global Crop Productivity. *Plant Physiology Review*. https://doi.org/10.1104/pp.112.208298.

Long SP, Ainsworce EA, Leakey AEB, Nosberger J, Ort DR (2006) Food for Thought: Lower-Than-Expected Crop yield Stimulation with Rising CO_2 Concentrations. *Science* 312:1918–1921.

Nordhause DN (1993) Reflections on the Economics of Climatic Change. *Journal of Economic Perspectives* 7(4):12.

Parry M (1990) *Climatic Change and World Agriculture*. London: Earthscan, pp. 24–36.

Sato T, Kimura F, Kitoh A (2006) Projection of global warming onto regional precipitation over Mongolia using a regional climate model. *Journal of Hydrology* 333(1):144–154.

Taub RD (2010) Effects of Rising Atmospheric Concentrations of Carbon Dioxide on Plant. *Nature Education Knowledge* 3(10):21.

Tsujii H (1977) Effect of Climatic Fluctuation on Rice Production in Continental Thailand. In: Takahashi K, Yoshino M (eds) *Climatic Change and Food Production*, pp. 167–79. Tokyo: University of Tokyo Press.

Tsujii H (1986) An Economic Analysis of Rice Insurance in Japan. In: Hazel P, Pomareda C, Valdes A (eds) *Crop Insurance for Agricultural Development: Issues and Experience* pp. 143–155. Baltimore: Johns Hopkins University Press.

Tsujii H (1987) Effects of Temperature Variations on Japanese Rice Market and Their Implications for Policy. In: Parry ML, Carter TR, Konjin NT (eds) *The Impact of Climatic Variations on Agriculture. Volume 1. Assessments in Cool Temperate and Cold Regions*. Dordrecht: Reidel.

Tsujii H, Yoshino MM (1988) The Effects of Climatic Variations on Agriculture in Japan. In: Parry ML, Carter TR, Konjin NT (eds) *The Impact of Climatic Variations on Agriculture. Volume 1. Assessments in Cool Temperate and Cold Regions*, Part VI, Dordrecht, The Netherland: Kluwer Academic Publishers, for International Institute of Applied Systems Analysis (IIASA) at Vienna and United Nations Environment Program, pp. 725–863.

Uchijima Z, Seino H (1987) The Effect of Climatic Variation on Latitudinal Shift of Plant Growth Potential. In: Parry ML, Carter TR, Konjin NT (eds) *The Impact of Climatic Variations on Agriculture. Volume 1. Assessments in Cool Temperate and Cold Regions*. Dordrecht: Reidel.

Tsujii H (2005) An Economic and Institutional Analysis of the Impacts of Climatic Change on Agriculture and Farm Economy in Eastern Mediterranean and Central Anatolia Regions in Turkey. In: Tsujii H (ed) *An Interim Report of the Socio-Economic Sub-group of ICCAP*. Kyoto, Japan: The Research Institute for Humanity and Nature.

USDA (2012) *Foreign Agricultural Service. Grain Report*. Washington, D.C.: USDA, 3.

Chapter 17
Response of Farm Households to Climate Change with Social Customs of Female Labour Participation in the Mediterranean Region of Turkey

Takeshi Maru and Motoi Kusadokoro

Abstract The aims of this paper are to simulate how farm households react to the changes in agricultural production caused by climate change and to examine how social customs affect the changes in agricultural production in rural areas of the Mediterranean Turkey. Based on the field survey results, a household model that includes the social customs that express dis-utility due to divergence from socially-determined female labour participation in crop production is constructed. Additionally, computable general equilibrium (CGE) simulation analysis that enables quantitative evaluation of the effects of climate change is conducted. In this paper, the A2 global warming scenario is assumed and the following results are obtained: (a) the farm profit will be increased through the enhanced crop yield under the sub-scenario that does not assume any CO_2-effect; (b) the farm profit will be decreased through the reduced crop yield under the sub-scenario that assumes doubling CO_2-effect; and (c) social customs slightly restrain agricultural production by restricting female labour participation in farm cultivation.

Keywords Climate change · Female labour participation · Social customs

17.1 Introduction

Agriculture is an industry whereby people manage natural environments and resources to obtain output from plants and animals; i.e. agriculture is a field of interactions between humans and nature. Climate change affects natural environ-

T. Maru, Assistant Professor, Hitotsubashi University, Institute of Economic Research, Naka 2-1, Kunitachi, Tokyo 186-8603, Japan; e-mail: marl@ier.hit-u.ac.jp.

M. Kusadokoro, Assistant Professor, Tokyo University of Agriculture and Technology, Institute of Agriculture, Saiwai-cho 3-5-8, Fuchu, Tokyo, Japan; e-mail: motoi_k@cc.tuat.ac.jp.

© Springer Nature Switzerland AG 2019   375
T. Watanabe et al. (eds.), *Climate Change Impacts on Basin Agro-ecosystems*,
The Anthropocene: Politik—Economics—Society—Science 18,
https://doi.org/10.1007/978-3-030-01036-2_17

ments, such as temperature, precipitation, water availability, humidity, pests, diseases, etc. Agriculture will apparently be one of the industries which is most vulnerable to climate change in the future. Evaluating the effects of climate change on agriculture is an urgent agenda, because it is already becoming a reality, especially in the most vulnerable part of the world.

Many agronomic studies which explore the dependence of plant and animal productivities on natural environments have projected and quantified the potential impacts of climate change on crop and animal productivities (Gornall et al. 2010; Kurukulasuriya/Rosenthal 2003). However, lack of awareness about the possibility of adaptations to climate change may lead to a significant bias in the projections made by these studies (Mendelsohn et al. 1994). Even if high temperature and low precipitation in one area reduce the yields of some major crops, new varieties suitable to the new climatic conditions might be developed by local research institutions. This would entail the localisation of the high-yield varieties that led the Green Revolution to success and facilitated the dramatic increase in wheat and rice across the world in the latter half of the 20th century. Even in the absence of technological innovations, farmers could change the timing of planting and the cropping system and introduce some new crops to mitigate the adverse effects of climate change. Furthermore, if the CO_2 concentration could result in an increase in the yields of some crops, farmers might be able to find more beneficial cropping systems than those currently used. If these adaptations are not taken into consideration, the studies might overestimate or underestimate the damage or benefits of climate change.

A growing body of interdisciplinary literature has tried to take into account adaptations to climate change. These studies can be roughly classified into two groups based on their approaches to quantify the effects of climate change. The first group adopts simulation-based approaches. These studies generally use the computable general equilibrium (CGE) model or a related methodology as the key components to take into account the response of the human beings, including consumers, producers, and governments, and to evaluate the effects of climate change in terms of welfare (Darwin et al. 1995; Stern 2007; Robinson et al. 2012). An obvious benefit of taking this approach is that it can define the types and abilities of the adaptations which can be adopted by each agent in response to anticipated or actual climate change. However, since the model cannot capture the effects of adaptations which are not defined in the model, more assumptions and models that are more complex are required to account for different types of adaptation.

The second group evaluates the effects of climate change based on econometric analyses, whereby some dependent variables, such as yield, land rent, and agricultural profit, are regressed on the factors which will be affected by climate change, as well as farm structures, social structures, etc. (Mendelsohn et al. 1994; Schlenker et al. 2005; Deschenes/Greenstone 2007). In order to derive the econometric model, these studies generally assume that producers and landowners exhibit profit max-imisation or rent maximisation behaviours. If the econometric model is well con-structed and the statistical data used for the estimation includes sufficient variations to control the responses of producers to climate change, the evaluation based on the econometric model will extract the pure effects of climate change on agriculture. This would be due to the confounding effects of climate change and the responses of the farmers that could mitigate them. Since this approach does not need to define specific adaptation behaviours, broader types of adaptation might be captured than the former approach. However, careful treatment of the econometric models and the data is necessary for fair evaluations of the effects of climate change (Deschenes/Greenstone 2007).

Small and family farmers, whose objective might not be the profit maximisation that is generally explicitly or implicitly assumed in the existing literature, produce a substantial part of food in the world. Local institutions and social customs in the rural areas may regulate the distribution of the natural resources among farmers and the allocation of natural and human resources within farmers (e.g. community-based property rights on land, and separate property rights between the husband and wife in a family). When we evaluate the effects of climate change at global or nationwide levels, little attention to these factors might be justifiable because of the complexity of modelling and the difficulty of obtaining sufficient data. Accounting for the characteristics of producers and the social and cultural aspects will be one of the contributions of the local level studies which will focus on the effects of climate change in specific regions.

In this paper, we use data obtained from a farm household and village survey in the Adana prefecture to investigate the effects of climate change on labour allo-cation in farm households using an agricultural household model framework with social customs and CGE modelling. The framework of the model in this paper is based on the model developed by Maru (2014) to evaluate the effects of Turkey's accession to the European Union on Turkish agriculture. In Sect. 17.2 the general situation of the study site is explained and in Sect. 17.3 the essence of a household model with social customs is explained. A series of CGE simulation analyses is conducted in Sects. 17.4 and 17.5 offers the concluding remarks.

17.2 Study Site and Social Customs

17.2.1 Agricultural Production in Adana

The Adana prefecture is located in the Mediterranean region, central south of Turkey, with two rivers flowing through the town of Adana into the Mediterranean Sea. The Çukurova plain has fertile soil in the lower parts of the basin, and the

Table 17.1 Agricultural production in Adana prefecture. *Source* Turkish Statistical Institute (2007)

	Production value (million TL)		Rank of Adana in Turkey	Share (%)	Production value per land (thousand TL/ha)	
	Adana	Turkey			Adana	Turkey
Crop production	1,699	45,680	5	3.7	2.85	1.72
Fruits	469	12,712	10	3.7	10.58	4.67
Vegetables	386	11,494	6	3.4	15.43	14.27
Other field crops	844	21,474	3	3.9	1.63	1.19
Wheat	318	7,605	3	4.2	1.02	0.82
Maize	227	1,060	1	21.4	2.29	1.94
Livestock production	168	15,574	35	1.1	–	–
Milk	100	6,595	25	1.5	–	–

Note Turkish Lira (TL) is expressed in present currency notation

middle/upper reach is a hilly area leading into the Anatolian highlands. Irrigation facilities are well-developed in the lower basin, and the proportion of the irrigation-introduced villages is high, whereas irrigation facilities are not well-developed in the middle/upper reach.

The Adana prefecture consists of 12,651 km^2 land area, of which 4,776 km^2 is used for farmland (State Institute of Statistics 2004). This land size used for agriculture is the largest in the Mediterranean region of the country. As shown in Table 17.1, Adana is the fifth largest prefecture in terms of value of crop production in Turkey. Furthermore, Adana is the only prefecture that ranked in top ten positions for vegetables, fruits and other field crops. More specifically, Adana is the third largest producer of wheat and the top producer of maize in Turkey. The crop yield of the prefecture is also higher than average in Turkey. All this makes it one of the most important crop production areas in Turkey. By contrast, Adana is only the 35th largest in terms of the value of its livestock production and, in this category, can be regarded as an average prefecture in Turkey.

Our study site was the Çukurova plain, where many irrigation-introduced villages exist (hereafter, irrigated area), and wheat, maize, citrus, cotton and vegetables are mainly cultivated. According to Kusadokoro/Maru (2007), maize, as a second crop, has a roughly 57% share of the planted area. Wheat and citrus follow with about 35% and 11% shares respectively. The reason for the popularity of maize in this area is its good balance in yield and labour demand, in addition to the aptitude of maize for flourishing in this area. Compared to wheat, its yield is high, and to vegetables and fruits, its labour demand is low because the tasks entailed by maize production are mechanised. In this area, some farmers manage more than

50 ha of land, and the average managed land size of this area is large compared to the average of Turkey. These farmers can take risks in production and produce more profitable crops, such as citrus and vegetables. However, this increases the labour requirements. The labourers needed during the harvesting period of these crops have previously been supplied by south-east Anatolia, but in recent years these labourers have been using their skills to develop agriculture in their own area. Consequently, farmers of the study area seem to produce mainly maize, taking into consideration the balance between profitability and labour requirements.

As for animal husbandry, the area was highly populated with sheep and goat flocks together with cattle in the past. However, due to the specialisation in crop production, which is more profitable than animal husbandry, the area of grazing land decreased and the cost of animal feed increased. According to Maru (2010), there exists a large productivity difference between crop and livestock production in the irrigated area of Adana, and farmers tend to stop keeping livestock as the managed land size gets larger. Consequently, only some of the farmers keep a few cattle in barns today.

17.2.2 Social Customs in the Study Site

Some existing literature has pointed out the labour division based on social customs in the agricultural production of Turkey. Morvaridi's (1993) analysis of the labour division in the eastern part of Turkey indicated that the honour of a male household member could be compromised when he did so-called women's work, even if the household was poor and the female could not work because of illness or injury. Hoshiyama (2003), who conducted a sociological survey in the Çukurova plain, showed that the labour type is customarily determined by sex. For example, cultivating vegetables and picking cotton are being considered to be the work of women. In these studies, factors such as religion, economic circumstances and conditions of location/technologies for suitability of agricultural production are introduced as the bases of social customs. In other words, social customs can be considered as a complex of such factors that reflect the characteristics of the society, and attaining an equilibrium status for reaching a standard/norm society that regulates the sharing of labour.

The male members of the study site mainly work on crop production, such as ploughing, planting, fertilising and, in some places, animal husbandry. Female members primarily do housework and animal husbandry, such as barn-cleaning and feeding. Additionally, a few women do supplementary work, undertaking various tasks, such as planting vegetable seedlings, ploughing and irrigating, alongside male family members when the household manages a small area of land. Thus, strict sanctions on labour division seem not to exist in this part of the country.

Meanwhile, concerning the other years, farmer surveys showed that, for men, the main labour on their own farm was crop production, while for women it was animal husbandry. Furthermore, in the village survey conducted in 2013, some significant statements were obtained in some cases, stating that female members needed to have the permission of male household heads to work outside. In addition, female members were not permitted to work outside the household except for a few days, mainly due to urgent household needs. The answer given for this by the household heads was that the female members had to give top priority to housework and animal husbandry. Moreover, some of the male household heads refused to answer this question. These answers and responses imply that farmers believe there exist different and well-defined types of work which males and females should do. As a matter of fact, they think letting women work outside is a sensitive problem related to the honour of the family. Although this outcome might differ to some extent according to the characteristics of the villages or size of the farms, it can be said as a whole that there exists a labour participation constraint which prevents women working in crop production outside the residential area due to social customs that keep them in their houses.

Standard household models may also partly explain the above situation. Let's assume that there is no labour market for female members, and farmers have a strong preference for non-tradable goods; these goods can be produced by domestic work only, and the technology might be more suitable for female members. In that case, an increase in the household income may increase the demand for housework and leisure, and thus decrease female labour participation in crop production.

However, women might have aptitudes for sensitive work, such as cotton-picking, and for attentive field management practices. The contemporary relief expected to occur by the decrease in the time allocated to housework due to technological progress in recent years might give female members more time for fieldwork. Nevertheless, farmers seldom allow female family members to participate in field cultivation, which implies that social customs regulating female labour participation cause inefficient resource allocations. Consequently, if female family members participate in field cultivation and if housework is properly shared among household members, agricultural production may in turn increase the utility of the household.

While south-east Turkey has traditionally been the source of outside labour, its development via governmental projects has made it difficult for the farmers in the study area to employ workers from that region. Under these circumstances, the restrictive social customs could have negative effects on crop production due to the insufficient labour supplied by female members. In this context, some counter-measures will be required if the impact of this restriction would create problems regarding the analyses of the climate change impact on farmers in the study site.

Finally, it must be noted that the study site does not represent the whole rural area of Turkey and the results obtained in this paper cannot be automatically applied to all other parts of the country.

17.3   Concept of Social Customs in the Model

In this section, the concept of social customs in the model is explained, based on the information mentioned above. In the study site and other similar areas, the decision of the household concerning the labour input level by economic and manpower aspects is different from the decision of society. If society imposes some sanctions on the household, the utility level of the household might decrease. Therefore, in this paper, social customs that directly decrease the utility level by deviating from the socially determined standard input level are introduced. The utility function of the household is modified as

$$U = \prod Z^{\alpha_z} - R$$

where U represents utility function, Z represents various goods and services consumed and R is the social customs restriction expressed as below:

$$R = \frac{\theta}{2}\left(t_f^C - \bar{R}\right)^2$$

t_f^C is the amount of labour participation of female members in crop production, $\bar{R}$ is the standard input level of t_f^C ruled by social customs, and θ is a parameter that shows the strength of the social customs. Here, $t_f^C > \bar{R} \geq 0$ is assumed to be based on the observed results in the study site. The more t_f^C is determined apart from $\bar{R}$, the more the utility level decreases, and the larger θ becomes, the more strongly the level decreases. In other words, when θ is large, the disapproval of society strongly decreases the utility level even though deviation from $\bar{R}$ is low. In societies where it is regarded as shameful for female family members to work outside ($\bar{R} = 0$), even small amounts of working outside on crop production by female members make the household utility level decline.

On the other hand, when θ is not so large, more deviations can be accepted. In this case, the following pattern would occur. It is considered acceptable for women to spend one or two days working on crop production activities with male family members if there is an unexpected drop in the labour force (e.g. a permanent employee is too ill to work for a couple of days). However, if female labour participation is scheduled for more than a few days, this means the household makes its female members work outside even though the work is substitutable with employed labour, causing a greater decrease in the utility level. Furthermore, if female labour participation is scheduled permanently, female members have to put working outside above housework and animal husbandry, which are considered to be important tasks for women. This drives the household to recognise its own behaviour as shameful.

By including restrictions imposed by social customs, the utility maximisation problem changes and the following equation is obtained by partially differentiating the maximising function with t_f^C:

$$p^C \frac{\partial Q^C}{\partial t_f^C} = p_f^* + \frac{\theta\left(t_f^C - \bar{R}\right)}{\mu_\pi}$$

and the shadow price of female labour on crop production p_f^{C*} contains new parts arising from the social customs restriction in addition to the normal shadow price of female labour, which is composed of marginal utility of female time consumption μ_{t_f} and marginal utility of income μ_π and expressed as $p_f^* = \mu_{t_f}/\mu_\pi$.

t_f^C in the second term of the right-hand side of the equation makes marginal valuation of the female labour upward from left to right. The positively larger the degree of response to social customs θ becomes, the larger the slope of shadow price of female labour on crop production. By contrast, the shadow price decreases wholly if the standard level $\bar{R}$ is positively large.

Female labour in crop production is generally considered unfavourable in the study site, therefore we set $\bar{R}$ at 0. If θ does not have a large effect on the marginal utility of income μ_π and marginal utility of time consumption (expressed as μ_{t_f}), the shadow price p_f^{C*} increases totally and t_f^C gets closer to 0 in the case of strict customs.

Even though the marginal product of female labour increases because of the increase of inputs or the rise of productivity, the amount of female labour on crop production may not increase compared to the case of no social customs if parts of the second term of the shadow price other than t_f^C increase. By contrast, the increase of female labour participation in crop production may approach the normal shadow price if $\theta \to 0$.

If social customs obstruct female labour participation in crop production, females have to spend more time on leisure or work more on livestock production, which has low productivity compared to crop production. This means the decline of μ_{t_f}. Additionally, μ_π increases due to the income decrease caused by the increase in payments for employees and leisure consumption. This means a drop in the normal shadow price p_f^*, indicating that social customs suppress the shadow price of females as a whole.

17.4   Simulation

17.4.1   Basic Settings

Here, the village-level CGE simulation analyses based on the household survey data are conducted. Based on the model in Sect. 17.3, the extent of the effect of climate change on households between the present case and the absence of social customs is simulated. Before proceeding to the explanations of setting details, we must note that there are lots of other possible policy/circumstance changes that might affect simulation strategy and results. For example, Turkey's accession to the EU might change market conditions and therefore affect the effects of climate change on household behaviour. A technological breakthrough in agricultural production or other industries might also affect the impacts of climate change. However, it is difficult to consider all the possible other changes since the increase in the number of targeted changes makes the number of simulation processes increase drastically due to the arising need for distinguishing which targeted change has what kind of effects and to what extent. Therefore, we do not include all the possible policy/circumstance changes and instead focus on the effects of climate change and social customs. This is because existing literature on the evaluation of the effects of climate change has mainly considered the market factors and paid little attention to social customs, even though social customs are one of important factors that could affect the behaviour of the households in the area.

Data is basically collected from the results of the farm household survey conducted in the irrigated villages of the lower basin of Adana. The sample size is 34 and is somewhat small because it is difficult to include the whole data concerning households, such as production statuses, value of assets, and family structure. However, this is common to other village-level CGE studies such as Dyer et al. (2006) and Taylor et al. (1999).

With an assumption of linear homogeneity, factor shares – each factor input value divided by the produced value – are used for exponents of Cobb-Douglas production functions. To calculate the factor share of female labour, the shadow price of female labour is needed. It is desirable to estimate the shadow price econometrically. However, the number of farmers managing animal husbandry and farmers where female members participate in labour are limited; therefore it is necessary to adopt other methods. Maru (2010), who surveyed the Adana prefecture, showed that there were no significant differences between the marginal productivity of own farm labour and market wages for both male and female members. Thus, this paper assumes that the initial value of the male wage and of the aggregated shadow price of the female labour supply in crop and livestock production is equal to market wages, taking into consideration the differences in the type of work.

As for the restrictions imposed by social customs, the shadow price of female labour for crop and livestock production is affected by the social customs at the time the database was constructed. Hence, it is necessary to set up values for the

Table 17.2 Shadow prices of female member in the base equilibrium (unit: TL/day). *Source* The authors

	$\theta = 0$	$\theta = 0.2$	$\theta = 0.4$
	Value	Value	Value
Overall	18.202	16.946	15.825
Crop production	18.202	24.508	30.137
Livestock production and others	18.202	16.929	15.793

parameters $\bar{R}$ and θ at the time. For $\bar{R}$, $\bar{R} = 0$ is assumed since there should be no female labour participation in crop production as a norm. For θ, it is desirable to define the value of θ by using econometric methods. However, the method is so difficult that θ is set in a convenient way. Thus, three cases are considered: these are no social customs restriction case ($\theta = 0$), weak restriction case ($\theta = 0.2$: shadow price of crop production is around 1.4 times higher than that of livestock) and strong restriction case ($\theta = 0.4$: shadow price of crop production is around 1.9 times higher than that of livestock).

For each value of θ, the other parameters are separately calibrated so that the base equilibrium solution can replicate the current conditions of resource allocation, production, and consumption. Shadow prices of female labour in the base equilibrium are shown in Table 17.2. Analyses are conducted by using the General Algebraic Modeling System (GAMS) software.

17.4.2 Simulation Results – Changes with Climate Change

The A2 scenario is adopted for simulation with two sub-scenarios constructed for this study. One of these is the sub-scenario without the increase of CO_2 in 2070–2079 (hereafter, 2070–2079) and the other is with the increase of CO_2 to the double in 2070–2079 (hereafter, 2070–2079 & $2 \times CO_2$). Ideally, we should utilise full information regarding climate-related technological changes in every crop. Otherwise, it might cause some bias since other crops may have different degrees of productivity changes and more opportunities to utilise female labour. However, due to the difficulty in obtaining the same information about every crop, we use simulation results for wheat and maize provided by the other subgroup of ICCAP (Ben-Asher et al. 2018) for productivity (yield) changes in crop production caused by climate change. This can still be considered an acceptable assumption since wheat and maize are the most important crops and mostly sown in this area. We primarily calculate the average productivity change rate of wheat and maize from change rates obtained by the SWAP and by the DSSAT models. Next, we calculate the weighted average of productivity change rate of the crop sector from the average productivity change rates of wheat and maize according to the ratio of the production amount of each crop at the base equilibrium. The calculated productivity changes in crop production are as follows; around 1.7% increase in the

Table 17.3 Simulation results: changes with climate change (Scenario: 2070–2079). *Source* The authors

	$\theta = 0$		$\theta = 0.2$		$\theta = 0.4$	
	Value	Percent change[a]	Value	Percent change	Value	Percent change
Shadow prices of female labour (TL/day)						
Overall	18.316	0.6	17.058	0.7	15.935	0.7
Crop production	18.316	0.6	24.848	1.4	30.638	1.7
Livestock production and others	18.316	0.6	17.041	0.7	15.902	0.7
Time usage of female member (day)						
Crop production	2.016	3.2	2.001	2.5	1.996	2.2
Livestock production	6.311	−34.8	6.360	−34.3	6.402	−33.8
Leisure	460.832	0.7	460.799	0.7	460.762	0.7
Profits (TL)						
Overall agricultural production	11,043.520	3.4	11,041.494	3.4	11,039.770	3.4
Crop production	12,383.043	4.9	12,369.611	4.9	12,357.680	4.9
Livestock production	−1,339.523	18.8	−1,328.117	19.1	−1,317.910	19.3
Equivalent variation	434.583	–	434.249	–	433.933	–

[a]This is the percentage change from the base equilibrium

case of sub-scenario 2070–2079, and around 1.0% decrease in the case of sub-scenario 2070–2079 & 2 × CO_2. The reason why crop productivity decreases in this sub-scenario is that the yield of maize decreases because of the shortened growing duration. Unfortunately, we could not obtain exact simulation results for livestock production in our field survey, although the farmer questionnaires revealed a probable milk yield decrease with the rising temperature. Therefore, we assume a 10% decrease in the productivity of livestock production due to the rise in temperature in both sub-scenarios. Finally, social customs restriction and other exogenous variables are assumed to be at the same level as the base equilibrium level. Tables 17.3 and 17.4 show the simulation results of changes, along with climate change for each case of θ.

First, we will see the results of sub-scenario 2070–2079 (Table 17.3). The rise in crop productivity and drop in livestock productivity makes the farmer concentrate on crop production, and shadow prices of female labour increase (about 1–2% from the base equilibrium) via increase in profits.

Table 17.4 Simulation results: changes with climate change (Scenario: 2070–2079 & 2 × CO_2).
Source The authors

	$\theta = 0$		$\theta = 0.2$		$\theta = 0.4$	
	Value	Percent change[a]	Value	Percent change	Value	Percent change
Shadow prices of female labour (TL/day)						
Overall	17.497	−3.9	16.294	−3.8	15.220	−3.8
Crop production	17.497	−3.9	23.784	−3.0	29.350	−2.6
Livestock production and others	17.497	−3.9	16.278	−3.9	15.189	−3.8
Time usage of female member (day)						
Crop production	1.995	2.1	1.976	1.2	1.968	0.8
Livestock production	6.949	−28.2	6.990	−27.8	7.026	−27.4
Leisure	460.217	0.6	460.194	0.6	460.166	0.6
Profits (TL)						
Overall agricultural production	10,218.708	−4.3	10,217.195	−4.3	10,215.921	−4.3
Crop production	11,537.282	−2.3	11,524.570	−2.3	11,513.283	−2.3
Livestock production	−1,318.575	16.9	−1,307.375	17.2	−1,297.362	17.5
Equivalent variation	−639.214	–	−639.188	–	−639.191	–

[a]This is the percentage change from the base equilibrium

With regard to the restrictions imposed by social customs, the results show that there exist differences between the shadow price of female labour for crop production and livestock production. The amount of female labour shifts from animal husbandry to crop production, but the shift is smaller in the socially restricted case.

Specialisation in crop production increases the total profit (about 3% from the base equilibrium). However, the increased amount of profit from crop production is weakened in the case of a higher θ. This indicates that restrictions imposed by social customs might reduce the income-augmenting effect of improvement in crop productivity caused by climate change and suppress agricultural production. However, there is little difference between the percentage change from the base equilibrium in the case of restriction and that of non-restriction (0.008 percentage point). This might mean that the differences in settings of base equilibrium also affect profit. Additionally, it must be noted that the diminution caused by restrictions imposed by social customs is not large compared to the whole growth caused by climate change.

The equivalent variation (the measure in evaluating the change in utility with income change by using the current price system) decreases when the effect of social custom restrictions is valid and strong. However, the change is small compared to the whole value caused by climate change, and this relatively small change might be due to the settings of the base equilibrium. It is necessary to judge the restrictions imposed by social customs after conducting another simulation that challenges the effects of the disappearance of social customs.

The results under the sub-scenario 2070–2079 & 2 × CO_2 (Table 17.4) reveal the decrease in productivity of both crops and livestock and consequently in the shadow prices. The total profit also decreases by about 4% from the base equilibrium, and the value is smaller in the case of the higher θ. Female labour in crop production increases in spite of a drop in productivity in crop production. This can be explained by a rise in marginal utility of income due to the income decrease. The effect of doubling CO_2 on agricultural production amounts to around a 7% decrease in the total profit from the base equilibrium in the case of no social restrictions. As with the results in Table 17.3, however, it should be noted that the effect of social customs is not large compared to the whole changes caused by climate change, and we have to wait to judge the effect of social customs until another simulation is conducted.

17.4.3 Simulation Results – Changes with Climate Change and the Disappearance of Social Customs

The objective of the simulation conducted here is to discover how the disappearance of social customs after climate change affects female labour participation in crop production and farm economy. It is assumed that the value of θ in each case changes to 0, which means that the shadow price of female crop labour becomes a normal shadow price p_f^*, after the simulation conducted in the previous subsection. Therefore, changes from the results obtained in the previous subsection show the effect of the disappearance of social custom restrictions. The obtained results appear in Tables 17.5 and 17.6.

Here, we see the results of sub-scenario 2070–2079 (Table 17.5) and of sub-scenario 2070–2079 & 2 × CO_2 (Table 17.6) jointly. In female labour, the disappearance of the social custom restrictions induces more participation in crop production and creates less participation in animal husbandry and leisure consumption. Under the sub-scenario 2070–2079, with the strict social customs ($\theta = 0.4$), the value of percentage change from the base equilibrium is 2.2 in Table 17.3 and increases to 96.2 in Table 17.5, suggesting that the effect of the increase in crop productivity caused by climate change is enhanced by the disappearance of the social customs. Under sub-scenario 2070–2079 & 2 × CO_2, female

Table 17.5 Simulation results: changes with climate change and the disappearance of social customs (Scenario: 2070–2079). *Source* The authors

	$\theta = 0$		$\theta = 0.2$		$\theta = 0.4$	
	Value	Percent change[a]	Value	Percent change	Value	Percent change
Shadow prices of female labour (*TL/day*)						
Overall	18.316	0.6	17.082	0.8	15.981	1.0
Crop production	18.316	0.6	17.082	−30.3	15.981	−47.0
Livestock production and others	18.316	0.6	17.082	0.9	15.981	1.2
Time usage of female member (day)						
Crop production	2.016	3.2	2.913	49.1	3.831	96.2
Livestock production	6.311	−34.8	6.338	−34.5	6.358	−34.3
Leisure	460.832	0.7	459.909	0.5	458.971	0.3
Profits (*TL*)						
Overall agricultural production	11,043.520	3.4	11,050.399	3.5	11,058.825	3.5
Crop production	12,383.043	4.9	12,378.936	4.9	12,377.580	5.0
Livestock production	−1,339.523	18.8	−1,328.537	19.1	−1,318.754	19.4
Equivalent variation	434.583	–	445.581	–	458.329	–

[a]This is the percentage change from the base equilibrium

labour participation also increases despite the decrease in crop productivity. This indicates the possibility that the disappearance of social customs makes households decide to increase female labour participation in crop cultivation to compensate for the negative effect of climate change on crop productivity and hence household income. Furthermore, the effect of the disappearance of the restrictions imposed by social customs on female labour participation increases if the stronger social customs restriction is supposed in both sub-scenarios.

Also, with the disappearance of the restrictions imposed by social customs, the total profit increases and exceeds the total profit in the non-restricted case. For example, in case of $\theta = 0.4$, the value of the overall agricultural production is 11,040 TL in Table 17.3 and increased to 11,059 TL in Table 17.5, whereas, in case of $\theta = 0$, the figures are 11,044 TL in Tables 17.3 and 17.5. It is difficult to simply conclude that the disappearance of restrictions imposed by social customs makes agricultural production more efficient. However, it is confirmed that the increase in female labour participation in crop production leads to the enlargement of agricultural production and its profit.

Table 17.6 Simulation results: changes with climate change and the disappearance of social customs (Scenario: 2070–2079 & $2 \times CO_2$). *Source* The authors

	$\theta = 0$		$\theta = 0.2$		$\theta = 0.4$	
	Value	Percent change[a]	Value	Percent change	Value	Percent change
Shadow prices of female labour (TL/day)						
Overall	17.497	−3.9	16.317	−3.7	15.264	−3.5
Crop production	17.497	−3.9	16.317	−33.4	15.264	−49.4
Livestock production and others	17.497	−3.9	16.317	−3.6	15.264	−3.4
Time usage of female member (day)						
Crop production	1.995	2.1	2.881	47.5	3.790	94.1
Livestock production	6.949	−28.2	6.967	−28.0	6.978	−27.9
Leisure	460.217	0.6	459.312	0.4	458.392	0.2
Profits (TL)						
Overall agricultural production	10,218.708	−4.3	10,225.633	−4.3	10,233.965	−4.2
Crop production	11,537.282	−2.3	11,533.445	−2.2	11,532.203	−2.1
Livestock production	−1,318.575	16.9	−1,307.812	17.3	−1,298.238	17.6
Equivalent variation	−639.214	–	−627.957	–	−615.069	–

[a]This is the percentage change from the base equilibrium

The results also show that the total shadow price of female members slightly increases compared with the results in Tables 17.3 and 17.4. This can be considered as a result of the relatively large increase in marginal utility of time consumption due to a decrease of leisure consumption compared to the marginal utility of income caused by the increase in goods consumption along with profit growth.

As for the equivalent variation, this value increases when the restrictions imposed by social customs disappear (in the case of $\theta = 0.4$, the increases are 5.6% and 3.8% compared with the results in Table 17.3 and Table 17.4 respectively). Thus, it can be concluded that the effects of increasing goods consumption on utility is larger than those of the decreasing leisure consumption due to the increase in female labour supply in crop production.

However, the changes caused by the disappearance of social customs are relatively small compared to the total increase caused by climate change. It can be said that the disappearance of social customs regarding female labour participation in irrigated areas of the Adana prefecture does not have a strong impact compared to the effects of improvement in production environments from economic points of view, even though it is an important issue for gender studies.

17.5 Conclusion

The restrictions imposed by social customs are evaluated in this paper and are proven to affect climate change impacts on female labour participation in agricultural production.

Climate change substantially affects agricultural production and farm profits. Under the sub-scenario 2070–2079, the effect of the productivity increase in crop production exceeds that of the productivity decrease in livestock production and the profit of overall agricultural production increases by around 3%. Under the sub-scenario 2070–2079 & $2 \times CO_2$, the productivity in both crop and livestock production decrease and therefore the profit of overall agricultural production decreases by around 4%. The effect of doubling CO_2 is calculated to be 7% of the decrease in the profit of the overall agricultural production.

The shadow price of female crop labour increases in cases where restrictions imposed by social customs are valid. Additionally, female crop labour, profits, and equivalent variation increase in the case of the disappearance of social custom restrictions. The disappearance of restrictions seems to promote technical changes towards further utilisation of the labour participation of female family members.

However, the obtained results also show that the effect of the disappearance of social custom restrictions is not large compared to that of changes in production circumstances caused by the climate change. From this viewpoint, policies that improve agricultural production circumstances become more important. On that basis, it is considered useful to ensure opportunities for female family members to work by promoting improvement of livestock production circumstances that have a particular affinity with social customs.

Nevertheless, the results indicate that the disappearance of social customs can let climate change give momentum to the increase in female labour participation. In the simulations of this paper, no change is assumed in terms of production technology. If technological adjustment is allowed, farmers can make decisions about crop production and respond to changes in production caused by climate change without needing to consider social customs. As a result, female labour participation can be encouraged and agricultural production can change and be more efficient in terms of both production technology and farm economy. This indicates that it is also important for agricultural policy to loosen social custom restrictions if they exist and to support female members indirectly with housework-mitigating means.

Acknowledgements This chapter is a partial contribution to The Research Project on the Impact of Climate Change on Agricultural Production System in Arid Areas in the Research Institute for Humanity and Nature. Also, the authors received financial support from the Japan Society for the Promotion of Science under Grant-in-Aid for Scientific Research [number 25850158, 26850147]; and the Asahi Glass Foundation under Research Encouragement Grant for Humanities and Social Sciences. The authors are grateful to Dr. Atsuyuki Asami, Dr. Seiichi Fukui, Dr. Hiroshi Tsujii and Dr. Tsugihiro Watanabe for their helpful comments to the early version of this chapter, and also Dr. Onur Erkan and Dr. Ufuk Gültekin for their kind help in the field survey (in alphabetical order). Finally, the authors appreciate the contribution of the editors of this book and the reviewers for improvements of this chapter. (Since the authors don't know the names of reviewers and all editors, these acknowledgements are perforce anonymous.)

References

Ben-Asher J, Yano T, Aydın M, Garcia y Garcia A (2018) Enhanced Growth Rate and Reduced Water Demand of Crop due to Climate Change in the Eastern Mediterranean Region. In: Watanabe et al. (ed) *Climate Change Impacts on Basin Agro-ecosystem*. Springer.

Darwin R, Tsigas ME, Lewandrowski J, Raneses A (1995) *World Agriculture and Climate Change: Economic Adaptations*. Agricultural Economic Report 703. U.S. Department of Agriculture, Economic Research Service; Washington, DC.

Deschenes O, Greenstone M (2007) The Economic Impacts of Climate Change: Evidence from Agricultural Output and Random Fluctuations in Weather. *American Economic Review* 97 (1):354–385.

Dyer GA, Boucher S, Taylor JE (2006) Subsistence Response to Market Shocks. *American Journal of Agricultural Economics* 88(2):279–291.

Gornall J, Betts R, Burke E, Clark R, Camp J, Willett K, Wiltshire A (2010) Implications of Climate Change for Agricultural Productivity in the Early Twenty-first Century. *Philosophical Transactions of the Royal Society B* 365(1554):2973–2989.

Hoshiyama S (2003) Female Inferiority and Gender Division of Labor: "Ayip" ("Shamefulness") in Turkish Rural Society (in Japanese). *Forum of International Development Studies* 24:95–111.

Kurukulasuriya P, Rosenthal S (2003) *Climate Change and Agriculture: A Review of Impacts and Adaptations*. World Bank; Washington, DC.

Kusadokoro M, Maru T (2007) The Features of Agriculture in Adana Prefecture: From the Result of Farm Survey. In The Research Project on the Impact of Climate Changes on Agricultural Production System in Arid Areas. *The Final Report of ICCAP*. Research Institute for Humanity and Nature; Kyoto, pp 257–264.

Maru T (2010) Agricultural Productivity and the Constraints on Female Labor Supply in Turkey (in Japanese). *Journal of Rural Problems* 46(1):148–153.

Maru T (2014) Economic Study on Effects of EU Accession on Agriculture and Farm Economy in Adana Prefecture of Turkey. In Consideration of Social Customs Restriction on Female Labor (in Japanese). PhD thesis, Kyoto University, Kyoto.

Mendelsohn R, Nordhaus WD, Shaw D (1994) The Impact of Global Warming on Agriculture: A Ricardian Analysis. *American Economic Review* 84(4):753–771.

Morvaridi B (1993) Gender and Household Resource Management in Agriculture: Cash Crops in Kars. In Stirling P (ed) *Culture and Economy: Changes in Turkish Villages*. The Eothen Press; Huntingdon, pp 80–94.

Robinson S, Willenbockel D, Strzepek K (2012) A Dynamic General Equilibrium Analysis of Adaptation to Climate Change in Ethiopia. *Review of Development Economics* 16(3):489–502.

Schlenker W, Hanemann WM, Fisher AC (2005) Will U.S. Agriculture Really Benefit from Global Warming? Accounting for Irrigation in the Hedonic Approach. *American Economic Review* 95(1):395–406.

State Institute of Statistics (2004) 2001 *General Agricultural Census: Village Information*. Ankara.

Stern N (2007) *The Economics of Climate Change: The Stern Review*. Cambridge University Press; Cambridge.

Taylor JE, Yunez-Naude A, Hampton S (1999) Agricultural Policy Reforms and Village Economies: A Computable General-Equilibrium Analysis from Mexico. *Journal of Policy Modeling* 21(4):453–480.

Turkish Statistical Institute (2007) *Agricultural Structure: Production, Price, Value 2004*. Ankara.

Chapter 18
Cost Impact of Climate Change on Agricultural Production in Turkey

Nejat Erk, Sinan Fikret Erk and İnanç Güney

Abstract Agriculture is a highly sensitive field in terms of climate change, as farming activities are strongly influenced by climatic conditions. Nonetheless, agriculture is one of the major contributing factors to increases in the greenhouse gases in the atmosphere. With this in mind, agriculture offers effective solutions to climate change by emissions and by sequestering carbon. This work explores and justifies comprehensive research being done in the area of the climate change cost impact on agriculture. the study examines three major pieces of research in the area of climate change cost impact on agriculture. Each methodology is unique in terms of the modelling used. Striking major findings seem to complement each other, sharing similar sensitivity figures. As commonly underlined, the biggest drawback is derived from models optimising their own outcome, given the restrictions and assumptions being made. Our modest proposal is to adopt holistic methods, with special emphasis on the ecosystem to see more realistic outcomes of climate change on agriculture in terms of costs born.

Keywords CGE models · Ecosystem · Holistic cost calculation

18.1 Introduction

Unlike most changes, climatic changes are global in scope and in most cases irreversible if we do not take the necessary action at an appropriate time. This in return negatively influences prosperity and the well-being of nations. In an extreme case, climate change has the power to deteriorate existing competences and can lead to wars between nations. Substitutability of fossil energy seems to be far easier than

N. Erk, Retired Professor, Çukurova University, Department of Economics, Adana 01330, Turkey; e-mail: erk@cu.edu.tr.

S. F. Erk, Ph.D. Student, Çukurova University, Department of Economics, Adana, Turkey; e-mail: inanfikreterk@gmail.com.

İ. Güney, Lecturer/Researcher, Çukurova University, Vocational School, Adana, Turkey; e-mail: iguney@cu.edu.tr.

© Springer Nature Switzerland AG 2019   393
T. Watanabe et al. (eds.), *Climate Change Impacts on Basin Agro-ecosystems*,
The Anthropocene: Politik—Economics—Society—Science 18,
https://doi.org/10.1007/978-3-030-01036-2_18

the substitutability of drinking water. Thus a reduction in the quality or quantity of pure drinking water could be a major source of conflict, and create neighbouring restrictions for outgoing river flows as the result of climatic change conditions. The worst outcome is that the relatively low-income regions that cannot afford to pay for adaptation policies will be further hit by low Gross Domestic Product (GDP) growth, low investment, high unemployment rates and other social problems. This, in brief, will affect the existing value of assets for all income levels. One other difficulty with climate change is that it is not a given that deterioration will be linear. Most econometric studies back non-linear trends.

Although they are just a minority, some business societies have started taking action in response to the environmental risks posed by climate change, but the percentage is at very low levels. One warning about risk calculations simply shows that by sectors and by regions there is great variability in risk assessment.

In the case of agriculture, cost variation is the highest because all climate-related impacts occur under the sun or blue skies. Reliability and accuracy of data have a vital impact on calculations, thus there seems to be a very fast-growing market for precise data to be used in risk assessment. One factor that stipulates cost or risk variation that is hard to predict is when the financial regulatory actions will come into force and with what scope. In the future we are expecting more unanimous intergovernmental timing in international regulations via reciprocity rules.

There is a competitive opportunity for firms with good calculations about the future prospects of a low carbon economy. The bias will be moving away from agricultural or manufacturing activities with high carbon economies. As for today, countries with high carbon emissions try to solve the problem by shifting high-carbon economic activities abroad. But, as in the case of the US, you can shift production of high carbon content to China but you cannot eliminate the risk of getting high carbon content through the wind over the Pacific Ocean. On top of that, as climate change impacts are far more commonly felt, country and regional advantages will be minimum.

After setting the general scope of cost analysis towards climate change, our key interest will be the agricultural sector. Initially this research will start with the measurement of the cost impact for climate change in agriculture, followed by international experiences of costs born in different countries and the study will end with detailed research done in Turkey. An intermediary international focus will give us a chance to make comparisons and benchmark the required roadmap.

Autumn 2015 was a turning-point for investors keen to mitigate climate change cost impacts. From 2016 onwards international interest rates have been expected to increase. Not only in emerging markets, but also among high-income countries, higher interest rates will deter investments towards climate change. There are still analysts who believe that climate change impacts will be restricted to specific zones, thus global efforts could be misallocation of scarce resources. One fact to be underlined is that collective action with parallel regulations is a must to mitigate the costs born by future generations. Among economists, there is a consensus on the pricing of carbon to truly understand the cost of climate change in agriculture. In the case of market failure, which is reflected in non-appropriate climate change costs, collective action will be hindered. Regulation at all levels should focus on removing market failures, which simply means that costs won't be paid by farmers, businesses or other institutions which create the climate change. This rule should apply within countries as well as among countries. Climate change can affect asset management by putting assets at risk of direct value loss from events like storms and floods, or by indirectly affecting asset returns in terms of large variation in temperature, humidity and beyond.

18.2 Climate Change Ecosystem and the Cost Dimension

Looking at countries around the world, we see that most countries are aware of the costs of climate change. Thus, no responses to environmental issues shows non-awareness on climate change issues. Some countries, like Finland, try to take all necessary actions to mitigate climate change, while Portugal shows no systematic action towards eliminating climate change outcomes. The ecosystem approach looks at the world from a global perspective and tries to understand the intricate relationships that exist between living resources, habitats and residents of the area. The approach verifies the linkages between variables like plants, trees, animals, fish, birds, microorganisms, water, soil and people living in the area. It also emphasises that micro-level approaches miss the link that exists and interacts globally. In terms of evidence-gathering, there is very little research done in terms of ecosystem dynamics around the World. There are modelling attempts which try to calculate the price dynamics of a unit change in plants, trees, animals, fish, birds, microorganisms, water, soil and people living in the area. But most empirical work, like the UK NAS MONARCH (Modelling Natural Resource Responses to Climate Change) programme 48, tries to assess the impact of climate change on wildlife in a particular area, such as the UK.

The ecosystem approach simply shows that reducing or mitigating CO_2 emission should begin by questioning global activities that lead to higher emissions. CO_2 emission is not the cause but the result of the way of living and production. As in the above cited case, there is a lack of global dynamics under the name of ecosystem policies towards partial optimisation. If a country reduces chemical fertiliser use to improve climate change conditions, one has to ask, what alternative fertiliser use will do to other living resources, to people and to all elements in the ecosystem. Additionally, use of chemical fertilisers could also create benefits not captured in the first level interaction. The interest towards ecosystem accounting comes from several emerging demands for integrating information on environmental sustainability and human well-being. The method simply creates a checklist to capture market and non-market areas of change from the base period for the global environment. Some other studies split the impacts of climate change into two headings (as underlined by the UNEP 2009): change impact on the environment and the social structure, which in no way reflects the interaction between plants, trees, animals, fish, birds, microorganisms, water, soil and people living in the area.

REPORTS

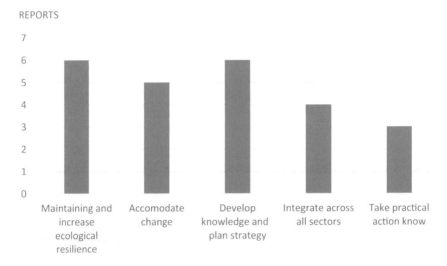

Fig. 18.1 Number of countries mentioning the different types of adaptation actions/measures in reports (their NAS) (total number of reports analysed: six). *Source* The authors

This in no way restricts alternative measures created to capture climate change and adaptation policies.

Figure 18.1 simply reflects the differences of understandings used for climate change and adaptation policies. Thus policies towards reducing carbon emission can reduce the negative impacts of climate change which may not capture indirect interactions between ecosystem elements. One solution might be offering carbon storage area availability as a remedy towards CO_2 emission reductions. Looking at systems strategically simply means selecting priorities, which could miss the dynamic interactions between ecosystem elements. These observations in no way attempt to undermine the valuable contributions of carbon emission reduction studies or other mitigation studies which have a more holistic view. As underlined in UNEP (2009) research, "Climate stabilisation can only be achieved by balancing emission sources (human and natural) and the global ecosystems' sink capacity. The protection and management of the world's ecosystems offers a highly cost-effective multiple 'Win' mechanism for mitigation by enhancing sink capacity and protects the essential life-supporting ecosystem services that will enable societal adaptation to climate change" (UNEP 2009). The need to include ecosystems management was likewise included in the Conference of the Parties COP 15 (COP 15 Agenda).

The same report also offers policy options to be adapted towards ecosystem approach solutions:

- Protection and maintenance of ecosystem services.
- Progress towards poverty alleviation.
- Enhanced food and water security.

- Buffering against extreme events.
- Support for disaster risk reduction.

An ecosystems approach is a strategy for the integrated management of land, water and living resources that promotes conservation and sustainable use in an equitable way (The United Nations Convention on Biological Diversity 2000).

18.3 Holistic Approaches to Climate Change

Another holistic study related to Turkey has been monitored by the IPCC (Şen 2013). The models used include ECHAM5 (Lohmann et al. 2007), the General Circulation Model (GCM) and the Regional Climate Model (ICTP-RegCM3). The study employs multiple scenarios involving demographic, economic, sociological and technology factors. A2 is the base scenario adapting upper bound changes. The model seems to be similar to RCP 8.5, which also provides upper bounds. Another model used in the study, GCM, uses 20th century observations to predict the coming century yields taking into account climate changes on Earth. A major weakness of the model is in the area of predicting precipitation. But, in general, GCM has high predictive power if outcomes are restricted to a single output. The GCM study developed specific scenarios for Turkey which downscaled higher resolution. The data used is derived from the Turkish State Hydraulic Works, the General Directory of Meteorology, and the General Directory of Forestry and Turkish Statistical Institute.

Apart from the IPCC study, ECHAM5 (Lohmann et al. 2007), which is a global circulation model, yields high-resolution outputs which were shared by UNFCCC in 2007. More research along similar lines has been undertaken by UNDP Turkey using the model of CCSM3 of the National Center for Atmospheric Research USA. In a recent study the National Communication of Turkey on climate change used UNFCCC and CCSM3 outputs. Another study by UNDP and State Hydraulic Works explored flood risks in the Seyhan Basin. More comprehensive research by Çukurova University and the Research Institute of Humanity and Nature (RIHN) team assessed the impacts of climate change on the water resources of the Seyhan River Basin in Turkey. Historical changes in the climate of Turkey are underlined in the IPCC report (Şen 2013), where temperatures are predicted to increase throughout the country together with the expected changes given below:

- Precipitation will be increased in the north-eastern parts;
- Mountain glaciers have been retreating (about 10 m/year);
- Timing of the peak discharges is shifting to earlier days;
- Sea level has been rising at different magnitudes in the surrounding seas of Turkey;
- The number of natural hazards seems to rise with increasing temperatures.

Within the same context, the same report has the following projections for the period 2041–2070 (Sen 2013):

- Temperatures will increase ubiquitously in all seasons, but the increases will be higher in summer than in winter;
- Precipitation will decrease in the southern parts of Turkey. It may slightly increase in the north-eastern parts;
- Wind potential will increase in the north-western parts of Turkey. It may decrease in the eastern parts;
- Solar radiation will increase across the country, but the increases will be larger in the western parts;
- Sea level rise is expected to impact the low-lying areas of the river deltas and coastal cities;
- The changes in the climate parameters will likely increase the water stress;
- The landslide risk is high at the north-eastern parts of Turkey. The projected increase in precipitation for this area could enhance the frequency and intensity of the landslides;
- Overall, the intensity and duration of droughts and hot spells could increase in response to increasing temperatures and decreasing precipitation in Turkey.

A more striking outcome of the report is the section which focuses on the climate change impact on the Turkish economy. If we focus on impacts on agriculture, climate change brings predictions with pessimistic tones, showing an adverse impact on agriculture. In brief, decrease in rainfall will threaten the food security of Turkey (Fig. 18.2).

A holistic approach concerning the climate change impacts on the Turkish economy over-emphasises the existing risks like water scarcity, high urbanisation

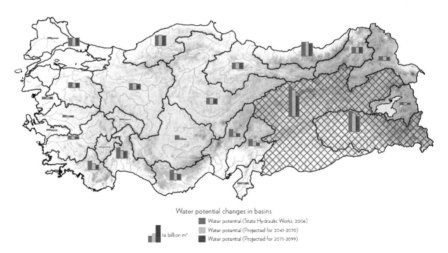

Fig. 18.2 Water potential changes in the basins of Turkey. Ornamented areas show the highest water potential changes. *Source* The authors

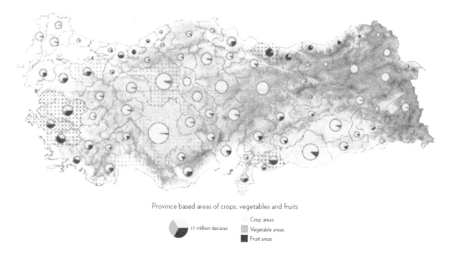

Province based areas of crops, vegetables and fruits

17 million decares

Crop areas
Vegetable areas
Fruit areas

Fig. 18.3 Areas of Crops, vegetables and fruits in the 81 provinces in Turkey. Ornamented areas with red dots show a higher distribution of fruit areas, whereas the yellow-dotted areas show a higher distribution of crop areas. *Source* The authors

and population increases (Fig. 18.3). Due to the country's geographical location, more heatwaves, hot days and nights should be expected. These trends seem to negatively affect the quality of life in Turkey.

18.4   Cost of Climate Change in Agriculture in Turkey

This section offers a specific look at climate change impact on agriculture. In this respect, we will cover four alternative approaches. As commonly emphasised, economic models maximise the outcome given the assumptions or restrictions we impose. From the analysis so far, we understand that Turkey is a water-stressed country, and climate change is going to worsen water shortage.

18.4.1   Computable General Equilibrium Approach to Climatic Change

The first model is by Dudu/Çakmak (2011). It employs a computable general equilibrium model (CGE) to assess crop water requirement for the period 2010–2099. The advantage of the model is that, in calculating the crop water requirement, it takes into account forward and backward leakages. The CGE model covers linkages between agriculture and other sectors at a 12 NUTS1 regions level. An analytical framework is given below (Fig. 18.4).

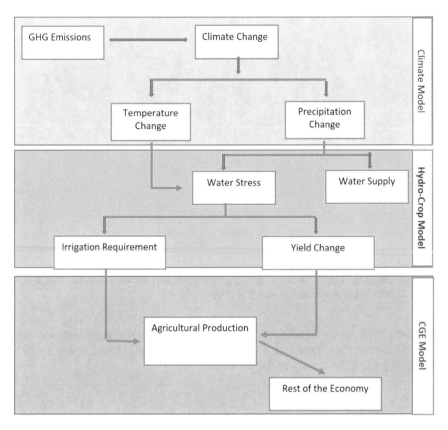

Fig. 18.4 Analytical CGE model. *Source* The authors

Findings show that Turkey still has time to take the necessary actions towards plant water requirements, because climate change impact will start from 2030 onwards. Although shocks are directly on agriculture, climate change has a negative impact on all sectors. Most agricultural production involves intermediate inputs for all other sectors, thus there will be forward and backward leakages due to competition between factor demands. This transmission process will mostly be via change in relative prices. If we examine the climate change impact in three different time intervals, 2010–2035 (P1), 2035–2060 (P2), 2060–2099 (P3), agriculture and food production will be influenced as follows. Table 18.1 shows the production and consumption figures in terms of quantities. The 2010–2035 period does not bear significant costs in Turkey for the agricultural sector. Prices slightly decrease, while wages in water and rain-fed land show a decrease. The decline in water requirement causes the price of water to fall during the climate change period of 2035–2060, causing agricultural production and consumption to decrease. The value added in the irrigated land decreases, trade deficit for agriculture drastically increases, followed by a food production decrease and labour wage decrease. As stated earlier, the major impact of climate change is felt during the 2060–2099 period. Agricultural production decreases by 5.12%, whereas the agricultural trade deficit increases by 76.32%, which in turn decreases agricultural production by 1.66%. For the second and third terms, all production suffers from production decreases (Table 18.1).

As production falls during the second and third terms, imports increase, leading to a trade deficit. So far analysis of both Turkish agriculture as a whole and specific regions has shown that diverse features deepen or dampen climate change impacts on agricultural and food production. The Mediterranean region seems to be affected most by changing climatic conditions. The following map shows agro-food production impacts for three consecutive periods (Table 18.2).

The 2066–2090 period witnesses severe agro-food production declines. The impact of climate change seems to be greater in the Mediterranean, Marmara and in the Western parts of Central Anatolia. South-eastern Anatolia seems to be disadvantaged for all three terms. For the rest of Turkey, although regions are negatively influenced, the impact is still negligible. The Aegean zone seems to improve its water use, thus agro-food production seems to be sustainable.

Figure 18.5 shows the pessimistic GDP changes for some regions of Turkey.

As a preliminary conclusion, we can state that productivity of operations change at any level will reduce the cost of climate change. Thus, all CGE predictions assume existing productivity levels not changing for the extended three time periods. The second study in the coming section will specifically focus on draught impact, which is an extension of the first model.

18.4.2   Economic Impact of Drought on Turkish Agriculture

The study conducted by Dellal/McCarl (2010) reveals that, as we have previously captured in the CGE model, climate change influences agricultural production, agricultural prices, cost of agricultural production, changes in the crop mix due to changing agricultural returns, agricultural exports and imports, food security and safety. Dellal/McCarl (2010) obtained the following observations for the impacts of drought on plant and livestock production as well as economy and food. This in no way eliminates the existing risk perception among Turkish farmers (Aydoğdu/ Yenigün 2016).

18.4.2.1   Plant Production

- Annual and perennial crop losses
- Damage to crop quality
- Income loss for farmers due to reduced crop yields

Table 18.1 CGE iteration results. *Source* The authors

			Base level	% Change		
				P1	P2	P3
Agriculture	Mark	Prod.	107,560	0.36	-1.69	-5.12
		Cons.	64,939	0.19	-0.15	-3.31
		Prices	1.00	-0.7	2.58	7.30
	Empl	Labour	5,018	0.08	1.54	4.52
		Irr. Land	5,261	0.78	-3.96	-13.92
		Rf. Land	16,708	0.21	1.40	3.49
		Capital	55,017	0.03	1.20	3.23
		Water	1,935	0.00	0.00	0.00
	Wage	Labour	7.68	0.00	-0.45	-1.69
		Irr. Land	0.28	1.24	0.96	0.47
		R. Land	0.33	-0.32	-0.71	-1.23
		Capital	1.09	0.00	0.00	0.00
		Water	1.00	-1.56	8.04	26.63
	Trade	Import	9,117	0.02	5.86	15.60
		Export	5,759	3.53	-5.76	-19.80
		Deficit	-3,358	-6.00	25.78	76.32
Food production	Marketing	Prod.	30,330	0.11	-1.14	-3.32
		Cons	92,422	0.08	-0.71	-2.12
		Prices	1.00	-0.06	0.64	2.05
	Employment	Labour	687	-0.04	-0.55	-1.66
		Capital	21,218	0.14	-1.37	-3.99
		Water	131	0.03	0.15	0.40
	Wage	Labour	13.07	0.11	-1.11	-3.13
		Capital	1.00	0.00	0.00	0.00
		Water	1.00	0.17	-1.98	-5.74
	Trade	Import	5,416	0.04	1.40	3.74
		Export	9,310	0.41	-3.89	-10.97
		Deficit	3,893	0.94	-11.25	-31.43

Table 18.2 GDP change with alternative time periods (Dudu/Çakmak 2011). *Source* Dudu/Çakmak (2011)

	Base level million TL.	% Change								
		2010–2035			2035–2060			2060–2100		
		Min.	Avg.	Max.	Min.	Avg.	Max.	Min.	Avg.	Max.
TR1	212,394	−11.13	0.13	10.44	−13.80	−2.50	10.45	−16.49	−6.90	2.67
TR2	41,916	−3.94	−0.14	4.21	−2.20	0.09	2.04	−3.95	−0.04	2.66
TR3	117,556	−7.71	−0.15	6.58	−9.65	−1.87	7.36	−11.60	−4.98	1.36
TR4	87,828	−7.86	0.10	7.25	−9.82	−1.71	7.39	−11.81	−4.92	1.73
TR5	89,146	−8.28	0.09	7.64	−10.32	−1.81	7.73	−12.52	−5.16	1.78
TR6	100,333	−5.74	0.10	5.35	−7.17	−1.18	5.54	−8.88	−3.61	1.30
TR7	38,343	−2.35	0.15	3.63	−2.32	0.17	2.81	−3.38	−0.45	1.45
TR8	46,688	−4.32	0.06	3.93	−4.70	−0.98	3.07	−5.57	−2.30	2.77
TR9	29,798	−2.80	0.05	2.82	−3.72	−0.80	2.26	−4.45	−1.92	0.19
TRA	21,083	−2.37	0.75	5.55	−3.46	0.69	5.44	−3.53	2.17	7.93
TRB	35,165	−1.02	0.88	3.41	−2.49	1.54	3.78	−1.10	2.94	6.05
TRC	70,180	−1.35	0.00	1.03	−1.434	−0.22	1.10	−2.01	−0.80	0.81
Turkey	890,431	−6.18	0.10	5.73	−7.66	−1.33	5.76	−9.28	−3.81	1.61

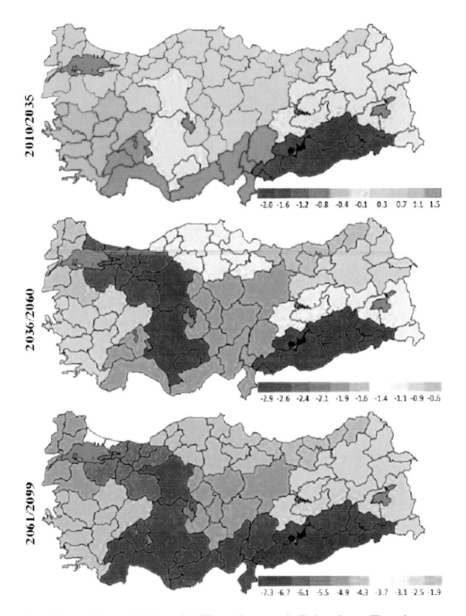

Fig. 18.5 Agro food production under different time spans in Turkey. *Source* The authors

- Reduced productivity of cropland
- Inspect infestation plant disease
- Increased irrigation costs
- Cost of new or supplemental water resource development (wells, dams, pipelines).

18.4.2.2 Livestock Production

- High cost/unavailability of feed for livestock
- Increased feed transportation costs
- High livestock mortality rates
- Disruption of reproduction cycles (delayed breeding, more miscarriages)
- Decreased stock weights
- Reproduced productivity of rangeland
- Reproduced production
- Forced reduction of foundation stock
- Limitation of public lands to grazing
- High cost/unavailability of water for livestock
- Cost of new or supplemental water resource development (wells, dams, pipelines)
- Increased predation
- Range fires.

18.4.2.3 Economic Impact

- Income losses
- Loss to industries directly dependent on agricultural production
- Decreased land prices
- Unemployment from drought-related declines in production
- Strain on financial institutions (foreclosures, more credit risk, capital shortfalls)
- Reduction of economic development
- Fewer agricultural producers (due to bankruptcies, new occupations)
- Rural population loss.

18.4.2.4 Food Impact

- Decrease food production
- Increase in food prices
- Increased importation of food (higher costs)
- Decreased exportation of food.

18.4.2.5 Observations and Findings

As seen from the observed findings, livestock and economic impacts are positive extensions of the model used. Given the findings, we should expect five major reflections as a result of climate change for the case of Turkey (Dellal/McCarl 2010):

- It is expected that by 2050 mean temperature will increase by 1.5 °C
- Precipitation will decrease by 1.5 mm/day
- Projected largest temperature change is expected to be 4.1 °C in Central Anatolia, and 3–4 °C in Mediterranean and Aegean regions
- The least temperature change is expected to be in the Black Sea region (1.6–2.5 °C).
- The greatest precipitation change will be in the Mediterranean region (March–April), and the least change will be in south-eastern Anatolia.

Dellal and McCarl employed the Turkish Agriculture Sector Model (TARSEM) for the base year 2008. The model makes predictions related to production, consumption, import and export, price, and production cost data and demand elasticity calculations. The study is restricted to five crops and seven geographic regions. The diagram below shows the expected agricultural production change (Dellal/McCarl 2010).

The TARSEM model calculates that barley, corn and cotton imports will increase and wheat, sunflower and corn exports will decrease (Fig. 18.6). The study shows that agricultural production will decrease and in turn the agricultural prices will increase and, consequently, consumers will be in a worst condition and producers will be better off due to the increase in agricultural product prices. The overall outcome will lead to a welfare loss.

18.4.3 Climate: Observations, Projections and Impacts

The last research related to climate change impact on agriculture has been conducted by the Meteorological (Met) Office in the UK (Huhne/Slingo 2011). Among the four alternative research findings covered, the Huhne/Slingo (2011) research is the most comprehensive approach towards the cost of climate change on agriculture. This research not only assesses agricultural output changes, but also focuses on food security, water stress and drought and flooding in the coastal regions.

The study starts with certain observations related to the Turkish economy and the warming trend since 1960. The general trend is the decrease in the number of cool nights and increase in the number of warm nights since 1960. Precipitation decreased in winter in the west and increased in autumn in the north.

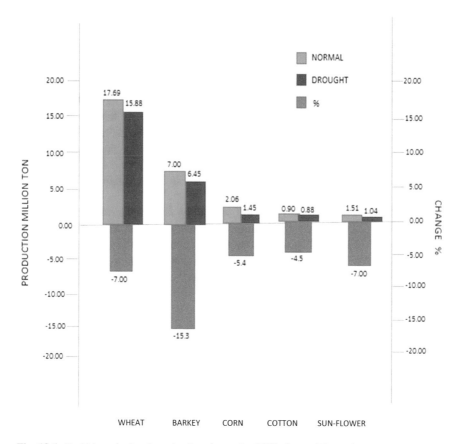

Fig. 18.6 Turkish agricultural production change by 2050. *Source* The authors

The A1B emissions scenario projected temperature increases over Turkey of around 2.5–3 °C in the north, 3–3.5 °C in central and south-western regions, and 3.5–4.0 °C in the east.

Turkey is projected to experience decreases in precipitation, in common with the wider Mediterranean and majority of the Middle East. Decreases in precipitation of over 20% are projected in the south of the country, with strong agreement across the CMIP3 ensemble. Smaller changes of between 0 and 10% are projected towards the north, but with more moderate agreement between the CMIP3 models.

Climate change impact projections are taken from the impacts sector of the AVOID (Avoiding Dangerous Climate Change Program) program (Huhne/Slingo 2011).

18.4.3.1 Climate Change Impact Projections

Under this heading alternative cost elements will be discussed and assessed for the Turkish agriculture sector.

Crop Yields. By 2030, approximately 75–95% of current croplands will experience declining suitability in both scenarios of the model. Among the declining agricultural crops, maize seems to have the largest loss. As known, maize production is vital for animal feed as well as vegetable oil. Predictions show that the maize production decrease is also consistent with the regional productions. Compared with the Iglesias/Rosenzweig (2009) findings, we expect a relatively small decline in agricultural production and mostly in the production of wheat. This is followed by rice production increases.

The AVOID Study focuses on land suitability for calculation, instead of the crops itself. Predictions show that among the crop raising land, a very small portion, such as 2–13% is expected to show improvement within scenario periods. By 2030, approximately 75–95% of current croplands will show declining suitability for both scenarios employed. Research also shows a very strong causality between the degree of climate change and land suitability for cultivation.

The A1F1, A2a, A2b, A2c, B1a, B2a and B2b are scenarios under different assumptions. The time span for the predictions are 2020, 2050 and 2080. The wheat and maize column shows the percentage change in output with respect to the base period.

The model assumes no adaptation efforts, thus costs related to mitigation have not been reflected in the costs of climate change. 2080 impacts seem to be larger with respect to the base time period, and negative impacts increase at an increasing rate (Table 18.3).

Research conducted by Giannakopoulos et al. (2005, 2009) shows that some summer crops (C4) and tuber crops show a decline with climate change, while cereal production increases for alternative scenarios (Fig. 18.7).

Food Security. Food security covers complex interactions between food availability, socio-economic, policy and health factors like undernourishment. When climate change interacts with population dynamics, national policies may need to change, or global food security will be strongly influenced. Recent research shows that, for the coming forty years, overall, Turkey will not face food insecurity, with the exception of marine fisheries (Cheung et al. 2010).

Water Stress and Drought. Water stress is a situation where water stores are not replenished at a sufficient rate to match water demand and consumption. In this respect, Turkey continuously faces water stress and drought in several zones. The map below has been developed by Vörösmarty et al. (2010). Human water security (HWS) threat seems to be at its limit for most regions. Population increases and drought increases with climate change seem to worsen the performance of agricultural crop production.

The AVOID results for Turkey show similarity with the Global Climate Model 21 GCMs exposed to an increase in water stress (45%) due to climate change under the mitigation scenario (30%).

Table 18.3 Changes in wheat and maize production under alternative scenarios. *Source* Iglesias/Rosenzweig (2009)

Scenario	Year	Wheat	Maize
A1FI	2020	2.69	−3.07
	2050	7.36	−3.70
	2080	5.79	−0.75
A2a	2020	3.64	−0.80
	2050	6.94	−3.47
	2080	9.90	−1.63
A2b	2020	2.14	−2.02
	2050	7.27	−3.84
	2080	10.83	−1.36
A2c	2020	1.60	−2.67
	2050	6.85	−4.12
	2080	12.27	−2.01
B1a	2020	−0.82	−3.27
	2050	3.19	−4.77
	2080	4.84	−5.43
B2a	2020	2.18	−4.43
	2050	3.11	−5.17
	2080	3.68	−3.68
B2b	2020	1.18	−3.92
	2050	3.27	−5.35
	2080	7.01	−3.64

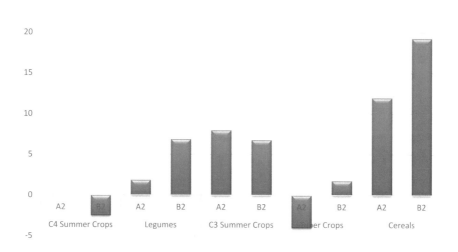

Fig. 18.7 Impact of climate change on crop productivity for different types of crops in Turkey. *Source* The authors

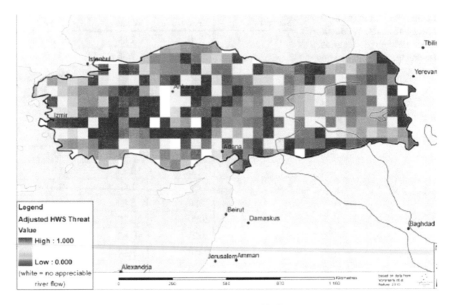

Fig. 18.8 HWS for Turkey. *Source* Vörösmarty et al. (2010)

International research shows that HWS will increase in frequency and magnitude with climate change, especially in the southern regions of Turkey (Fig. 18.8).

Fluvial Flooding. Fluvial flooding involves flow in rivers either exceeding the capacity of the river channel or breaking through the riverbanks, and so inundating the floodplain. Using 21 GCM extended with AVOID simulations shows that the flood risk should be expected to decline with climate change scenarios for the coming thirty years.

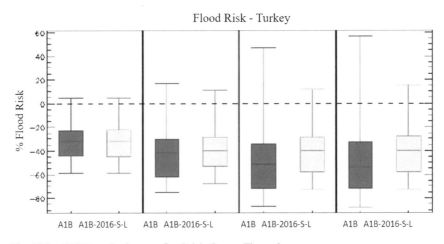

Fig. 18.9 AVOID projections on flood risk. *Source* The authors

The AVOID program also show that the flood risk in Turkey could decrease with climate change throughout the 21st century (Fig. 18.9).

Coastal Regions. Research conducted on coastal regions shows that with climate change Turkey will witness a sea level rise (SLR) for all regions. Predictions are more pessimistic for the Mediterranean and Aegean coasts. Kuleli (2009) explored the population, settlements, land use, contribution to national agricultural production and taxes within 0–10 m elevation of national level, as an indicator of risk to SLR. The study found that approximately 7,319 km^2 of land area lies below 10 m elevation in Turkey, and is hence highly vulnerable to SLR. Twenty-eight coastal cities, 191 districts and 181 villages or towns are located below the 10 m elevation. Kuleli (2009) estimated that the population in Turkey at risk of SLR was around 428,000 along the Mediterranean coast, 208,000 along the Aegean coast, 842,000 in the Marmara region and 201,000 along the Black Sea coast (Huhne/Slingo 2011).

18.5 Conclusion

This study aims to cover the direct and indirect costs of climate change on agriculture. Focusing on the major research findings in the field shows that holistic measurements are not easy to come by at an implementation level. Maybe the most difficult part of assessing climate change impact on agriculture is in terms of quality of life. Awareness that water will always be a scarce factor has its stress factor which is not very easy to assess. We believe a global cost structure for the agricultural sector influenced by climate change can be grouped as follows:

A. Amount of crop losses
B. Export earnings losses
C. Consumption losses due to restricted output
D. Price increase impacts leading to a consumer welfare loss
E. Food security considerations
F. Water stress and drought
G. Fluvial flooding
H. Coastal region costs
I. Costs related to CO_2 content in agricultural exports
J. Mitigation costs.

This primary list reflects the magnitude of costs related to climate change which affect agriculture. The main observation that we have made throughout the analysis is that the social, demographic, economic and geographical location of a country or region will also define the different cost factors for assessment. Although Turkey seems to have sufficient time to adapt to climate change, agriculture is an area that needs special attention due to its impact on industry and service sectors as well as on quality of life. Potential climatic change conditions is a function of international regulations and illegal misuse. Thus, all climate change trends rely on these two assumptions. Sample product production changes can be extended towards other national agricultural outputs, which is beyond the scope of our study. The scope change in agricultural production technology also needs to be explored, which changes the definition of fields to be cultivated. As stated earlier, climate change mitigation could be accomplished via three complimentary tools. Reduction of CO_2 in terms of demand, agricultural strategy changes towards crop mix and others, and via technological changes that reduces CO_2 emission. The future is in front of us and the choice will be ours.

References

Aydoğdu MH, Yenigün K (2016) Farmers' Risk Perception towards Climate Change: A Case of the GAP, Şanlıurfa Region, Turkey. *Sustainability* 8(8):806.

Cheung WWL, Lam VWY, Sarmiento JL, Kearney K, Watson REG, Zeller D, Pauly D (2010) Large-Scale Redistribution of Maximum Fisheries Catch Potential in the Global Ocean Under Climate Change. *Global Change Biology* 16:24–35.

Dellal I, McCarl BA (2010) The economic impacts of drought on agriculture: The case of Turkey, *Options Mediterraneennes: Serie A. Seminaires Méditerraneens*; no 95 169–174.

Dudu H, Çakmak EH (2011) *Regional Impact of the Climate Change: A CGE Analysis for Turkey*, The Economic Research Forum (ERF) Working Paper 644 October 2011.

Giannakopoulos C, Bindi M, Moriondo M, Lesager P, Tin T (2005) *Climate Change Impacts in the Mediterranean Resulting from A 2c Global Temperature Rise*. WWF Report. Gland, Switzerland: WWF.

Giannakopoulos C, Le Sager P, Bindi M, Moriondo M, Kostopoulou E, Goodess CM (2009) *Climatic Changes and Associated Impacts in the Mediterranean Resulting from A 2 °C Global Warming*. *Global Planet Change* 68:209–224.

Huhne C, Slingo J (2011) Climate: Observations, Projections and Impacts. Turkey: Printed by the Met Office.

Iglesias A, Rosenzweig C (2009) *Effects of Climate Change on Global Food Production under Special Report on Emissions Scenarios (SRES) Emissions and Socioeconomic Scenarios: Data from a Crop Modeling Study*. Palisades, NY: Socioeconomic Data and Applications Center (SEDAC), Columbia University.

Kuleli T (2009) Quantitative analysis of shoreline changes at the Mediterranean Coast in Turkey, *Environmental Monitoring and Assessment* ISSN: 0167–6369 (Print) 1573-2959 (Online), Springer Link Date: Tuesday, June 30, https://doi.org/10.1007/s10661-009-1057-8.

Lohmann U, Stier P, Hoose C, Ferrachat S, Kloster S, Roeckner E, Zhang J (2007) *Cloud Microphysics and Aerosol Indirect Effects in The Global Climate Model*, Published by Copernicus Publications on Behalf of the European Geosciences Union, ECHAM5-HAM. *Atmospheric Chemistry and Physics* 7:3425–3446.

Şen ÖL (2013) *A Holistic View of Climate Change and its Impacts in Turkey*. İstanbul: İstanbul Policy Center, 7 December 2013.

The United Nations Convention on Biological Diversity (2000); at: http://www.cbd.int/ (5 January 2015).

UNEP (2009) The need to include ecosystems management as part of the COP 15 Agenda, Research Brief; at: http://www.macaulay.ac.uk/copenhagen/documents/UNEP-CC-EM-12-page-brief.pdf, 1–12.

Vörösmarty CJ, Mcintyre PB, Gessner MO, Dudgeon D, Prusevich A, Green P, Glidden S, Bunn SE, Sullivan CA, Liermann CR, Davies PM (2010) Global Threats to Human Water Security and River Biodiversity. *Nature* 467:555–561.

Correction to: Enhanced Growth Rate and Reduced Water Demand of Crop Due to Climate Change in the Eastern Mediterranean Region

Jiftah Ben-Asher, Tomohisa Yano, Mehmet Aydın and Axel Garcia y Garcia

Correction to:
Chapter 13 in: T. Watanabe et al. (eds.), *Climate Change Impacts on Basin Agro-ecosystems*, The Anthropocene: Politik—Economics—Society—Science 18, https://doi.org/10.1007/978-3-030-01036-2_13

The original version of the book was inadvertently published with incorrect author name tagging in Chapter 13, which has been now corrected as "First Name: Axel" and "Last Name: Garcia y Garcia". The correction chapter and the book have been updated with the change.

The updated version of this chapter can be found at
https://doi.org/10.1007/978-3-030-01036-2_13

About the Editors

Tsugihiro Watanabe (Japan), born in 1952 in Tochigi Japan. Professor of the Graduate School of Global Environmental Studies, Kyoto University, Japan. The President of the Japan Society of Hydrology and Water Resources. He has about 40 years' research experience related to irrigation management, rural hydrological regime and climate change impacts on basin hydrology and agriculture. He had served as the president of the Japanese Society of Irrigation, Drainage and Rural Engineering, and the International Society of Paddy and Water Environment Engineering. He has been the chair of the Working Group on Global Climate Change and Agricultural Water Management of International Commission of Irrigation and Drainage.

Address: Prof. Dr. Tsugihiro Watanabe, Graduate School of Global Environmental Studies, Kyoto University, Yoshida-Honmachi, Sakyo, Kyoto, 606-8501 Japan.
Email: nabe@kais.kyoto-u.ac.jp.
Website: http://www2.ges.kyoto-u.ac.jp/en/lab/regional-planning.

© Springer Nature Switzerland AG 2019
T. Watanabe et al. (eds.), *Climate Change Impacts on Basin Agro-ecosystems*,
The Anthropocene: Politik—Economics—Society—Science 18,
https://doi.org/10.1007/978-3-030-01036-2

 *Selim Kapur* (Turkey), born in 1946 in Ankara, Turkey. Retired Professor of Soil Science and Plant Nutrition at the Faculty of Agriculture and Chairman of the Dept. of Archaeometry at the University of Çukurova, Adana, Turkey. He is the Scientific Comm. Member of the European Soil Bureau Network of the EU-JRC-IES in Ispra, Milan, Italy. He is the former Secretary of the International Working Group of Land Degradation and Desertification. He has acted as the Science and Technology Correspondant of Turkey for the UNCCD and he organised the First International Meeting on Land Degradation and Desertification in 1996 in Adana, Turkey. He led projects supported by NATO, EU, JSPS, TUBITAK, Mitsui Environmental Fund, FAO and GEF concerning soil protection and environmental issues. He has authored numerous papers in cited and authored/edited books on land degradation and desertification, archaeometry and micromorphology.

Address: Prof. Dr. Selim Kapur, University of Çukurova, Department of Soil Science and Archaeometry, 01330, Adana, Turkey.
Email: kapurs@gmail.com.
Website: www.cu.edu.tr.

 *Mehmet Aydın* (Turkey), born in 1955 in Kars, Turkey. Retired Professor of Soil Physics. He founded the Department of Soil Science at Mustafa Kemal University, Hatay, Turkey, in 1995. He carried out/coordinated many national/international research projects. He developed a mathematical model (acronym: E-DiGOR) for predicting runoff, drainage and actual evaporation from bare soils. He was competitively appointed as a Visiting Professor in ALRC of Tottori University, Japan, between 2000 and 2003 (two years in total). He served as a Professor in the Department of Biological Environment of Kangwon National University of South Korea from 2011 to 2014. He has authored 172 publications.

Address: Prof. Dr. Mehmet Aydın, Mustafa Kemal University, Department of Soil Science, Faculty of Agriculture, Antakya, Turkey.
Email: maydin08@yahoo.com.

Rıza Kanber (Turkey), born in 1944 in Erzurum. Emeritus Professor of Agricultural Structures and Irrigation at the Faculty of Agriculture University of Çukurova, Adana, Turkey. He experienced on farm irrigation system (sprinkler, drip, and gravity systems) design and management, water resources development, monitoring and evaluation of irrigation projects, evaluation and assessment of effectiveness of the design and management of the dams and their irrigation system network. He has authored numerous papers on Participatory Irrigation Management, irrigation water requirements of crops, Irrigation scheduling, quality of irrigation water, drainage system design, improvement of saline and alkaline soil.

Address: Prof. Dr. Rıza Kanber, Çukurova University, Department of Agricultural Structures and Irrigation Engineering, Adana, Turkey.
Email: kanber@cu.edu.

Erhan Akça (Turkey), born in 1969 in Adana, Turkey. Professor Dr. at Adıyaman University School of Technical Sciences, Adıyaman, Turkey. His expertise is in soil mineralogy, archaeometry and natural resource management. He has written and contributed to more than 110 scientific publications (papers, chapters, books) on soil science, archaeometry, land use and desertification. He has participated in projects supported by NATO, EU, JSPS, TÜBİTAK, Mitsui Environmental Fund, FAO and GEF. He is currently scientific board member of the Ministry of Forestry and Water Works, General Directorate of Combating Desertification and Erosion Control of Turkey.

Address: Prof. Dr. Erhan Akça, Adıyaman University, School of Technical Sciences, 02040, Adıyaman, Turkey.
Email: eakca@adiyaman.edu.tr.
Website: www.adiyaman.edu.tr.

About the Authors

Akça, Erhan (Turkey), see above as coeditor.

Address: Adıyaman University, School of Technical Sciences, Adıyaman, Turkey.
Email: eakca@adiyaman.edu.tr.
Website: www.adiyaman.edu.tr.

Alpert, Pinhas (Israel) served as the Head of the Porter School of Environmental Studies, Tel-Aviv University, till 2013, and as the Head of the Department of Geophysics. He coauthored more than 300 articles. His research focuses on atmospheric dynamics, climate, aerosol dynamics and climate change. Alpert served as co-director of the GLOWA-Jordan River BMBF/MOS project to study the water vulnerability in the East Mediterranean and as the Israel representative to the IPCC TAR WG1. A novel method for monitoring rainfall employing the cellular network data was proposed by Alpert and colleagues.

Address: Tel Aviv University, The Dept. of Geophysics, Exact Science Faculty, Tel Aviv, Israel.
Email: pinhas@post.tau.ac.il.
Website: http://www.tau.ac.il/~pinhas/.

Alphan, Hakan (Turkey), born in 1972 in Adana, Turkey. Professor of Landscape Architecture and Chairman of the department at the Faculty of Agriculture University of Çukurova, Adana. Author/co-author of numerous scientific publications, including peer-reviewed journal papers, book chapters and papers in conference proceedings concerning his major fields of study on remote sensing and GIS, landscape-level environmental monitoring, change detection, landscape pattern and environmental processes.

© Springer Nature Switzerland AG 2019   417
T. Watanabe et al. (eds.), *Climate Change Impacts on Basin Agro-ecosystems*,
The Anthropocene: Politik—Economics—Society—Science 18,
https://doi.org/10.1007/978-3-030-01036-2

Address: Çukurova University, Department of Landscape Architecture, Adana, 01330, Turkey.
Email: alphan@cu.edu.tr.
Website: http://aves.cu.edu.tr/alphan/.

Ando, Makoto (Japan), Associate Professor, University of Kyoto, Division of Forest Biosphere, Field Science Education and Research Center. He has been involved in the following projects: study on the regeneration of boreal forest in Canada in 1996, study on the regeneration of coniferous forest in British Columbia in 1997, technical cooperation on the improvement of the vegetation in inland semi-arid region of Rio Grande do Norte state in Brazil in 1999, study on the ecological characteristics of salt-tolerant plants and the soil amendment in the salinised area of the basin of the Yellow River in China in 2000, study on the ecological characteristics of salt-tolerant plants and the possibility of soil amendment in semi-arid and arid regions.

Address: Kyoto University, Field Science Education and Research Center, Sakyo-ku, Kyoto 606-8501, Japan.
Email: ando@kais.kyoto-u.ac.jp.

Aydın, Mehmet (Turkey), see above as coeditor.

Address: Mustafa Kemal University, Department of Soil Science, Faculty of Agriculture, Antakya, Turkey.
Email: maydin08@yahoo.com.

Barutçular, Celaleddin (Turkey) was born in 1964 in Dortyol/Turkey. He graduated from the Department of Field Crops, Faculty of Agriculture, University of Uludag in 1987. Between 1990 and 1999, he obtained his M.Sc. and Ph.D. degree from the Department of Field Crops of Applied Science Institute of the Çukurova University in Adana/Turkey. He has been studying abiotic stress physiology on cereals since his M.Sc. degree. He has been a full Professor at Çukurova University, Department of Field Crops, Faculty of Agriculture, Adana, Turkey since 2017.

Address: Çukurova University, Department of Field Crops, Faculty of Agriculture, Adana, Turkey.
Email: cebar@cu.edu.tr.
Website: http://aves.cu.edu.tr/cebar/.

Ben-Asher, Jiftah (Israel), Retired Professor of Soil Physics at Ben-Gurion University of the Negev, Israel. His Ph.D., entitled *Energy and Water Balance in Soil partially wetted by a point source*, was the world's first Ph.D. on drip irrigation. He founded the Katif R&D Center for coastal deserts development, which he still heads. As a partner in the Agroecology Group he developed an algorithm to determine crops' water and NPK deficiencies using cellular phones. The technology is now installed in an application that is used by growers in China under the auspices of Israel-China AgriOT, China Ltd. company.

Address: Ben-Gurion University Agroecology Group, The Katif R&D Center, Ministry of Science and Technology, Sedot Negev Regional Council 85200, Israel.
Email: benasher@bgu.ac.il.

Berberoğlu, Suha (Turkey) is head of the Remote Sensing and GIS department at the Institute of Basic and Applied Sciences and a lecturer in Landscape Architecture. He has 21 years of broad-based field and laboratory environmental experience in the application of environmental monitoring and management. Dr. Berberoğlu is also involved in various national and international projects, such as biodiversity mapping, change detection, hydrological modelling and biomass estimates. His current field of research is GIS and satellite remote sensing for driving ecosystem models and conservation in the Mediterranean region. He has been using geo-spatial technologies to model NPP, basin hydrology and land use/cover change. He is the founder of the CU Landscape Architecture Remote Sensing and GIS Lab, which is dedicated to providing innovative, state-of-the-art monitoring of environment using geospatial technologies (GIS, GPS and Remote Sensing).

Address: Çukurova University, Landscape Architecture Department, Adana, 01330, Turkey.
Email: suha@cu.edu.tr.
Website: http://aves.cu.edu.tr/suha/.

Çilek, Ahmet (Turkey), received his Ph.D. in Landscape Architecture from Çukurova University in 2017. He worked at the Aerial Application Technology Research Unit, Agricultural Research Services in USDA for six months with TÜBİTAK 2214 – a scholarship for Ph.D. research abroad. He has been involved in many national and international research projects focusing on different aspects of forest productivity, erosion modelling and carbon cycles research in the frame of integrated water and land resources management at the Eastern Mediterranean part of Turkey.

Address: Çukurova University, Landscape Architecture Department, Adana, 01330, Turkey.
Email: acilek@cu.edu.tr.

Coşkun, Ziya (Turkey), born in 1964 in Yozgat, Turkey. Ph.D., Agriculture Faculty of Çukurova University. He has worked as Agricultural Economist, Planning Section, 6th Regional Directorate of State Hydraulic Works since 1985. He is an expert on agricultural economy, water rights and expropriation, and has carried out national and international research projects.

Address: Planning Section, 6th Regional Directorate of State Hydraulic Works, 01120, Adana, Turkey.
Email: ziyacoskun2008@gmail.com.

Donma, Sevgi (Turkey) was born in 1958 in Adana, Turkey and obtained her Ph.D. from the Agriculture Faculty of Çukurova University. She has worked as an engineer in the Operations and Maintenance Department of the 6th Regional Directorate of State Hydraulic Works since 2006. She has also attended the Land Drainage Course at the International Institute for Land Reclamation and Improvement Centre at Wageningen in Holland (Netherlands Fellowship Programme) in 1997 for four months; a short advanced course on Water Supply and Demand Management (Malta, 2000); international workshops on the Mediterranean Dimension of Pilot River Basins (PRB) linking rural development and land degradation mitigation with river basin management plans, (Italy, 2004); and the Workshop on Agriculture and Environment in the Balkans and Turkey: Networking Regional Experience (Belgium, 2006). She has carried out national and international research projects and currently works as coordinator of the Land Consolidation Department of 6th Regional Directorate of State Hydraulic Works.

Address: State Hydraulics Works, 6th Regional Directorate, Adana, Turkey.
Email: sevgi60@yahoo.com.

Dönmez, Cenk (Turkey) is a researcher and lecturer in the Landscape Architecture Department, Çukurova University. His main research interest is using process-based ecosystem modelling to understand the complex interactions between land-surface and atmosphere at different spatial scales with emphasis on ecosystem modelling in combination with hydrological and decision tree models and remote sensing data. He has been involved in different national and international research projects with various aspects of remote sensing and GIS techniques and process-based hydrological modelling applications for eleven years. He has international experience at Boston University, FSU-Jena, University of California Santa Barbara and Goethe University, having received support from the Turkish Research Council (TÜBİTAK), European Union and German Academic Exchange Service (DAAD).

Address: Çukurova University, Landscape Architecture Department, Adana, 01330, Turkey.
Email: cdonmez@cu.edu.tr.

Erk, Nejat, Prof. Dr. (Turkey), born in 1948 in Ankara, Turkey is Professor of Development and International Economics. His career has been spent in managerial roles as Department Head, Dean, Vice Rector, institute Principal and Senate member. He is a member of the European University Association (EUA) and has served on higher education evaluation committees nationally and internationally. He worked at Çukurova University and Maryland University as well as many guest instructor positions at universities around Europe. He earned various awards on banking, globalisation, employability, international economics, entrepreneurship, development and customs impact on economies. His focus has shifted to interdisciplinary research and higher education, accreditation. He has participated in the Bologna Process as well as being EUA's quality assurance expert. He has worked on various TÜBİTAK, EUA, European Commission, European Institute, European Union, Chambers of Commerce, and Turkish Ministry of Development projects. Through his work at such institutions he has earned various awards and worked on a wide variety of projects, including various projects on global warming, land degradation and their impact on the public.

Address: Çukurova University, Faculty of Economics and Administrative Sciences, Department of Economics, Adana, Turkey.
Email: erk@cu.edu.tr.

Evrendilek, Fatih (Turkey), received his MA in Energy & Environmental Policy from the University of Delaware and his Ph.D. in Environmental Science from the Ohio State University, and currently works for the Department of Environmental Engineering of Abant Izzet Baysal University (Turkey) as the founding chair. He is (co-)author of over 100 SCI-E articles and co-author of a university textbook on environmental science entitled *The Environment: Science, Issues, and Solutions*. As an ecosystem scientist and big data analyst, his research interests include eddy covariance, wavelet denoising, data-driven & process-based modelling, and GIS & remote sensing analyses of spatiotemporal ecosystem dynamics in a changing climate and environment.

Address: Abant İzzet Baysal University, Department of Environmental Engineering, Bolu, 14052, Turkey.
Email: fevrendilek@ibu.edu.tr.

Fikret Erk, Sinan (Turkey), born in 1981 in Ankara, Turkey, is a Ph.D. student of Marketing. Throughout his academic career he has taught courses in Kosovo. His research mainly focuses on marketing, cultural dimensions and different attributes of consumer preference. He is a member of the European Young Researchers network and on the Scientific Committee of the Euro-Mediterranean Researchers network. He has worked on the effects of carbon on rural and plain areas as well as the culture and external factors effect on marketing attributes.

Address: Çukurova University, Institute of Social Sciences, Marketing, Adana, Turkey.
Email: sinanfikreterk@gmail.com.

Fujihara, Yoichi (Japan) was born in 1977 and raised in Shimane Prefecture. He is Associate Professor of Bioresources and Environmental Sciences at Ishikawa Prefectural University in Japan. He teaches courses on hydrology and agricultural water resources. He gained extensive experience of rainfall-runoff modelling, snow hydrology and remote sensing as a researcher at the Disaster Prevention Research Institute of Kyoto University (2004–2005), the Research Institute for Humanity and

Nature (2005–2008) and the Japan International Research Centre for Agricultural Sciences (2008–2012).

Address: Ishikawa Prefectural University, Bioresources and Environmental Sciences, 921-8836, Japan.
Email: yfuji@ishikawa-pu.ac.jp.
Website: http://water.ishikawa-pu.ac.jp/.

Garcia y Garcia, Axel (USA) received his Ph.D. in Crop Sciences (2002) from the University of Sao Paulo, Brazil. Since 2014 he has been an Assistant Professor in the Deparment of Agronomy and Plant Genetics at the University of Minnesota. He has worked at the University of Georgia—Griffin Campus on crop-modelling applications and at the University of Wyoming as an Assistant Professor of Irrigated Crops. His research focuses on sustainable cropping systems, emphasising site-specific management, primarily in corn-soybean rotation. Cover crops, crops' water and nitrogen use, environmental sustainability and crop-modelling are his main research areas of interest.

Address: Department of Agronomy and Plant Genetics, University of Minnesota, Southwest Research and Outreach Center, Lamberton, MN, United States.
Email: axel@umn.edu.
Website: https://swroc.cfans.umn.edu/about/faculty-staff/axel-garcia-y-garcia.

Gültekin, Ufuk (Turkey) is Assistant Professor of Agricultural Economics at Çukurova University's Faculty of Agriculture in Turkey, where he is also Vice-Dean. He trained at the Development Study Center in Israel (1996–1997) and received a postgraduate diploma on Integrated Rural Regional Development Planning. His M.Sc. and Ph.D. are from Çukurova University, Faculty of Agriculture. He has worked on several projects and served as a Visiting Professor at Tokyo University of Agriculture and Technology in 2015, where he taught a one-semester module on General Aspects of Regional Development.

Address: Çukurova University, Department of Agricultural Economics, Adana, 01330 Turkey.
Email: ugultekin@gmail.com.
Website: http://aves.cu.edu.tr/ugultek/.

Güney, O. İnanç, Dr. (Turkey), was born in 1979 in Adana, Turkey. He is Lecturer/researcher of Agro-Food Economics and Marketing at the Vocational School of Adana, University of Çukurova, Adana, Turkey. He works on Agricultural and Food Markets and Environmental Economics. He is on the consultant board of Istituto di Servizi per il Mercato Agricolo Alimentare (ISMEA) and has led projects supported by the EU, TÜBİTAK and the University of Çukurova. He has participated in several conferences and workshops, and attended training courses on the topic of food and agricultural marketing and economics, mainly in Europe.

Address: Çukurova University, Adana Vocational School, Beyazevler, 01170, Adana, Turkey.
Email: iguney@cu.edu.tr.
Website: http://aves.cu.edu.tr/iguney/.

Harmanci, Didem (Turkey), born in 1975 in Adana, Turkey, is Forest Engineer at the Regional Directorate of Forestry in Adana. She gained her M.Sc. degree in Landscape Planning from the University of Çukurova. Her research subject was investigating land allocation pattern in a protected area located by the Mediterranean coast and influenced by agricultural activities.

Address: Adana Regional Directorate of Forestry, Adana, 01120, Turkey.
Email: didemharmanci@ogm.gov.tr.

Hoshikawa, Keisuke (Japan), born in 1975 in Fukui, Japan, is Associate Professor at the Department of Environmental and Civil Engineering at Toyama Prefectural University. He is engaged in research activities related to land and water management using numerical simulation models, remote sensing and spatial information analysis.

Address: Faculty of Engineering, Toyama Prefectural University, 5180 Kurokawa, Imizu, Toyama 939-0398, Japan.
Email: hoshi@pu-toyama.ac.jp.

Jin, Fengjun (China), born in 1969. Having graduated from the Department of Atmospheric Science at Nanjing University of Information Science & Technology, he gained his Ph.D. at the Department of Geophysics and Planetary Science at Tel Aviv University in Israel. His research interests are climate change, assessment of meteorological disasters and atmospheric water vapour budget. He is now working in Xiamen Meteorological Bureau, Xiamen City, China. He has published papers on global warming, atmospheric water and moisture budget in the Mediterranean and the USA.

Address: Xiamen Meteorological Disaster Prevention Technology Centre, Qixiangtai Road, 85#, Huli District, Xiamen City, Fujian Province, China.
Email: jfj9999@hotmail.com.

Kanber, Rıza (Turkey), see above as coeditor.

Address: Çukurova University, Department of Agricultural Structures and Irrigation Engineering, Adana, Turkey.
Email: kanber@cu.edu.

Kapur, Burçak (Turkey) was born in 1978 in Adana/Turkey. He graduated from the Department of Soil Science, Faculty of Agriculture, University of Çukurova in 1996. In 2000–2002 he obtained his M.Sc. degree from the International Centre for Advanced Mediterranean Agronomic Studies in Bari/Italy. He completed his Ph.D.

thesis, entitled *Enhanced CO₂ and Climate Change Effects on Wheat Production and Adaptation Strategies in the Lower Seyhan Plain* in the Department of Agricultural Structures and Irrigation, University of Çukurova. He has been Assistant Professor Dr. in the Department of Agricultural Structures and Irrigation, Çukurova University since 2014.

Address: Çukurova University, Department of Agricultural Structures and Irrigation Engineering, 01330, Adana, Turkey.
Email: bkapur@cu.edu.tr.
Website: http://aves.cu.edu.tr/bkapur/kimlik.

Kapur, Selim (Turkey), see above as coeditor.

Address: Çukurova University, Department of Soil Science and Plant Nutrition, 01330, Adana, Turkey.
Email: kapurs@cu.edu.tr.

Kitoh, Akio (Japan), born in 1953 in Osaka, Japan, is Director of the Office of Climate and Environmental Research Promotion, Japan Meteorological Business Support Center. He engaged climate modelling and future climate projections at MRI/JMA. He acted was a lead author of the 2nd through 5th Assessment Report of IPCC WGI. He has authored more than 200 papers on climate variability and climate change.

Address: 1-1 Nagamine, Tsukuba, Ibaraki 305-0052, Japan.
Email: kitoh@jmbsc.or.jp.
Website: www.jmbsc.or.jp/tougou/EN/index.html.

Koç, Müjde (Turkey) was born in 1947 in Erzurum/Turkey. She graduated from the Department of Field Crops, University of Atatürk, Erzurum in 1970. Between 1973 and 1982 she obtained her M.Sc. and Ph.D. degrees from Study and Diploma, Agriculture, Swiss Federal Institute of Technology, Zürich, Switzerland and the Institute Plant Science, Swiss Federal Institute of Technology, Zürich, Switzerland. She is retired Professor at Çukurova University, Department of Field Crops, Faculty of Agriculture, Adana, Turkey.

Address: Çukurova University, Department of Field Crops, Faculty of Agriculture, Adana, Turkey.
Email: mkoc@cu.edu.tr.

Kojiri, Toshiharu (Japan). The late Professor Toshiharu Kojiri served as head of the Water Resources Research Center at Kyoto University. He was a Council Member of the International Association for Hydro-Environment Engineering and Research (IAHR), Chair of the IAHR Working Group on Climate Change and closely involved in launching the IAHR Japan Chapter, of which he was President. He passed away on 2 November 2011.

Address: Kyoto University, Disaster Prevention Research Institute, Uji, Japan.

Koluman (Darcan), Nazan (Turkey), born in 1969 in Mardin, Turkey, is Professor of Small Ruminant Breeding and Production in the Animal Science Department of Çukurova University. She is a board member of the International Goat Association and a consultant to the Food and Agriculture Organization of the United Nations. She has led projects supported by the EU, UN, TÜBİTAK and the Ministry of Food, Agriculture and Animal Husbandry in Turkey concerning animal science and environmental issues. She has authored numerous papers and chapters on animal science and environment. She has organised various national and international meetings. She is founder of Koluman-Gen Research & Consultancy Company. She has been certified as a mentor for entrepreneurs and has acted as investment consultant for some national and international companies.

Address: Çukurova University, Agricultural Faculty, Department of Animal Science, Saricam, Adana 01330, Turkey.
Email: nazankoluman@gmail.com.
Website: http://aves.cu.edu.tr/ndarcan/.

Krishnamurti, Tiruvalam Natarajan (USA), known as TNK or Krish (10 January 1932–7 February 2018), was an American meteorologist. His birthplace was Chennai, India. He earned his doctorate at the University of Chicago in 1959. He taught at the University of California, Los Angeles and was Emeritus Professor at Florida State University, where before his retirement he was the Lawton Distinguished Professor of Meteorology. He was awarded the Carl-Gustaf Rossby Research Medal in 1985 by the American Meteorological Society, the International Meteorological Organization Prize by the World Meteorological Organization in 1996, and the Sir Gilbert Walker Gold Medal by the Indian Meteorological Society in 2012. A symposium was held in his honour by the American Meteorological Society in New Orleans in 2012.

Address: Florida State University, Department of Earth, Ocean and Atmospheric Science, Tallahassee, FL, United States.

Kumar, Vinay (USA) is a research scientist at Texas A&M University, Corpus Christi, USA. He has more than twenty years' experience of weather and climate research, multi-model ensemble schemes, downscaling, extreme rainfall prediction, droughts, MJO, thunderstorms, lightning, monsoonal heat and Arctic ice melt, and modelling using WRF. He holds a Ph.D. degree in Meteorology from IITM, affiliated to Pune Univesity, India. He has authored thirty-five scientific research works, including four book chapters. He is on the editoral board of many online journals. His principal research interests are monsoonal studies; extreme rainfall events; tropical weather and climate; and thunderstorms.

Address: Department of Physical and Environmental Sciences, Texas A&M University, Corpus Christi, TX 78412, United States.
Email: vinay.kumar@tamucc.edu.
Website: https://sites.google.com/site/neeluvinaysite/.

Kume, Takashi (Japan), born in 1973 in Aichi Prefecture, Japan, is Associate Professor at the Graduate School of Agriculture, Ehime University, Japan. He has studied soil and water management in arid regions from the viewpoint of soil hydrology. He has started to conduct transdisciplinary study based on the concepts of resilience and adaptive governance to promote and enhance farmers' activities in Turkey, Thailand and Indonesia.

Address: Graduate school of Agriculture, Ehime University, 790-8566, 3-5-7, Tarumi, Matsuyama city, Ehime, Japan.
Email: kume@ehime-u.ac.j.
Website: http://yoran.office.ehime-u.ac.jp/profile/en.834808d0a9e78141.html.

Kusadokoro, Motoi (Japan), born in 1979 in Gunma, Japan, has been Senior Assistant Professor at the Institute of Agriculture, Tokyo University of Agriculture and Technology, Japan, since 2016 and was previously Assistant Professor at the same institution between 2011 and 2016. He was Postdoctoral Researcher at the Institute of Economic Research, Hitotsubashi University, in 2011. His expertise is in agricultural economics, development economics, economic history and applied econometrics. He has authored twenty-four publications.

Address: Tokyo University of Agriculture and Technology, Division of Studies in Sustainable and Symbiotic Society, 3-5-8 Sawaicho, Fuchu, Tokyo 183-8509, Japan.
Email: motoi_k@cc.tuat.ac.jp.

Kutlu, Hasan Rüştü (Turkey) was born in 1963 in Turkey and is Professor of Feeds and Animal Nutrition in the Department of the Animal Science of Çukurova University. He lectures and conducts research on feeds, feed technology and animal nutrition. He has headed the university's international affairs office, and also the Department of Animal Science. He was Acting Dean of the Agricultural Faculty of Çukurova University and head of the Research Grant Committee of Agriculture, Forest, Veterinary and Animal Sciences of TÜBİTAK. He heads the Turkish Society of Animal Science. He is the expert member of the Turkish Delegation of EU FP7, on Theme 2, "Food, Agriculture and Fisheries, and Biotechnology". He is a member of the World's Poultry Science Association. He is the Turkish representative of the European Federation for Animal Science (EAAP). He has had 78 international and 132 national publications and four books on feeds and animal nutrition published. He is also a member of the editorial board of four scientific journals on feeds and animal science. He is married and has a son.

Address: Çukurova University, Agricultural Faculty, Department of Animal Science, Saricam, Adana 01330, Turkey.
Email: hrkutlu@gmail.com.
Website: http://aves.cu.edu.tr/hrk/.

Maru, Takeshi (Japan), born in Japan, is Assistant Professor at Bunri University of Hospitality and Associate Research Scholar at Hitotsubashi University from 2018. He served as Postdoctoral Researcher, Research Associate and Assistant Professor at the Institute of Economic Research, Hitotsubashi University from 2011 to 2018, and as Visiting Researcher at Georg-August-Universität Göttingen from 2017 to 2018. He has authored papers on economic development and rural society in Turkey and Japan.

Address: Hitotsubashi University, Institute of Economic Research, 2-1 Naka, Kunitachi, Tokyo 186-8603, Japan.
Email: marl@ier.hit-u.ac.jp.

Nagano, Takanori (Japan), born in 1970 in Tokyo, Japan, is Associate Professor at the Graduate School of Agricultural Science, Kobe University. He has participated in many interdisciplinary research projects on climate change, integrated water resources management and regional planning.

Address: Kobe University, Graduate School of Agricultural Science, 1-1 Rokkodai, Nada-ku, Kobe, Japan.
Email: naganot@ruby.kobe-u.ac.jp.
Website: http://www.ans.kobe-u.ac.jp/itp/english/kyouin/nagano.html.

Özekici, Bülent (Turkey), born in 1960 in Adana, is Professor of Agricultural Structures and Irrigation at the Faculty of Agriculture University of Çukurova, Adana, Turkey. He completed his Ph.D. thesis at the University of North Carolina State University. He has experience of sprinkler, drip, and gravity systems design and management, water resources development, monitoring and evaluation of irrigation projects, evaluation and assessment of the effectiveness of the design and management of dams and their irrigation system network. He leads many drip irrigation projects.

Address: Çukurova University, Department of Agricultural Structures and Irrigation Engineering, Faculty of Agriculture, Adana, Turkey.
Email: ozekici@cu.edu.tr.
Website: http://aves.cu.edu.tr/ozekici/.

Sano, Junji (Japan), born in 1952 in Hokkaido, northern Japan, graduated from the Faculty of Agriculture, Hokkaido University (Ph.D.). He is Professor of Forest Ecology and Ecosystem Management Laboratory, Faculty of Agriculture, Tottori University, western Japan. He is a forest ecologist concerning forest dynamics and management, especially in primeval forests dominated by *Quercus* spp. and *Castanopsis* spp. He has acted as the chairperson of the Committee of Bamboo Forest and Environmental Assessment of Tottori prefecture. His research sites were Turkey, Mexico, Russia, China and Japan. He has some qualifications as a tree doctor, technician, nature interpreter, and therapist. He will establish the Institute of Forest Ecology and Ecosystem Management, focusing on forest ecology and environmental education, at Fujisawa city and its branch at Tottori city in 2018.

Address: Forest Ecology and Ecosystem Management Laboratory, Faculty of Agriculture, Tottori University. Tottori, 680-8553, Japan.
Email: jsano@muses.tottori-u.ac.jp.

Simonovic, Slobodan P. (Canada), born in 1949 in Belgrade, Yugoslavia, is Professor of the Department of Civil and Environmental Engineering at the University of Western Ontario, London, Canada and Director of Engineering Studies at the Institute for Catastrophic Loss Reduction. His primary research interest focuses on the application of systems approach, and development of the decision support tools for management of complex water and environmental systems. He has received a number of awards for excellence in teaching, research, and outreach.

Address: The University of Western Ontario, Department of Civil and Environmental Engineering, 6NA 5B9.
Email: ssimonovic@eng.uwo.ca.
Website: https://www.eng.uwo.ca/civil/faculty/simonovic_s/.

Tamai, Shigenobu (Japan) completed his doctoral thesis, *Stand Structure and Light Conditions in the Stand*, at Kyoto University in 1974. Professor Emeritus of Tottori University. Division of Environmental Conservation, Arid Land Research Center, Tottori University. His research sites were China, Thailand, Brazil and Turkey. His work experience includes: 1. Conservation of mangrove in southern Thailand; 2. Ecological characteristics of halophytes in Inner Mongolia; 3. Revegetation of Caatinga in north-east Brazil; 4. Impact of climate changes on stand structure and growth of plants in the south-eastern Mediterranean region of Turkey.

Address: Tottori University, Arid Land Research Center, 1390, Hamasaka, Tottori, 680-0001, Japan.
Email: tamai@alrc.tottori-u.ac.jp.

Tanaka, Kenji (Japan) was born in 1969 in Kusatsu, Japan. Associate Professor of Water Resources Research Center at the Disaster Prevention Research Institute at Kyoto University, Uji, Japan. He has been working on better representation of land surface processes, land-atmosphere interaction, impact assessment of climate change on water resources. He was a board member of the Japan Society of Hydrology and Water Resources (JSHWR) and Japan Geoscience Union (JpGU), and Secretary of the Asia Pacific Association of Hydrology and Water Resources (APHW).

Address: Kyoto University, Disaster Prevention Research Institute, Water Resources Research Center, 6110011.
Email: tanaka.kenji.6u@kyoto-u.ac.jp.
Website: http://rwes.dpri.kyoto-u.ac.jp/~tanaka/.

Tezcan, Levent (Turkey), born in 1965 in Ankara, is Assistant Professor of Hydrogeological Engineering at the Faculty of Engineering, Hacettepe University, Ankara. He is also staff member of the International Research Center for Karst Water Resources (UKAM). He joined several research projects and has authored numerous journal and conference papers on construction management, geotechnical engineering, and hydrogeology.

Address: Hacettepe University (UKAM), International Research Center for Karst Water Resources, Beytepe, 06800, Ankara, Turkey.
Email: tezcan@hacettepe.edu.tr.
Website: http://yunus.hacettepe.edu.tr/~tezcan/.

Topaloğlu, Fatih (Turkey) was born in 1969 in Çankırı, Turkey. He graduated from the Department of Agricultural Engineering, Faculty of Agriculture, University of Çukurova, Adana, Turkey in 1990. In 1993–1995 he obtained his M.Sc. degree, which was entitled *Effect of Tillage Systems on Soil Hydraulic Properties Through Ponded Infiltration Tests*, from the International Centre for Advanced Mediterranean Agronomic Studies in Bari, Italy. He completed his Ph.D. thesis, entitled *Determining an Appropriate Method for Estimating Flood Magnitude and Frequencies in the Rivers of the Seyhan Basin*, in the Department of Agricultural Structures and Irrigation, University of Çukurova (1995–1999). He worked as a researcher on the projects 'Partial Root Drying: A Sustainable Irrigation System for Efficient Water Use without Reducing Fruit Yield' and 'Impact of Climate Changes on Agricultural Production System in Arid Areas'. He has been Prof. Dr. in the Department of Agricultural Structures and Irrigation since 2013.

Address: Çukurova University, Faculty of Agriculture, Department of Agricultural Structures and Irrigation, 01330, Adana, Turkey.
Email: topaloglu@cu.edu.tr.
Website: http://aves.cu.edu.tr/topaloglu/.

Tsujii, Hiroshi (USA/Japan) was born in 1941 in Kyoto, Japan. Professor and Professor Emeritus of Kyoto University, Ph.D., University of Illinois, USA in the field of agricultural economics. President of Agricultural Economics Association in Kansai, Japan. He has published many books and papers on the relationships between climate and weather and agricultural production, and on the international rice market and rice policy in Japan, Asian countries, and in the world in English and Japanese.

Address: University of Illinois, USA, Professor Emeritus, Kyoto University, Agricultural Economics, 104-1 Higashianshincho, Okamedani, Fukakusa, Fushimi, Kyoto, Japan.
Email: tsujii1809press@yahoo.co.jp.

Umetsu, Chieko (Japan) is Professor of Resource and Environmental Economics at the Graduate School of Agriculture, Kyoto University. She was a project member of Impact of Climate Changes on Agricultural Production System in Arid Areas (ICCAP) at the Research Institute for Humanity and Nature (RIHN), Kyoto during 2002–2007. She was a faculty member at Kobe University and Nagasaki University before moving to Kyoto University in 2016. She has research experience in arid and semi-arid regions, including Zambia, Turkey and India. Her work has been published in *American Journal of Agricultural Economics*, *Journal of Environmental Economics and Management* and *Journal of Economic Dynamics and Control*.

Address: Kyoto University, Regional Environmental Economics, Graduate School of Agriculture, Kitashirakawa Oiwakecho, Sakyo-ku, Kyoto 606-8501, Japan.
Email: umetsu.chieko.5e@kyoto-u.ac.jp.
Website: http://www.reseco.kais.kyoto-u.ac.jp/en/faculty/#%e6%a2%85%e6%b4% a5-%e5%8d%83%e6%81%b5%e5%ad%90.

Ünlü, Mustafa (Turkey), born in 1969 in Bor, Turkey, is Professor in the Department of Agricultural Structure and Irrigation, Faculty of Agriculture, University of Çukurova, Adana, Turkey. He specialises in irrigation, evapotranspiration and micrometeorology topics and has led projects supported by the EU and TÜBİTAK. He has authored numerous papers on irrigation, evapotranspiration and micrometeorology for various crops.

Address: Çukurova University, Department of Agricultural Structures and Irrigation Engineering, Faculty of Agriculture, Adana, Turkey.
Email: munlu@cu.edu.tr.
Website: http://aves.cu.edu.tr/munlu/.

Ünlükaplan, Yüksel (Turkey), born in 1976 in Adana, Turkey, is Assistant Professor of Landscape Architecture at the Faculty of Agriculture, University of Çukurova, Adana. She has been involved in the projects supported by TÜBİTAK, as project assistant. She has contributed to several publications in the field of landscape planning, in particular the application of multivariate statistical techniques in landscape analysis.

Address: Çukurova University, Department of Landscape Architecture, Adana, 01330, Turkey.
Email: yizcan@cu.edu.tr.
Website: http://aves.cu.edu.tr/yizcan/.

Watanabe, Tsugihiro (Japan), see above as coeditor.

Address: Kyoto University, Regional Planning, Graduate School of Global Environmental Studies, Sakyo-ku, Kyoto 606-8501, Japan.
Email: nabe@kais.kyoto-u.ac.jp.
Website: https://www2.ges.kyoto-u.ac.jp/en/members/watanabe-tsugihiro/.

Yano, Tomohisa (Japan) was born in 1939 in Iizuka, Japan. Emeritus Professor. Director of Sand Dune Research Institute and Arid Land Research Center of Tottori University, Japan, between 1989 and 1994. Councillor of the Japanese Society of Soil Physics and Japanese Association for Arid Land Studies from 1992 to 1993. Director of the Japanese Society of Water, Land and Environmental Engineering from 1996 to 1997. He served as a Visiting Researcher in the Department of Land, Air and Water Resources, California University, USA from 1980 to 1981, and was Visiting Professor in the Faculty of Agriculture at Mustafa Kemal University in Turkey from 2004 to 2005. He has authored 117 publications.

Address: Tottori University, Arid Land Research Center, Tottori, Japan.
Email: yano@ant.bbiq.jp.

Yatagai, Akiyo (Japan) Professor, Hirosaki University, Japan, gained her Ph.D. at Tsukuba University in 1996. She has worked at the Research Institute for Sustainable Humanosphere (RISH), Kyoto University (June 2012–October 2013), and before that at the University of Tsukuba (June 2011–March 2012). She has also taught at Kobe University and Meiji University, Japan, and Vietnam-Japan University. Her research topics are climatology, hydrology, meteorology and statistical data analysis. She has authored more than 110 papers in English and Japanese.

Address: Hirosaki University, 3 Bunkyocho, Hirosaki, Aomori 036-8561, Japan.
Email: yatagai@hirosaki-u.ac.jp.

Yılmaz, K. Tulühan (Turkey), born in 1963 in İzmir, Turkey, is Professor of Landscape Architecture at the Faculty of Agriculture, University of Çukurova, Adana, Turkey. He has acted as the Chairman of the Dept. of Landscape Architecture. He is a former Steering Committee Member of the Small Grant Programme of GEF II, and a Board Member of the Turkish National Committee on Coastal Zone Management in Ankara. He has led research projects supported by TÜBİTAK, and been involved in several projects conducted by both national and international institutions. He has authored numerous journal and conference papers on vegetation dynamics, coastal zone management, biodiversity conservation and environmental issues.

Address: Çukurova University, Department of Landscape Architecture, Adana, 01330, Turkey.
Email: tuluhan@cu.edu.tr.
Website: http://aves.cu.edu.tr/tuluhan/.

Printed in the United States
By Bookmasters